T5-ARI-492

Ecological Studies

Analysis and Synthesis

Edited by

W. D. Billings, Durham (USA) F. Golley, Athens (USA)

O. L. Lange, Würzburg (FRG) J. S. Olson, Oak Ridge (USA)

Volume 21

V. Ya. Alexandrov

Cells, Molecules and Temperature

Conformational Flexibility of Macromolecules and Ecological Adaptation

Translated from the Russian by
V. A. Bernstam

With 74 Figures

Springer-Verlag Berlin Heidelberg New York 1977

Professor Dr. VLADIMIR YA. ALEXANDROV
Laboratory of Cytophysiology and Cytoecology
Komarov Botanical Institute
The USSR Academy of Sciences
Leningrad 197022/USSR

Dr. VICTOR A. BERNSTAM
Department of Cell Cultures Institute of Cytology
The USSR Academy of Sciences Leningrad 190121/USSR

For explanation of the cover motive see legend to Fig. 59 (p. 188)

ISBN 3-540-08026-0 Springer-Verlag Berlin Heidelberg New York
ISBN 0-387-08026-0 Springer-Verlag New York Heidelberg Berlin

Library of Congress Cataloging in Publication Data. Aleksandrov, Vladimir Iakovlevich. Cells, molecules, and temperature. (Ecological studies; v. 21). Translation of: Kletki, makromolekuly i temperatura. Bibliography: p. 1. Thermobiology. 2. Macromolecules. 3. Adaptation (Biology) I. Title. II. Series. (DNLM: 1. Adaptation, Physiological, 2. Cells, 3. Temperature. 4. Macromolecular systems. WI EC912K v. 21/QH581.2 A366k). QH516A4313. 574.8'76.76-55609.

Typesetting, printing, and binding: Brühlsche Universitätsdruckerei, Lahn-Gießen, 2131/3130—543210

Foreword

After the great achievements in the field of molecular foundations of genetics and protein synthesis, molecular biology undertook the successful deciphering of a number of other important biological problems. By this time ecology in its various branches was far enough advanced to tackle the problems arising at the level of molecular biology. The monograph of Professor Alexandrov, which takes as an example the adaptation of organisms to habitat temperatures, presents a vivid picture of this major ecological problem as viewed at the cellular and molecular levels.

As main theme of the book the author advances a hypothesis on a correlation between the level of conformational flexibility of protein molecules and the temperature ecology of a species, as a result of which the protein molecules are maintained in a semilabile state. This principle may also be applied to other factors of the environment which affect the level of flexibility of protein macromolecules. The principle of semistability is shown to be applicable also to the nucleic and fatty acids.

While temperature adaptations realized at the biocoenotic, social, organismal and organic levels are essentially different in different groups of organisms, the adaptation achieved at cellular and molecular levels proves to be of general biological applicability. For this reason, the book contains experimental observations which span the whole domain of thermoecology of plants, animals, and microorganisms, thus differing from the majority of other ecophysiological monographs, which usually touch on only one kingdom of the living world. The book serves moreover as an introduction for those who cannot read Russian papers to the vast investigations on relevant problems undertaken by Soviet scientists, which have either formerly not been known at all, or only to a few researchers partially familiar with the material. Over 500 publications by Russian authors are referred to in this volume.

The book is valuable for a wide field of biologists interested in modificational and genotypic adaptations of organisms at cellular and molecular levels, but will also be of great use to practical workers who are dealing with acclimation of organisms to new temperature environments. Students and newcomers in the field of adaptation and resistance of organisms will find many methodological approaches employed, illustrated by a wealth of experimental material critically discussed by the author.

Würzburg, April 1977 OTTO L. LANGE

Preface

Recent years have witnessed a considerable acceleration of the rate of research on various biological problems carried out both at the cellular and molecular levels. It is becoming more and more difficult to cope with the ever-increasing flow of data and to keep abreast with new developments in science. The earlier advice of my teacher, Academician Aleksei Alekseevich Zavarzin, suggesting that "... if a scientist finds pleasure in reading proofs of his own paper, it is time for him to quit science ..." appears even more pertinent at present. In this situation I ventured to summarize the evidence accumulated by my colleagues and myself, as well as by researchers in other laboratories, dealing with the problem of cellular and molecular mechanisms responsible for adaptation of organisms to the temperature of environment. After a quarter of a century of studies in this field of ecology, I felt it justified and necessary to evaluate the significance and value of the results obtained and to seek the most promising pathways for further investigation on the problem. A number of reasons might be advanced against this decision, however. A book of this kind, as any book in our field, would present, more or less accurately, only the state of the science at the time of writing. On the other hand, the incessant and ever accelerating accumulation of new data and ideas may transform this book in the near future into a faithful account of the fallacies of the recent past. Many a time I have been pleased to realize that I had been absolutely right in refraining from writing a book a few years earlier. In so doing I avoided publication of a number of ideas which turned out to be unsubstantiated. In addition, the book would have lacked the data and concepts that, to my mind, seem valuable at present. Moreover, writing of a book would have prevented me from conducting research for a considerable length of time. In spite of all of these considerations I decided to write this book, realizing that the disadvantages mentioned will hardly be invalidated as time goes on.

This book could not have been written unless a great deal of work on the problem had been performed by my dear students and collaborators: O.A. Ageeva, A.P. Andreyeva, T.A. Antropova, N.I. Arronet, K.A. Barabalchuk, V.A. Bernstam, N.L. Feldman, I.S. Gorban, L.S. Gubnitsky, E.I. Denko, I.E. Kamentseva, I.M. Kislyuk, M.F. Konstantinova, A.G. Lomagin, M.I. Lutova, H.G. Shukhtina, A.M. Shcherbakova, L.V. Ulyanova, M.D. Vaskovsky, V.N. Vitvitsky, V.I. Yakovleva, A. Yazkulyev and I.G. Zavadskaya. They also offered invaluable assistance in the preparation of this book. My heartfelt thanks are due to all of them.

The present monograph deals with animals, plants and microorganisms, cells, proteins, nucleic and fatty acids, various cellular functions and enzymatic processes, not to mention numerous other topics. Attempts to cover such a diverse and sundry material within the framework of a single book are justified by my willingness to use a great variety of data to substantiate one unifying idea. At this time of profound differentiation of biological disciplines, such tasks present great difficulties for an author who is likely to misinterpret material outside the scope of his particular field. Under these circumstances the only solution for me, as a cytologist, was the extensive use of friendly assistance offered by specialists in neighboring disciplines: taxonomy, ecology, genetics, biochemistry, biophysics, etc. I hope the fruitful discussions I have had with them, as well as their reading of appropriate chapters, has helped to diminish possible misinterpretations and fallacies. I am deeply grateful to all of them for their invaluable help. I refrain, however, from giving a list of their names since I am afraid this might suggest to the reader that I am inclined to make my consultants responsible for the possible errors in the book.

I would appreciate any comments that the readers of this book might wish to make.

Leningrad, April 1977 VLADIMIR YA. ALEXANDROV

Contents

Introduction

In living organisms all physico-chemical processes responsible for the functional activity of cells are, to a greater or lesser extent, dependent on temperature. This is due to the temperature dependence of thermodynamic and kinetic constants that determine directions and rates of chemical reactions, conformational transitions of biological macromolecules, phase transitions of lipids, changes in the structure of water, etc. Any temperature shift will invariably result in an alteration of a metabolic rate and distort original relationships between the reaction rates of numerous constituents. At the same time, normal functioning of cells is maintained within a wide range of temperature fluctuations in the environment (the biokinetic zone of the temperature scale).

The relative independence of the living process from changes in temperature is achieved through a variety of adaptive mechanisms that have developed in organisms during the entire biological evolution. These are based on various physical and chemical principles. In order to maintain a relatively normal level of living after a change in temperature has taken place, organisms make use of the regulation of biochemical processes by accelerating or suppressing the enzyme functioning, by enhancing or reducing an enzyme synthesis, by modifying properties of biological macromolecules, by triggering the processes effective in producing or absorbing heat, by employing thermoinsulation, by avoiding habitats with unfavorable temperature conditions, etc.

If cooling or overheating reaches the level surpassing adaptive capacities of organisms and becomes incompatible with their active life, some organisms move to the second line of defence of which they are capable, i.e. to a state of a more or less concealed life. They retain, however, the ability to return to normal life upon re-establishment of acceptable temperature conditions. At any given moment in the life of an organism at every stage of evolution, all living systems are forced to conform to temperature variations in the environment. Adaptation to the temperature factor involves all levels of the organization of a living being, ranging from the molecular to the biocenotic one so that an adaptive result may be attained at various levels.

Certain plants avoid excessive overheating and illumination by growing under the canopy of other plants. In this case, an adaptive effect is achieved at the biocenotic level. The surprising devices employed by bees in their attempts to cope with overheating and overcooling of the bee-hives may serve as an example of social adaptation. Lowering the temperature of an animal body through functioning of the sweat glands, similar to cooling of plants by transpiration, is an adaptation at the organismal level. Rearrangement of chloroplasts from a random (apostroph) to a profile (parastroph) distribution close to the cell wall under excessively intense illumination should be considered as an adaptation at the cellular level. Going deeper, the high thermostability of proteins, characteristic of thermophilic microorganisms, can be visualized as an adaptation at the molecular level of the organization of living systems.

Biological evolution has used all these contrivances to create animal, plant and microorganismal species which exhibit extremely diversified requirements for their temperature environments. As a result, all the climatic zones on our planet have been inhabited by living organisms. On land, the higher plants can grow on the Franz-Joseph Land where the maximal air temperature never exceeds 6–7° C, as well as in tropical deserts where the temperature maximum is around 50° C, while the soil temperature frequently rises above 70° C. The multiform world of animals that occupy the ultra-abyssal oceanic regions, exists at a constant temperature of about 2° C, whereas in the surface waters of the Arctic and Antarctic one finds organisms living at −1.8° C. On the other hand, the temperature preferred by reptiles living in African deserts is as high as 45° C, and that preferred by Arthropoda up to 49° C (Herter, 1943). Still wider is the temperature range for microorganisms (Skinner, 1968). Among these we can find the psychrophiles with a temperature optimum around 0° C and the maximum of around 15–20° C. Larkin and Stokes (1968) have referred to reports of microbial growth observed at −17.8° C, −20° C, and −34° C. These authors were able to detect bacterial growth at −10° C in a culture medium supplemented by anti-freeze compounds. At the other extreme are the thermophiles living at optimal temperatures ca. 70–80° C, the maximum being around 90° C.

Brock et al. (1971) have found sulfur bacteria that live in boiling waters in the Yellowstone National Park. In the salt lakes of the Borovsky Reservation in Khazakhstan, Zernov and Schmalhausen (1944) have detected flagellated algae (*Paramimonas* and *Dunaliella* spp.) that move vigorously at −7.75° C. Under experimental conditions these algae were seen moving even at −15° C. Active microbial life has also been detected in the salt lakes in the Antarctic where water freezes at −48° C (Meyer et al., 1962). Thermophilic sulfate-reducing bacteria were able to grow at +104° C under pressure of 1000 atm (ZoBell, 1958, quoted by Skinner 1968). Thus, active life is feasible at any temperature, provided water is maintained in a liquid state. The temperature range for concealed life is even wider. There are living organisms capable of surviving the temperature of liquid helium (−272° C) and those that withstand, in a desiccated state, heating up to 170° C.

Adaptation of organisms to the temperature of environment is an almost limitless subject. It is of prime importance to various branches of theoretical biology and is directly related to scores of practically important issues in agriculture, medicine and industry. Hence, it is apparent that a single author can cover within the limits of one book only a few facets of this subject, and not the field as a whole.

The subject matter of this book will be confined to a discussion of the problem in the context of cytoecology and molecular ecology.

The term "cytoecology" has been introduced into the literature in the title of a paper by Tischler (1937) *On some problems of cytotaxonomy and cytoecology*, dealing with the adaptive significance of natural polyploidy. In the same sense, the term "cytoecology" appears in the title of a paper by Hagerup (1941). Viktorov (1940), in his paper on this problem, wrote: "Cytoecology visualizes its goal in studies of those changes which occur in the structure and properties of the cell under the influence of environmental factors, as well as in those which analyze the significance of these changes for an organism and a species."

In 1950 Biebl wrote in his paper *Ökologie und Zellforschung* that there are grounds at present to speak about the "protoplasmic ecology", which studies relations between the properties of the cytoplasm and habitats. In an introduction to his monograph, Biebl (1962c) offers a somewhat modified definition: "... the term 'protoplasmic ecology', in a more or less restricted sense, stresses that here we are concerned with studies on only those ecologically determined properties of the cytoplasm, which can be observed even in single cells and can be studies primarily by cytophysiological methods" (p.4). Similar definition can be found in a paper by Biebl entitled *Protoplasmatische Ökologie* (Biebl, 1967b).

There are investigators who seem to understand the aims of cytoecology in a too broad sense. In fact, B. Ushakov states: "The aim of cytoecology, as a branch of ecology, is in elucidation of the role given to cells in creation and realization of supracellular adaptations (organismal, family-herd, populational, specific)" (1963a, p. 16). It should, however, be borne in mind that organisms adapt to environmental factors, employing all the devices available. An adaptive effect can be achieved at various levels of organization of the living. Since cells are the structural and functional elements of a multicellular organism, they take part in realization of every adaptation of the latter to the environment. Cooling of leaves by transpiration is accomplished through functioning of the root cells, the conductive system, and of the stomatal cells. If an animal avoids overheating by taking refuge in shadowed places, it resorts to an organismal adaptation, the realization of which involves stimulation of nerve cells and contraction of muscle fibers. When speaking of the aims of cytoecology, an advocation of studies concerning the participation of cells in organismal reactions, the adaptive significance of which becomes evident at the supracellular level of organization, is, in the opinion of the author, hardly justified. In this case, the scope of cytoecology would have been expanded so as the comprise the whole domain of cellular physiology and morphology. Every normal function and structure in the cell is of greater or lesser adaptive significance. Otherwise their appearance in phylogenesis would have been inexplicable. *It would be more reasonable to confine the sphere of cytoecology to studies on those forms of organismal adaptations to environmental factors, in which an adaptive effect is achieved already at the cellular level of organization of living systems.*

As will be shown later, a very important role is given also to such adaptations of organisms which are realized at the molecular level. This confirms the author's views on molecular ecology similar to those implied in the term "cytoecology" (Ushakov, 1963a; Alexandrov, 1965b). In both instances, it is the cellular and molecular basis of ecology of organisms that is the subject of investigation, but not the ecology of cells and/or molecules. This reservation, however, should be made, since, according to Paul Weiss, the terms "molecular ecology" and "cellular ecology" are currently used in essentially different senses. He writes: "... I have often characterized cell life as *molecular ecology*. And, in the same sense, the population of cells composing tissues and organs are integrated systems of *cellular ecology*" (Weiss, 1971, p. 184).

A discussion of the problem in question will be confined to cellular and molecular aspects. This limitation, however, is to a certain degree an artificial one, since delineation of the aims of cytoecology or molecular ecology, as any definition in biology, is relative. Modes of adaptation of organisms to environmental factors have been created in the course of evolution whithout any consideration of subsequent suitability of these for systematization by scientists. Take, for example, the cooling of the warmblooded organisms that suppresses oxidative phosphorylation in liver cells and in muscles so that the energy of electron and proton transport is channelled into heat generation (Neifakh, 1961; Skulachev, 1969). Such an adaptation might be assumed to belong to the cellular level of organization. In order to achieve any adaptation, however, more than one cell is required. A significantly larger cellular mass is needed in order that its temperature might be increased to a detectable degree. The dilemma of whether or not similar marginal adaptations should be included in the scope of cytoecology is essentially a matter of taste.

At present, cytoecology has firmly established itself as a field of study intermediate between cytology and ecology. To give the date of foundation of this branch of biology is a difficult task. The earliest intrusion of cytology in the solving of ecological problems occurred in the middle of the nineteenth century in studies dealing with plant frost resistance. The history of this event has been well described by Maksimov (1913). Credit for the first cytoecological experiments should be given, however, to the surprising research performed by Purkinje and Valentin. In their treatise *De phaenomeno generali et fundamentali motus vibrotorii continuiti* which appeared in 1835, i.e. four years prior to the formulation of the cell theory by Schwann, these authors compared the resistance of the ciliary cell beating to elevated and low temperatures in warm- and cold-blooded animals. Purkinje and Valentin found, among other things, that in the "tunic" of mammalia and birds, the ciliary beating stopped at 5° R, whereas in the turtle, in the mollusc *Unio pictorum* and *Anadonta cygnea* it can be observed even at 0° C.

In the present monograph, major attention will be given to the cellular and molecular adaptations to temperature by organisms found in a state of active life. The problem of winter and summer quiescence of plants and animals, encysting and spore formation and similar phenomena, despite the profound interest they present, will be dealt with to the extent determined by the necessity for developing certain general concepts. The boundary between active life and quiescence at low or elevated temperatures is also vague and ill-defined.

Presentation of the experimental material has, of necessity, involved differentiation between modificational and genotypic adaptations of cells and macromolecules to the temperature of environment. The modificational adaptations comprise those which result from immediate response of cells to shifts in ambient temperature. They are not related to changes in genomes. In this case, it is this very ability of cells to react in an adaptive fashion to a temperature shift that is genotypically determined. An adaptive reaction itself is revealed only when an appropriate shift in temperature, or in a factor substituting the thermal shift, occurs [1].

Genotypic adaptations are inherited ones, for whose appearance no provocative action of a temperature fluctuation is needed.

Attempts to distinguish modificational adaptations from genotypic ones are often confronted with major difficulties, in particular when an investigator deals with a genetically heterogeneous population of dividing cells: with cultures of unicellular organisms and of tissue cells. Consider, for instance, an adaptive increase in an average level of thermostability of a cellular population when the temperature is elevated. This might be a result of an elimination of thermolabile cells and of subsequent multiplication of the more heat-resistant cells, or it might be a consequence of a higher division rate of the cells featuring an increased heat resistance. In this situation a genotypic adaptation of a population takes place which is realized through multiplication of the cells, whose genomes endow them with enhanced thermostability. A temperature shift brings about a shift in the gene fund of the population. An apparently similar effect—an elevation of the average level of thermostability of a population—may be achieved also if, in response to a temperature elevation, every cell reacts by increasing its heat stability. A modificational adaptation is then said to have taken place. Reaction of a population might involve a combination of genotypic and modificational adaptations. Not always can one easily distinguish between the two in actual experiments.

There is another source of difficulties in analyzing the nature of temperature adaptations. As will be shown later, the onset of heat or autumn frosts induces shifts in the level of cellular thermoresistance. Such a shift could be regarded as a modificational reaction to a change in temperature. Some evidence is available, however, which indicates that heat resistance of plant cells may vary with changes in day-length. This reaction serves as an adaptation to seasonal changes in temperature so that the triggering agent is not the ambient temperature, but a more reliable seasonal indicator—the day-length. In this context, the latter parameter is, presumably, playing the role of a "conditioned stimulus". And, finally, temperature adaptations might be reflections of the internal—diurnal and seasonal—rhythms. They may be more or less independent of changes in the environment. In this case, the triggering agent remains unknown, whereas adaptations of this kind should nevertheless be regarded as modificational.

The existence of the coincident (Morgan, 1896; Baldwin, 1902; Kirpichnikov, 1935, 1947), stabilizing (Schmalhausen, 1940, 1946) selection, or, according to Waddington (1953), of the genetic assimilation, may account for a more or less complete substitution in the course of evolution of modificational adaptive reactions for the genetically established endogeneous adaptations. It is in this way that intermediate forms of adaptation have been created, which can hardly be

[1] In the cytoecological literature, the non-genotypic adaptations are often referred to as phenotypic or physiological adaptations. Since the genotypic mutations are also realized through the phenotype and with the aid of the physiological mechanisms, I prefer to use a more indifferent term, employed in genetics and the evolutionary science—modificational adaptations.

fitted into two sorting boxes: modificational and genotypic (Langridge, 1963a). A discussion on whether the long-lasting temperature modifications of *Protozoa* are of modificational or genotypic nature can serve as a vivid illustration of the complexity of the problem in question (see p. 21).

The trouble is often not in the lack of available knowledge of the phenomena themselves. The cause of the difficulties mentioned above lies deeper, and cannot always be eliminated. Order must be introduced into the research process and the various biological phenomena encountered, and appropriate definitions must be given. Each time, however, certain distortions enter into the picture of living nature, since definitions usually add an aspect of discontinuity to a sequence of naturally uninterrupted events. Admittedly, every phenomenon in living nature has appeared in evolution from something else, in the course of this evolution, brings forth something different. Intermediate links are often retained in this transformation. In giving definitions, a fraction of a process is usually selected, and thereby the limits of the fraction defined become rather relative. This must be tolerated because definitions facilitate presentation of material and exchange of ideas between scientists, and if they are not regarded as absolute truths, will not stimulate fruitless discussions. In this respect, we must agree with Engels (Marx and Engels, 1961), who wrote that definitions are of no importance for science because they always prove to be unsatisfactory. For practical purposes, however, they may often be useful and even indispensable and will do no harm, provided they are not expected to give more than they can express.

The first two chapters of this book deal with modificational and genotypic changes of the primary thermoresistance of cells occurring in response to fluctuations in ambient temperature. The reader may obtain the impression that too much space is given to the primary resistance of cells as compared with other aspects of cellular response to the temperature factor. However, this characteristic is indicative of an important property of protein macromolecules, namely, their conformational flexibility. Koshland was right in saying that "... this flexibility is a key feature of biological action of enzymes" (Koshland, 1964a, p. 722); in other words, a central feature of cellular metabolism. Conformational flexibility is temperature-dependent, and, therefore, plays an important role in adaptation of organisms to the temperature environment. Temperature adaptations of protein macromolecules are the subject of four subsequent chapters of this monograph. And, finally, two concluding chapters are devoted to a discussion of the involvement of nucleic acids as well as fatty acids—constituents of lipids—in temperature adaptations of organisms.

Discussion of this diverse experimental material brings us to the conclusion that for normal functioning of proteins and nucleic acids to occur their macromolecules must exhibit a definite level of conformational flexibility: macromolecules should be neither too rigid nor too labile. The same holds true also for the aggregate state of lipids: they should be neither too fluid, nor too solidified. Thereby, the general principle of semistability or semilability of biological macromolecules is advanced. This must be kept in mind when carrying out research on adaptation of organisms not only to temperature, but also to any other environmental factor that can affect the degree of stability of proteins, nucleic acids and lipids.

Presentation of an exhaustive literature review pertinent to the issues involved has not been the author's goal. Those problems which have been tackled experimentally both by myself and my collaborators are presented in greater detail. This should not be regarded as an overestimation of the significance of the author's own research. This is simply due to the fact that when dealing with material known only from the literature, the author feels himself less secure. Unfortunately, in many fields of biology one can draw from the relevant literature arguments in favor of almost any point of view, and, without practical experience, one can hardly distinguish the trustworthy from the unsubstantiated.

Chapter 1
Modificational Changes of the Primary Thermoresistance of Cells

Ancel and Vintemberger (1925) pointed out in their discussion concerning radiosensitivity of cells that the injurious effect of X-rays, estimated by microscopic technique, is a result of an interaction of three factors; namely, a radiation injury, factors promoting manifestation of the radiation injury, and factors restoring a normal condition. Much later, not knowing this work, the author drew similar conclusions as to the complexity of the process of injury and, consequently, the complexity of the concept of resistance (Alexandrov, 1952).

1.1 The Primary Thermoresistance of Cells

1.1.1 The Primary and General Thermostability of Cells

The injurious action of elevated temperatures may elicit the following:

1. Primary injury, i.e. an alteration of cellular components as a direct result of heating.
2. Destructive after-action—a complex of secondary alterations affecting cellular metabolism and directly or indirectly dependent on the primary site of injury.
3. Reactive increase in the stability of cellular components in response to heating—an adaptation.
4. Repair of the injury possibly taking place not only after heating is over, but also during heating.

To recapitulate, the processes occurring in the cell under conditions of injurious heat are of dual nature: some are destructive, whereas others maintain and restore the cell's integrity.

This dualism is often reflected in the shape of a curve which describes the dependence of a duration of any given life function upon the temperature of heating. Figure 1 shows the time during which protoplasmic streaming is maintained in epidermal cells of *Campanula persicifolia* leaves as a function of the temperature of injurious heating. If expressed in a semi-logarithmic scale, this curve reveals the existence of a simple dependence at temperatures from 46 to 42° C. Lowering the temperature by 2° C prolongs the time approximately two-

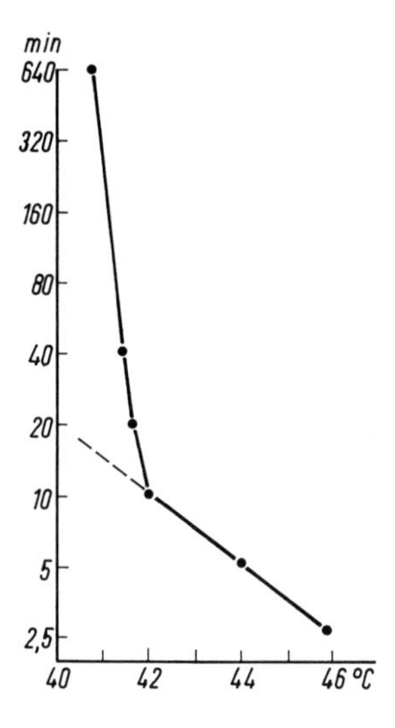

Fig. 1. Persistence of protoplasmic streaming in epidermal cells of *Campanula persicifolia* leaves as function of heating temperature (Alexandrov, 1963). *Abscissa:* temperature of heating; *ordinate:* time period during which protoplasmic streaming persisted (log. scale)

fold during which the protoplasmic streaming persists; in other words, the rate of development of an injury decreases twofold. This regularity changes, however, at temperatures below 42° C. Furthermore, even a minor lowering of temperature greatly prolongs the duration of the protoplasmic streaming. At 42° C the streaming of protoplasm stops in 10 min, at 41° C the streaming continues over 10 h. Similar curves displaying linear portions in the zone of intensive heating, and showing a more or less abrupt turn upwards at a break-point which comes on transition to a region of less intensive heating, have been obtained in the author's laboratory in scores of plant species. The location of the break in the curves for different plants is found at different points on the temperature scale and with different durations of heating. In *Campanula persicifolia* the break occurs at a temperature that is slightly lower than the temperature effective in stopping the protoplasmic streaming in 10 min. The curve for *Tradescantia* (Fig. 2, curve 1) turns upward only at that temperature which is lower than the temperature effective in stopping the streaming of protoplasm in 80 min. Similar plots have been described also for animal objects. Figure 3 shows the temperature-dependence of the time period during which the ciliary beat persists in the mollusc (*Lacuna divaricata*) larval cells (Arronet, 1963).

Occurrence of such breaks in the plots described is accounted for by the fact that at temperatures below the break, the cells become capable of resisting the injurious action of heat (for description of a similar dependence for other injurious agents see Clark (1933) and B. Ushakov (1959a). This is supported by direct observation during heating of the cell. Experiments have been carried out employing a specially designed heating chamber mounted on a microscope table. A piece of *Campanula persicifolia* leaf is placed in the chamber and the temperature of water in the chamber is gradually elevated. The same cells are being watched

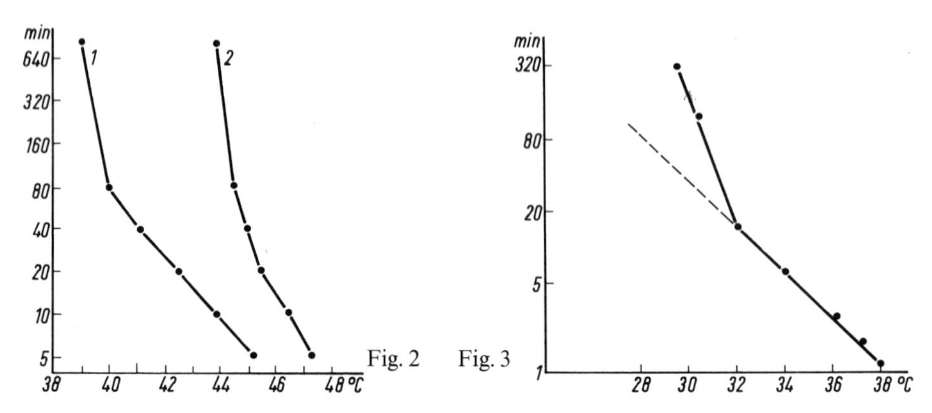

Fig. 2. Persistance of protoplasmic streaming in epidermal cells of *Tradescantia fluminensis* leaves as function of heating temperature (Alexandrov, 1963). *Abscissa:* temperature of heating; *ordinate:* time period during which the protoplasmic streaming persisted (log scale). *1:* control leaves; *2:* hardened leaves (18 h exposure to 36° C)

Fig. 3. Persistence of ciliary beat of mollusc *Lacuna divaricata* larvae as function of heating temperature (Arronet, 1963). *Abscissa:* temperature of heating; *ordinate:* time period during which ciliary beat persisted (log scale)

continuously through a light microscope. Early in heating the protoplasmic streaming, detected by translocations of spherosomes, accelerates. After 40 min of heating the chamber, when the object under study has been warmed to 41° C, movement of spherosomes slows down dramatically. Elevation of the temperature is discontinued and the temperature is maintained at 41–41.5° C (a break in the curve for thermostability of this object has been reported to occur at 42° C; Fig. 1). Ninety min from the beginning of the experiment, the movement of spherosomes ceases altogether. At 200 min, despite continued heating, many spherosomes start to move a little, and some are seen to cover short distances. Further progress resulted in a more vigorous flow of the protoplasm and after 6 h it was at a normal level. Had the heating been prolonged the streaming of protoplasm would have stopped once again and the cells would have eventually died (Alexandrov, 1963). Absolutely identical results have been obtained for the reed *Phragmites communis* cells (Alexandrov, 1956). Preobrazhenskaya and Shnol (1965) observed cessation of growth of the *Rhizopus nigricans* mycelium when the temperature was elevated from 35 to 40° C. In 2 h, however, growth resumed. Heating sundew leaves at 39° C for an hour, Favard (1963) detected pathological alterations in mitochondria and after 2–3 h of heating the chondriom morphology significantly improved.

These experiments demonstrate the ability of plant cells to repair an injury produced by elevated temperatures during heating.

At a temperature below the break-point in the curve of thermostability the suppression of protoplasmic streaming may not go as far as complete temporary cessation of the flow. In those cases, the resistance of the cell amounts to counteracting a certain initial suppression of the function involved. Repair of various functions during action of injurious agents has been given the name of "reparatory adaptation" (Alexandrov, 1965a). At less intensive heating the resistance of

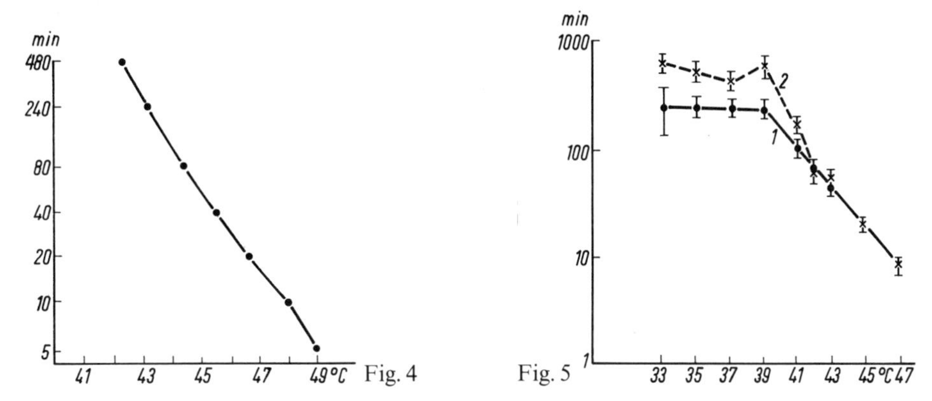

Fig. 4. Persistence of protoplasmic streaming in epidermal cells of *Echeveria secunda* leaves as function of heating temperature (Gorban, 1962). *Abscissa:* temperature of heating; *ordinate:* time period during which protoplasmic streaming persisted (log scale)

Fig. 5. Duration of excitability of albino rat muscles in non-aerated *(1)* and aerated *(2)* Ringer solutions as function of temperature (Skholl, 1962). *Abscissa:* temperature of heating; *ordinate:* time period during which excitability of muscles persisted at indicated temperatures (log scale)

the cell can cope with the continuing action of an injurious agent and support more or less prolonged activity of a function at a relatively normal level.

In some plant objects, no break in the curve of thermostability can be observed (Fig. 4). This may be explained as follows: in order to obtain the curves of thermostability of leaf cells, excised pieces of a leaf plate are exposed to heat. The cells in an excised leaf segment in a number of plant species, e.g. *Tradescantia* leaves, survive for more than two months, if the leaves are kept at room temperature in tap water changed periodically. In other plant species, the cells survive isolation poorly, and the thermostability curves for such cells often display no upward break. This is a result of the inability of the cells, which suffer from isolation, to preserve their metabolic activity at a level required for sustaining the function under study during the action of heat.

Animal cells survive isolation even less well and the curve of thermostability for these cells frequently shows a downward, not an upward break (B. Ushakov, 1965a). In that case, the cell function concerned is more rapidly destroyed at moderately high temperatures than would be expected from extrapolation of the right linear branch of the curve. Effects of conditions required for survival of animal cells on the shape of the thermostability curve are clearly seen from the experiments conducted by Skholl (1962, 1963a). Muscles isolated from albino rats were used to determine duration of the ability for excitation provoked by electrical stimulation. If the muscles were heated in Ringer solution without aeration, the curve revealed no signs of adaptation (Fig. 5, curve 1). Conversely, at 39–33° C the muscles died prematurely since at those temperatures and at the heat durations employed the adverse effects of isolation of the muscles from the organism seem to be enhanced. Similar experiments have been performed with aerated Ringer solution, which showed that an administration of oxygen does not affect the rate of the loss of muscle excitability at temperatures of 47–41° C. Exposure of the muscles to 41–39° C resulted, however, in apparent adaptation—the curve took an upward course (Fig. 5, curve 2). These experiments indicate that in order for the muscle fibers to resist the injurious action of heat actively, they must be able to maintain normal energy metabolism.

Thermostability of cells can be assessed by the extent of alteration of various functions; however different functions may vary in their heat resistance by a wide margin. Figures 6 and 7 illustrate the sequence of injury of various functions in the leaf cells of *Tradescantia* and the bluebell as the temperature of a 5-min heating is elevated. The beginning of a band corresponds to the temperature at which the first manifestations of damage of a given function begin to appear. The end of a band indicates the temperature of heating effective for complete suppression of the function concerned. Thus, cessation of the protoplasmic streaming in *Tradescantia* leaf cells occurs after heating the leaves for 5 min at 44° C, whereas a complete suppression of the cells' ability for plasmolysis requires heating for 5 min at 60° C. The cell cannot repair a thermal injury produced in highly thermostable functions, because suppression of these functions ensues only after such intensive heatings as are effective in destroying the reparatory mechanisms themselves. Therefore, a curve of thermostability displaying an upward break can be obtained in testing rather heat-sensitive functions. Reversibility of a heat-induced suppression of the capacity for plasmolysis can hardly, if ever, be observed (Alexandrov, 1955). Accordingly, the curve of thermostability of plasmolysis reveals no break and gradually turns downward, the more so the nearer we are to the limits of the control cell survival time at room temperature. Belehrádek and Melichar (1930) have also reported a plot of thermostability of plasmolysis in *Elodea* leaf cells that turns downward at milder heatings. Repair of respiration suppressed by heat treatment cannot be obtained in plant cells.

The finding that at temperatures above the point of discontinuity in a curve for cellular thermostability, the dependence of the effect upon heating temperature is represented by a straight line, when expressed in a semilog scale, disproves the possibility of an active resistance of the cell to the progression of a thermal injury. Had active resistance taken place, this would have manifested itself by lowering the temperature of heating and would have resulted, if expressed in a semi-log scale, in a deviation of the temperature dependence of the effect from a straight-line relationship.

An analysis of the curves of thermostability reveals that in registering the effect of heating immediately after intensive short-term exposures to heat, the results are essentially determined by the heat resistance of the different protoplasmic components, and the results are practically independent of the reparatory and/or adaptive capacities of the cells. At moderately high temperatures and with appropriately prolonged exposures, the outcome is determined by the interaction of the destructive effects of heat and constructive efforts of the cell. The same complicated result can be observed with intensive short-term heatings and registration of the effect produced after a more or less delayed time. Subsequent to Sachs (1864) it has been repeatedly shown that thermal injury in plant and animal cells can be reversed (Alexandrov, 1964b). That is why the state in which the cells are found several hours or days after exposure to heat is determined not only by

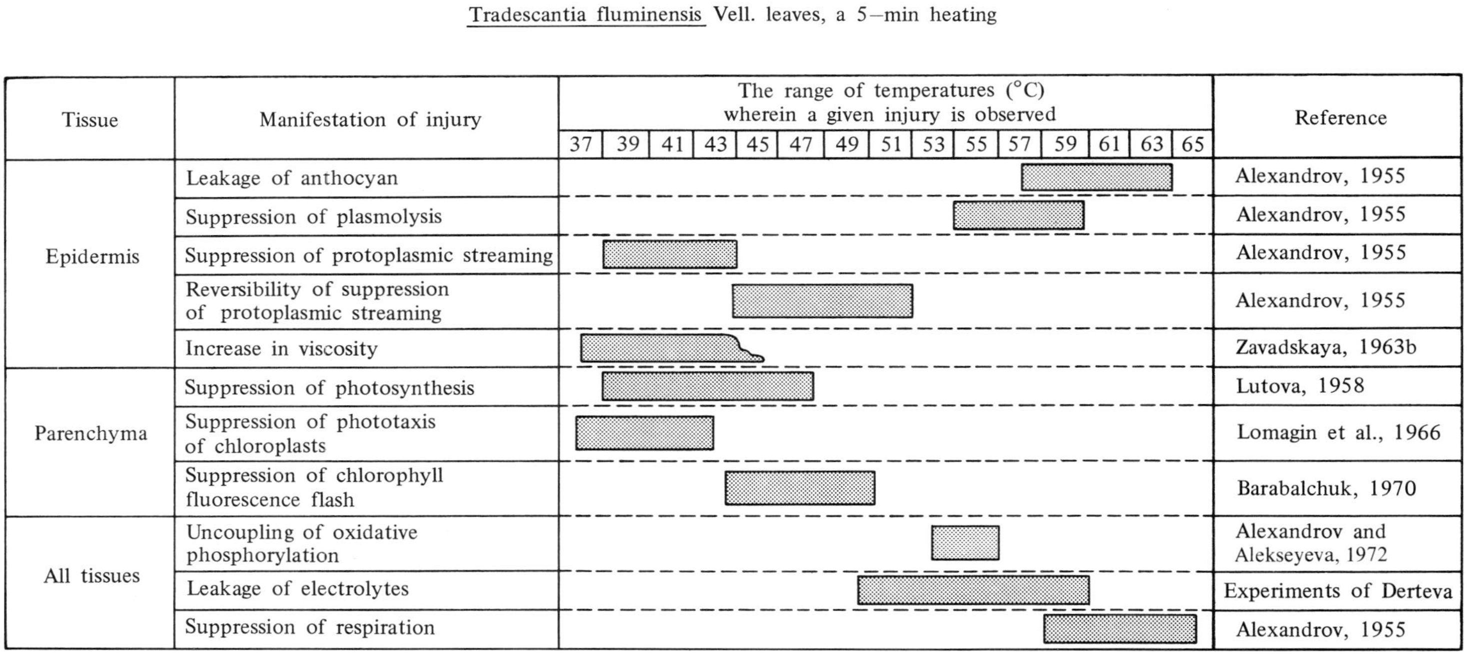

Tradescantia fluminensis Vell. leaves, a 5–min heating

Tissue	Manifestation of injury	The range of temperatures (°C) wherein a given injury is observed															Reference
		37	39	41	43	45	47	49	51	53	55	57	59	61	63	65	
Epidermis	Leakage of anthocyan																Alexandrov, 1955
	Suppression of plasmolysis																Alexandrov, 1955
	Suppression of protoplasmic streaming																Alexandrov, 1955
	Reversibility of suppression of protoplasmic streaming																Alexandrov, 1955
	Increase in viscosity																Zavadskaya, 1963b
Parenchyma	Suppression of photosynthesis																Lutova, 1958
	Suppression of phototaxis of chloroplasts																Lomagin et al., 1966
	Suppression of chlorophyll fluorescence flash																Barabalchuk, 1970
All tissues	Uncoupling of oxidative phosphorylation																Alexandrov and Alekseyeva, 1972
	Leakage of electrolytes																Experiments of Derteva
	Suppression of respiration																Alexandrov, 1955

Fig. 6. Sequence of suppression of cellular functions in a *Tradescantia fluminensis* leaf with increasing temperature of 5-min heating. Explanations in the text

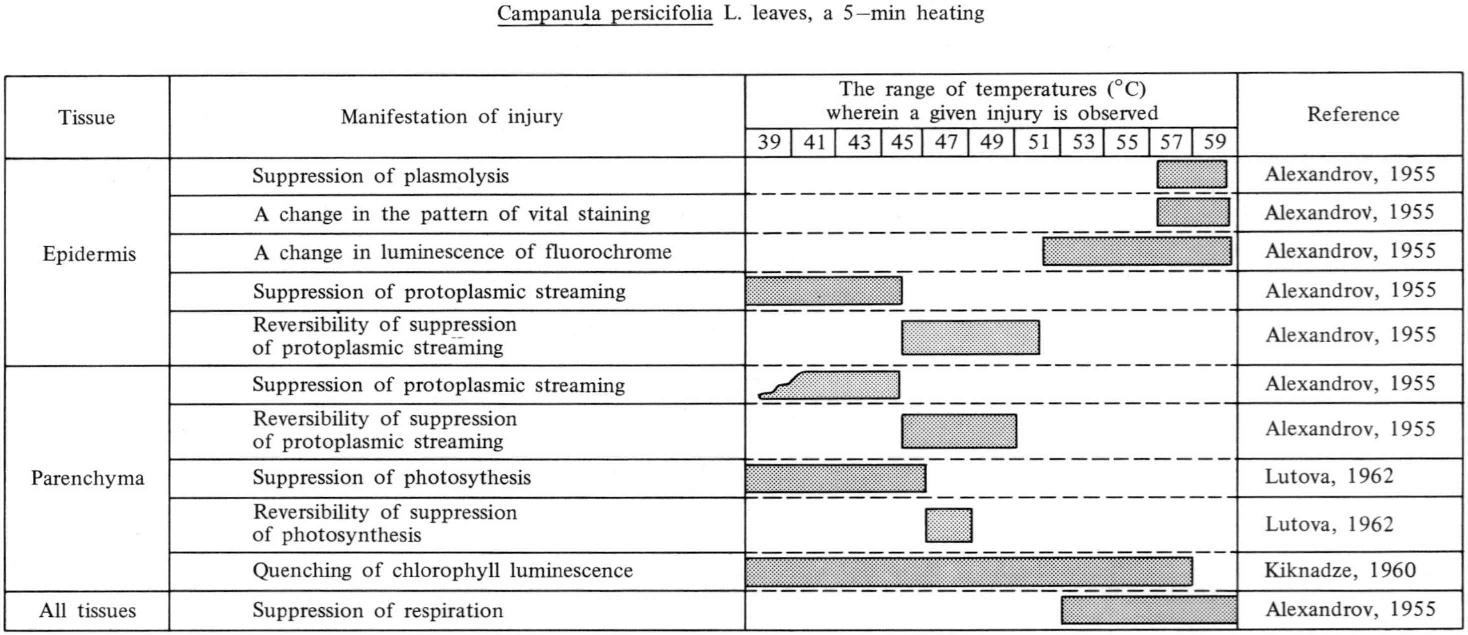

Campanula persicifolia L. leaves, a 5–min heating

Tissue	Manifestation of injury	The range of temperatures (°C) wherein a given injury is observed											Reference
		39	41	43	45	47	49	51	53	55	57	59	
Epidermis	Suppression of plasmolysis												Alexandrov, 1955
	A change in the pattern of vital staining												Alexandrov, 1955
	A change in luminescence of fluorochrome												Alexandrov, 1955
	Suppression of protoplasmic streaming												Alexandrov, 1955
	Reversibility of suppression of protoplasmic streaming												Alexandrov, 1955
Parenchyma	Suppression of protoplasmic streaming												Alexandrov, 1955
	Reversibility of suppression of protoplasmic streaming												Alexandrov, 1955
	Suppression of photosythesis												Lutova, 1962
	Reversibility of suppression of photosynthesis												Lutova, 1962
	Quenching of chlorophyll luminescence												Kiknadze, 1960
All tissues	Suppression of respiration												Alexandrov, 1955

Fig. 7. Sequence of suppression of cellular functions in a *Campanula persicifolia* leaf with increasing temperature of 5-min heating. Explanations in the text

Table 1. Capacity of algae *Chaetomorpha cannabiana* cells for plasmolysis estimated at various times after a 5-min heating. (From Biebl, 1969a)

Time elapsed after heating	% plasmolyzed cells heated at indicated temperatures				
	37° C	38° C	39° C	40° C	41° C
8 min	100	100	100	100	100
3 h	100	100	100	98	95
7 h	100	100	99	90	30
14 h	100	100	90	70	10
1 day	100	95	60	10	0
1.5 day	100	10	10	0	0
2 day	100	9	0	0	0
3 day	100	1	0	0	0
4 day	100	0	0	0	0

the extent of the primary thermal injury and the degree of the destructive after-action, but also by the ability of cells to eliminate the damage produced. If, on the other hand, an extensive thermal injury has been produced, and the cell is unable to repair it, the later the effect of injury is assessed the more dramatic it will be due to progressing development of the destructive after-action (Table 1, Fig. 15a and b).

It is obvious that the shorter the exposure and the sooner the effects brought about by the injury are estimated, the lesser will be the involvement of the destructive after-action and of the mechanisms of cellular homeostasis. Experiments dealing with the thermal effects on cells designed in this way primarily enable an assessment of the resistance of certain protoplasmic components to the immediate action of elevated temperatures. Accordingly, the test employed for evaluating the degree of thermal injury—let it be the protoplasmic streaming, photosynthesis, or plasmolysis, etc.—will restrict the information obtained to particular protoplasmic components. The heat stability of cells, assessed under conditions which minimize the interference of the destructive after-action and of homeostatic processes, is defined as the *primary thermostability*. (The term "primary thermostability" has been used by Lange and Lange (1963) in another sense). It would be more correct to speak of the primary thermoresistance of that life function only, which has been chosen as an indicator of the inflicted injury. In contrast to primary thermostability, that stability which is evaluated in prolonged experiments, where the result depends upon the complicated relationships between destructive and creative activities of the cell, should be defined as the *general cellular thermostability*. [The term "general thermostability" is employed by Christophersen and Precht (1952), and Precht et al. (1955) to designate an elevated stability of the cell against both heat and cold resulting from a decrease in the content of free water in the cytoplasm].

The right rectilinear branches of the curves depicted in Figures 1–3 represent the primary thermostability of cells estimated by the protoplasmic streaming test. It is fairly evident that by recording the time of cessation of protoplasmic streaming in the bluebell *Campanula* cells heated to 43 and 41° C, information is obtained about essentially different aspects of cellular resistance to heat.

Later it will be shown that confusion may arise through neglect of differences in primary and general thermostability. At this point only one example is given. If a comparison of thermostability of epidermal leaf sheath cells in two grasses—millet and orchard-grass—is made by estimating the temperature effective in stopping the protoplasmic streaming in 5-min heating, and by recording the moment of cessation of the streaming immediately after heat exposure, it can be seen that millet cells are the more stable ones, the difference between the two plants being 4.5° C. In the millet cells, protoplasmic streaming stops at 48.5° C, in *Dactylis glomerata* at 44° C. If, however, the heat stability of these cells is estimated by a reversible cessation of the protoplasmic streaming and is evaluated by microscopic technique over several days after 5-min heating, one may conclude that thermostability is equal in both cases, that is, the lowest temperature of heating effective for irreversible arrest of the protoplasmic streaming is 52° C (Alexandrov, 1956).

1.1.2 Biochemical Aspects of Primary Thermoresistance

Belehrádek (1935) was certainly right in saying that the mechanism of injury to cells by heat may differ depending upon intensity and duration of exposure. At this moment we are concerned with the primary alterations of a substrate which occur in the cells during short-term exposures to heat (minutes) and are effective in suppressing this or that cellular function. The whole body of evidence available at present indicates that the rapid thermal damage of cellular functions is due to the thermal denaturation of proteins responsible for maintenance of the function concerned. Here are the facts substantiating this conclusion:

1. Heating results in the appearance of manifestations of injury of the protoplasm similar to those which accompany thermal denaturation of protein solutions, i.e. lowering of dispersion, an increase in viscosity, an enhanced ability to absorb dyes (Nasonov and Alexandrov, 1940; B. Ushakov, 1965a).
2. The temperature coefficient for thermal injury of cells (Q_{10}), or the activation energy (E), displays as high values as the temperature coefficient for thermal denaturation of proteins, being in the order of hundreds and thousands (Belehrádek, 1935; Ushakov and Gasteva, 1953; Alexandrov, 1963; B. Ushakov, 1965a; Yazkulyev, 1969, and others).
3. Those agents and conditions which increase the resistance of proteins to heat confer enhanced resistance to heat also on cells (see Sect. 6.1.1.2).
4. If corresponding proteins derived from analogous cells of two related species are analyzed, provided a difference in heat resistance of the cells exists, one can, as a rule, also detect a corresponding difference in the resistance of proteins in vitro to the denaturing action of heat (see Sect. 3.2.2).

These data substantiate the contention that the primary thermostability of cells is an indirect indicator of resistance of cellular proteins to the denaturing action of heat. The primary thermostability of cells reveals the stability against heat of various proteins depending upon what particular cellular function has been chosen as an indicator of the thermal injury.

When comparing the primary thermostability of the same cellular functions in analogous cells taken from organisms belonging to two different species, or from organisms of the same species but found in different states, we may then with greater certainty conclude whether there is a difference in the thermostability of analogous proteins related to the function under study.

1.1.3 Methods for Evaluation of Primary Thermostability of Cells

Implicit in the definition of the primary thermostability is the contention that the methods adopted for evaluating the primary thermoresistance of cells should avoid prolonged exposure of cells to conditions favorable for active life after the end of a pulse-heat treatment. Many a botanist assesses the resistance of plants or plant parts to heat by measuring an area of macroscopically detectable necrotic spots. Since the appearance of these alterations requires time, an evaluation is usually carried out one to two weeks after cessation of heating. It is certainly evident that results of experiments of this kind will depend not only on the primary injury, but also on the degree of destructive after-action, and on the reparatory capacity of an organism or tissue.

Interference of the destructive after-action with the results of exposure to heat is seen in the experiments performed by Biebl (1969a) with the alga *Chaetomorpha cannabina*. The algae were exposed to heat for 5 min and afterwards, at various time intervals, the capacity of cells for plasmolysis was evaluated. The larger the time period after the end of a heat treatment, the lower were the number of cells able to undergo plasmolysis (Table 1). It has already been noted earlier that the capacity of cells for plasmolysis is a rather thermostable function. In order for this to be suppressed, one has to employ such intensive heating that it eventually destroys the reparatory apparatus of the cell. This accounts for the fact that in the cited experiments performed by Biebl, the state of the cells progressively worsened with time. Employing a similar experimental design, but taking the suppression of a thermolabile function—the protoplasmic streaming—as an indicator of the injury we may obtain an opposite result. Thus, immediately after the end of heating a *Campanula persicifolia* leaf for 5 min at 48° C, we observe cessation of the protoplasmic flow in all the cells inspected. No streaming is detected in 24 h. In two days, the streaming resumes in a minority of the cells, in three days the protoplasmic streaming can be seen in most cells, and, finally, in four days the protoplasm is in motion in all the cells (Alexandrov, 1955). This experiment clearly shows the decisive influence of the reparatory activity of cells on the result of thermal damage. Hence, it is evident that no evaluation of the primary thermostability can be made if the results are checked after an extended period following a heat shock. This holds true also for the experiments in which the resistance to heat of a fertilized embryo cell is estimated by a proportion of deformed embryos, etc.

When dealing with plant tissues, the best evaluation of primary thermostability is to test by protoplasmic streaming, suppression of the capacity for plasmolysis, leakage of antocyan or other substances from cells, a change in fluorescence of chloroplasts, distribution of vital dyes and luminescence of fluorochromes (see: Alexandrov, 1955; Kiknadze, 1960; Barabalchuk, 1970a; Alexandrov and Zavadskaya, 1971; Alexandrov and Dzhanumov, 1972). All of these methods can yield results in appropriate objects within 10–20 min or even faster. One should not employ section techniques when dealing with plant material: the wound injury may shift the level of thermostability of cells (Feldman, 1960). If difficulties arise in microscopic examinations of an intact tissue, then cut-out leaf pieces (not sections) should be used. The most convenient objects that can be tackled with the light microscopic technique are the epidermal leaf or petal cells. In some plants vital microscopy of epidermal cells is almost completely precluded due to the presence of a highly structured cuticle. This obstacle, interfering with light microscopy observations, can be overcome by applying a drop of vaseline or silicone oil to the object, which have refraction indices similar to those of the cuticle (Alexandrov, 1962). Often elimination of air from the intracellular spaces, achieved by infiltration, proves to be helpful in microscopy examination of plant tissues. The easiest way to perform this is to use a syringe. A plant object to be studied is placed in a syringe without piston. Some water is transferred to the syringe, the piston is then inserted, and, closing the outlet of the syringe with the finger, the piston is pulled out to produce negative pressure in the syringe (Alexandrov, 1954).

The primary thermostability of photosynthesis can be successfully estimated in studies on the effects of heat on incorporation of radioactive carbon, because pulse exposures (in the order of minutes) are sufficient. A monometric approach to assessment of photosynthesis and respiration requires prolonged exposures so that the result of heating may be complicated by repair and the destructive afteraction.

Working with animal material, one should take those cells which have an easily observable function. Among these cells are primarily the ciliary epithelium cells and muscle fibers. Thermal injury in the first case can be easily recognized by observing the cells with a light or phase-contrast microscope, and in the second case by application of electrodes to the tissue. The majority of animal tissue cells, however, tend to conceal from an investigator the state of their life. They may be provoked to give an answer with vital dyes, which make granular deposits in actively living cells, whereas in injured ones the cytoplasm is stained diffusely (Nasonov and Alexandrov 1940; Alexandrov, 1949). Among the best dyes—indicators of cell life—are neutral red, new methyl blue N and acridine orange. The pattern of distribution of the latter dye in the cell can be revealed with a luminescence microscope.

Both in animal and plant objects the primary thermoresistance can be investigated with the aid of various cytochemical approaches assessing activity of enzyme systems, provided the methods used do not involve prolonged exposure of cells to conditions compatible with active life of the cells under study.

1.2 The Level of Primary Thermostability of Cells and Habitat Temperature

1.2.1 Effects of Temperature Variations Within the Tolerant Zone on Primary Thermostability of Cells (the Temperature Adjustment)

The necessity for distinguishing shifts in the level of primary thermoresistance occurring during temperature fluctuations within the tolerant zone from those provoked by temperature changes outside this zone, is dictated by differences in the nature of these shifts. This difference can be revealed by comparing the reactions to variations in the environmental temperature manifested by plant and animal cells in organisms belonging to different levels of phylogenetic development.

Abundant literature is available concerning changes of thermostability of microorganisms affected by the growth temperature or by brief exposures to heat. Relevant data have been well summarized in a book by Loginova et al. (1966). There are two reasons, however, for my refraining from discussing these data in this subsection. First, many publications illustrate an extensive use of prolonged exposures of populations to different temperatures, so that it becomes practically impossible to distinguish a modificational aspect of adaptation from a genotypic one, related to selection of genotypes. Moreover, an evaluation of the level of thermostability after a pulse-test heating, is generally performed by observing the growth of the culture heated. Frequently, the criterion for assessing thermoresistance is the ability of microorganisms to survive at supraoptimal temperatures. The outcome of such an experiment is determined by the general thermostability of cells and difficulties arise in estimating the contribution made by primary thermostability. Thus, for instance, Bausum and Matney (1965) reported the resistance of *Bacillus licheniformis* and *B. subtilis* to pulse-heat shocks was greater, the higher the growth temperature had been. The degree of thermal injury of a bacterial suspension has been estimated by the number of colonies grown at a normal temperature. A series of reports demonstrate the dependence of the heat resistance of spores on the temperature at which they had been formed by bacteria. Lundgren (1967), for instance, found that if *B. cereus* forms spores at 20° C, a 10% suppression of the spore outgrowth requires heating for 5 min at 90° C, whereas an exposure for 13 min to 90° C is needed to achieve the same effect when the spore formation had occurred at 40° C. In that work too, the resistance has been estimated by culturing bacteria derived from spores.

1.2.1.1 Protozoa

With regard to evaluation of the primary thermostability, more unequivocal results have been obtained in *Protozoa*. A rather complete coverage of relevant data can be found in a monograph by Sukhanova (1968). Jollos (1913) seems to be the first to have detected a reversible uninherited shift of thermostability of *Paramecium caudatum* that depended on ambient temperature. Serebrovsky (1916) cultured *Paramecium caudatum* at 4.1, 18.5, 25, and 36° C. Thermostability was tested by an average duration of survival of the organisms at 40.5° C. The resistance to heat proved to be higher with more elevated cultivation temperatures. In the coldest culture, death ensued in 15 min, in the warmest one in 48 min. Selection cannot be ruled out in those experiments. Szabuniewicz (1929) worked with pure strains of *Paramecium caudatum*, which he cultured for several weeks or months at different temperatures ranging from 17 to 40° C. With an increase in cultivation temperature, the resistance of the infusoria to heat also increased. Similar findings have been reported for *Colpidium*. Upon transfer of a culture from 40° C to 16–18° C, the temperature maximum for life was found to become lower in five days. Hopkins (1937) cultivated amoebae at 15–20° C and 32° C. In

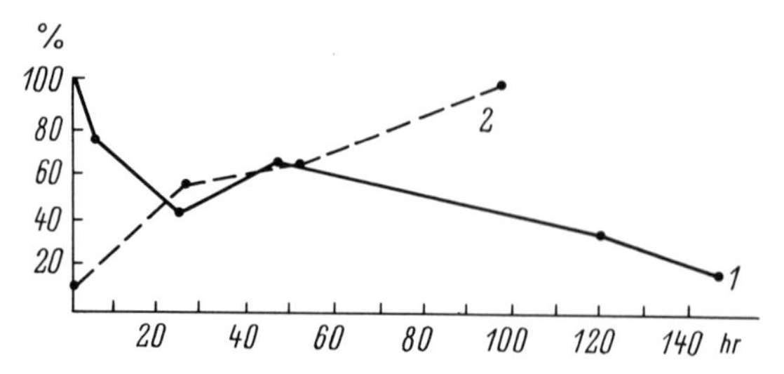

Fig. 8. Changes in heat stability of *Paramecium caudatum* (clone L) upon transfer of cells from 15–18° C to 5–6° C *(1)* and from 5–6° C to 17–18° C *(2)* (Polyansky, 1959). *Abscissa:* duration of exposure of infusoria to new temperature conditions; *ordinate:* survival time at 40° C (as percent of survived control cells kept at room temperature)

the first group, the organisms ceased to move when heated to 43° C, in the second one at 50° C.

The ability of cells to adjust their thermostability with ambient temperature, when the latter is changed within the limits of the tolerant zone, has been designated as the *temperature adjustment* (Alexandrov, 1956). This phenomenon has been extensively studied in *Protozoa* by Polyansky (Poljansky) and his collaborators (Polyansky and Orlova, 1948; Polyansky, 1957, 1959, 1963, 1973; Irlina, 1960, 1964; Sukhanova, 1963, Polyansky and Poznanskaya, 1964; Sukhanova and Poznanskaya, 1966; Polyansky et al., 1967; Polyansky and Irlina, 1973). An experimental design employed in the majority of their studies leaves no doubt as to the modificational nature of an adaptive response, whereas an approach to testing heat resistance ensures that it is the primary thermostability that has been evaluated. The findings obtained in the laboratory of Polyansky might be summarized as follows. The level of the primary thermoresistance of the *Protozoa* studied so far depends on ambient temperature at a given period in their life. Changes in thermostability are adequate to variations in ambient temperature within the whole range of the biokinetic zone. By exposing infusoria to higher or lower temperatures of cultivation, a shift in the heat stability of these organisms can be recognized as early as in several hours (Fig. 8). Transition to a new level of thermostability frequently follows a phase-like, undulating pattern. A new level is attained in several days. The magnitudes of the thermostability shifts, provoked by a transfer to a new temperature environment, may be significantly different in various clones of the same infusoria species.

Consequently, thermostability of *Protozoa* is not a constant parameter, that is, it is directly related to ambient temperature. Shifts of the heat resistance, if they occur, are readily reversible and may take place during the life of a single cell. A temperature adjustment of primary thermostability has been observed in the laboratory of Polyansky in the free-living fresh-water infusoria *Paramecium caudatum, Spirostomum minus, S. ambiguum, Oxitricha minor, Tetrahymena pyriformis.* Similar findings have been reported by Lozina-Lozinsky (1961) for *Paramecium multimicronucleatum* and by Vogel (1966) for the marine infusorium *Zoothamnium hiketes* (Fig. 9). Mikhalchenko (1958) and Sukhanova (1963) have shown that *Opalina ranarum*, a parasite found in frog cloaca, exhibits greater thermostability the higher the temperature at which the frogs had been kept or the culture of the

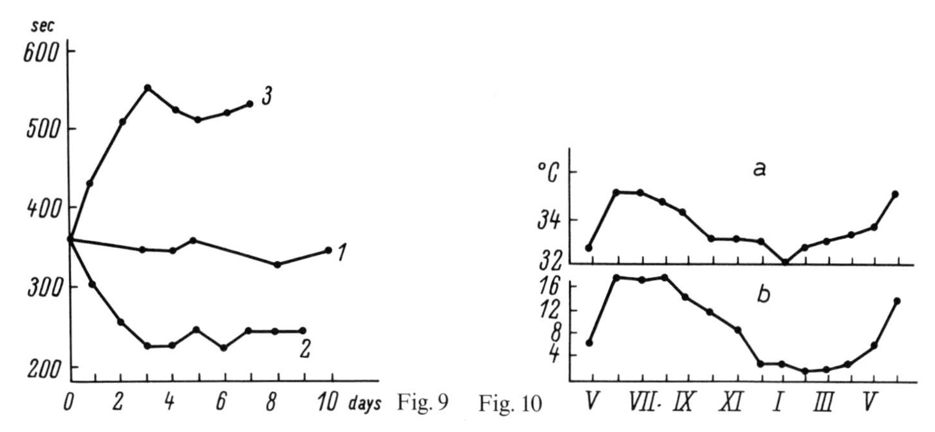

Fig. 9 Fig. 10

Fig. 9. Thermostability of *Zoothamnium hiketes* kept at 15° C *(1)* and that of cells transferred to 5° C *(2)* and 22° C *(3)* (Vogel, 1966). *Abscissa:* duration of exposure of the infusoria to indicated temperatures; *ordinate:* survival time at 37° C

Fig. 10. Seasonal variations of thermostability of *Zoothamnium hiketes (a)* as a function of water temperature variations *(b)* (Vogel, 1966). *Abscissa:* months; *ordinate: (a):* lethal temperature for infusoria; *(b):* water temperature

parasite maintained. Transfer of *Amoeba proteus* from room temperature to 28° C results in their resistance to the lethal temperature (41° C) gradually becoming higher, and, in a day, reaches a new level (Polyansky et al., 1967).

An adequate change in the level of the primary thermostability of *Protozoa* can be detected not only under laboratory conditions, but also in nature. This can be revealed by comparing thermostability of the organisms taken from natural habitats at various seasons (Fig. 10; Sukhanova, 1962a; Vogel, 1966).

In studies on the temperature adjustment of *Protozoa* it was found that a transfer of the organisms from higher temperatures to lower ones resulted in a rapid decrease in the heat stability of the cells. A trustworthy decrease could be observed within hours. Prolonged cultivation of the cells at an elevated temperature makes the enhanced thermostability less readily reversible. The latter persists for weeks and frequently for months after transfer to a lower temperature. During this period the cells can produce up to a hundred agamic generations. Transition from a readily reversible shift to a more or less stable level of thermostability during a prolonged exposure of cells to injurious agents was described for the first time by Jollos (1913), when he studied the effects of arsenic acid on *Paramecium caudatum.* He designated this transition as "a long-lasting modification". Such modifications of the level of heat resistance occurring during long-term cultivation of *Protozoa* at an elevated temperature have been obtained in infusoria (Jollos, 1921; Polyansky and Orlova, 1948; Polyansky, 1957; Crippa-Franceschi, 1970) and in *Heliozoa* (Jollos, 1934). Jollos believed that long-lasting modifications are not related to changes in the genotype, but result from rearrangements within the cytoplasm. This contention seemed to gain support from the findings that the long-lasting modifications disappeared after conjugation, and, moreover, they could be obtained in infusoria clones *(Colpoda steini)* devoid of micronucleus (Schuckmann and Piekarski, 1940). The problem of whether the long-lasting

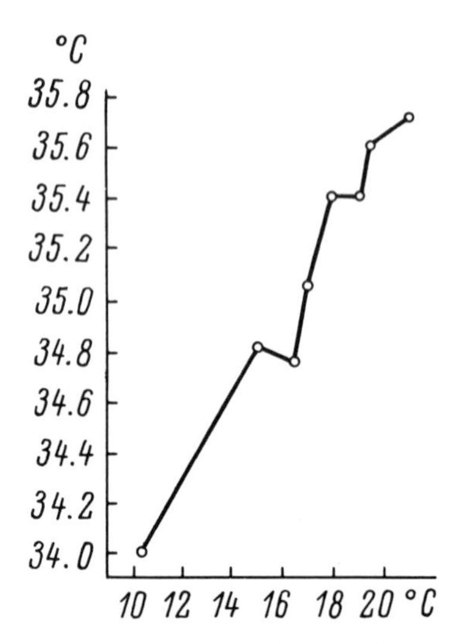

Fig. 11. Changes in thermostability of unicellular alga *Peridinium bipes* in nature (May–June) (Luknitskaya, 1963). *Abscissa:* water temperature; *ordinate:* temperature of a 5-min heating that stops movement of cells

modifications should be regarded as genotypic or non-genotypic adaptations later became rather confused and at present still remains unsolved. Crippa-Franceschi (1970) admits that long-lasting modifications are determined by unknown cytoplasmic extrachromosomal factors. Génermont (1970) regards long-lasting modifications as a consequence of an unapparent selection with respect to heterogeneity of macronuclei that is achieved at the expense of a difference between the division rates of different organisms in a clone. Jinks (1964) believes the long-lasting modifications represent an intermediate stage between phenocopies and extrachromosomal mutations.

1.2.1.2 Algae

Apart from *Protozoa*, the phenomenon of temperature adjustment has also been studied in detail in unicellular and multicellular algae, both in fresh-water and marine species (Feldman and Lutova, 1963). A correlation between thermostability of algal cells and environmental temperature has been repeatedly observed in nature. Prat (1954) showed that the blue-green alga *Oscillatoria princeps* taken from cold waters stops moving at 45° C, but algae of the same species growing in warm waters at about 56° C. It is difficult to say from these data whether the author has dealt with modificational differences or whether there were various ecological races in the water bodies that the studied. More unequivocal conclusions can be drawn from the finding reported by Luknitskaya (1963) concerning the dependence of the primary thermoresistance on the temperature of water in the case of the unicellular fresh-water alga *Peridinium bipes*. In her studies samples of the alga were taken from the same pond over a period from May to June (Fig. 11). Similar evidence has been provided by Schölm (1968) who

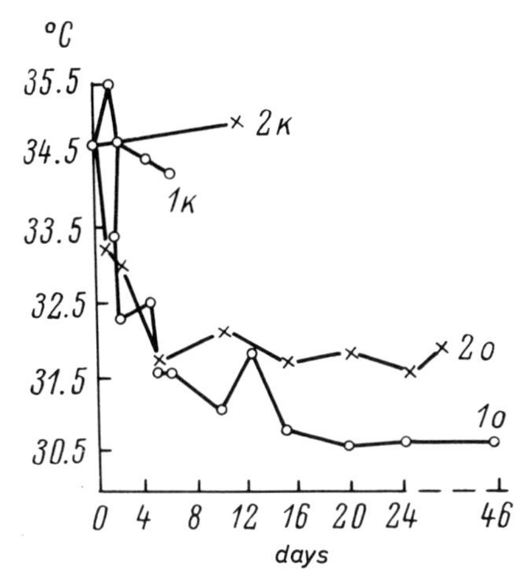

Fig. 12. Changes in thermostability of *Peridinium bipes* as a function of duration of exposure to different temperatures (Luknitskaya, 1963). *Abscissa:* time elapsed from start of experiment; *ordinate:* temperature of a 5-min heating that stops movement of cells. 1_0: culture kept at 2–3° C; 2_0: at 5–6° C; *1k* and *2k*: respective control cultures kept at 18–20° C

studied seasonal changes of thermostability of some fresh-water algae (*Rhizoclonium* sp., *Cladophora glomerata, Cl. fracta; Spirogyra* sp.). In order to suppress their capacity for plasmolysis a more intensive heat treatment was needed in summer than in winter. The difference for some species exceeded 10° C. Similar regularities have been observed also in marine algae: *Fucus vesiculosus, F. serratus, Ascophyllum nodosum* (Feldman et al., 1963; Feldman and Lutova, 1963), *Laminaria saccharina* (Lutova et al., 1968). The heat stability as estimated by changes in the pattern of vital staining of the algal cells proved to be 1–1.5° C higher in summer than in winter. In addition, *F. vesiculosus* and *Pelvetia canaliculata* showed higher thermostability during an ebb in tide pools than during a tide.

A possibility for the temperature adjustment of algae has been convincingly demonstrated in laboratory studies. A 5-min heating was required for suppression of movement of *P. bipes* taken for the experiment from room temperature (18–20° C) (Luknitskaya, 1963). After transfer of the cultures to a lower temperature, the heat stability of the cells grew increasingly less, the lower the ambient temperature had been (Fig. 12). In cells kept at 2–3° C, this decrease was as large as 4° C. Identical results were obtained by Luknitskaya in *Chlamydomonas eugametos* and *Spirogyra* sp. cells. In the latter case, thermostability was assessed by changes in fluorescence. When *Peridinium* and *Chlamydomonas* cells were transferred into cold, a statistically significant decrease in their heat stability could be detected in 4–6 h. Conversely, if the cells were returned from the cold to room temperature, an increase in their thermostability could already be observed within 2–4 h. The same holds true for marine multicellular algae examined under laboratory conditions (Fig. 13). Experimentally induced changes in thermostability produced by fluctuations of ambient temperature within the limits of the biokinetic zone have been reported for *Delessaria sanguinea, Phycodrys sinuosa* (Schwenke, 1959), *Fucus inflatus, Enteromorpha compressa, Porphyra* sp. (Lutova and Feldman, 1960),

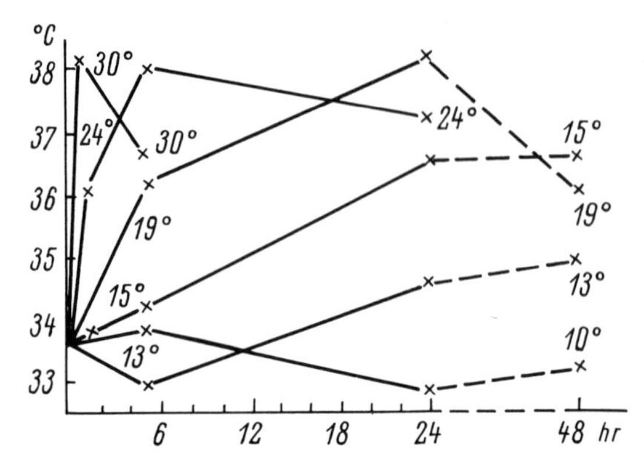

Fig. 13. Changes in thermostability of alga *Porphyra* sp. cells as a function of duration of exposure to different temperatures (Lutova and Feldman, 1960). *Abscissa:* time elapsed from beginning of experiment; *ordinate:* temperature of a 5-min heating effective in changing the normal pattern of staining by New Methylene Blue N into a necrotic one

F. vesiculosus, *F. serratus*, *Ascophyllum nodosum* (Feldman et al., 1963). In marine algae also, one can observe shifts in thermostability within hours after a change in ambient temperature.

These observations testify that unicellular animal organisms as well as unicellular and multicellular algae are able to shift the level of their primary thermostability rapidly when the ambient temperature is changed within the tolerant zone. Tissue cells of multicellular animals and plants, belonging to a higher stage of phylogenesis, display a fundamentally different response to temperature variations within the tolerant zone.

1.2.1.3 Animal Tissue Cells

At first glance the multitude of relevant facts impresses with its extreme diversity and contradictory nature. On closer examination, however, a number of definite regularities can be discerned.

Three major bodies of evidence are available on this topic. One covers comparison of analogous cells taken from organisms belonging to the same species, but living in different biotopes with contrasting temperature conditions. On the other hand, observations are available on changes in the primary thermostability of cells taking place in various seasons. And, finally, the third group of data comprises findings obtained under experimental conditions when the thermostability of cells from animals exposed to different temperatures has been studied.

The earliest findings provided information that can be included in the first group mentioned above. They were cited by the present author in 1952 and illustrated studies on the heat resistance of the ciliary epithelium of the frog *Rana ridibunda* taken from various habitats. It was found that the temperature of a 5-min heating effective in suppressing the ciliary beat was the same for frogs collected in Kharkov, Simferopol, Nalchik and for those taken in a pond here in Leningrad, into which they had been introduced 50 years prior to our experiments (the natural northern boundary of the area for distribution of *Rana ridi-*

bunda lies to the south of Leningrad). These data were not conclusive enough to establish the ability of animal tissue cells to change the level of their thermostability in response to variations in ambient temperature. One could admit that these shifts were reversible and that keeping frogs for a period of time at a constant temperature tended to equalize thermostability of their cells.

A series of publications appeared subsequently from the laboratory of B. Ushakov, which referred to field studies confirming the constancy of the level of thermostability of tissue cells in a given animal species exposed to different ambient temperatures. Thermostability of muscles isolated from the frog *Rana ridibunda* in the animals taken in the Moscow region, Pyatigorsk, Baku, Latvia, Astrakhan and Armenia proved to be the same (B. Ushakov, 1955, 1960b). Identical results of were reported also in studies of thermostability of spinal ganglia cells and cartilages from frogs in Latvia and Armenia (B. Ushakov, 1960b). Thermostability of the ciliary epithelium of *Rana ridibunda* from a hot spring in Zheleznovodsk proved to be the same as that of the ciliary epithelium from frogs collected in the localities listed above (S. Alexandrov et al., 1960). That was the case also for the spermatozoa of *Rana ridibunda* living in hot springs; the heat stability of the spermatozoa was identical to that of spermatozoa from frogs in the Moscow region, Erevan, Tashkent and Frunze (Svinkin, 1959, 1962a). Further studies reported by Glushankova and Chernokozheva (1971) revealed that in August there was no difference in the heat stability of muscles isolated from *Rana ridibunda* living in hot springs in Zheleznovodsk (water temperature: 30–33° C) and that for the same species of frogs taken from ponds in Kislovodsk (water temperature: 17–19° C). Nevertheless, during May–June, the muscles isolated from the Kislovodsk frogs proved to be less resistant to heat.

Furthermore, identical thermostability was found in muscles of two subspecies of the toad *Bufo bufo*: one from Leningrad, another from Sukhumi (the Caucasus) (B. Ushakov, 1955); of the lizards *Phrynocephalus helioscopus* taken from habitats differing in temperature conditions (B. Ushakov, 1962); of the lizards *Lacerta strigata* which live at altitudes of 800 and 2000 m above sea level; and of *Agama* from altitudes of 800 and 1300 m (B. Ushakov, 1958). In the Baikal Lake area, Kusakina (1962a) found no differences in thermostability of muscles from the fish *Crucian* and *Cobitis taenia sibirica*, which live in hot springs at temperatures of 30–34° C and at normal water temperatures. No difference could be detected in the thermostability of muscles and ciliary epithelium of various fresh-water and marine molluscs taken from biotopes characteristic of different temperature conditions. Thus, for instance, muscles of *Radix ovata*, taken from a hot spring (26–27° C) and from a depth of 20 m in the Baikal Lake, where the temperature is about 6–8° C, exhibited identical heat-stability (Dzhamusova and Shapiro, 1960). Ciliary epithelium of the mussel *Mytilus grayanus* from the Sea of Japan, where these organisms live at depths of 1–60 m and at temperatures differing by 8–10° C, was seen to arrest its ciliary beat at the same temperature of heating. Similar thermostability was recorded for the ciliary epithelium of ascidias and for the sea urchin muscles taken from organisms living at different depths in the sea (Zhirmunsky and Pisareva, 1960a, 1960b; Zhirmunsky, 1963). Again, similar data can be found concerning the ciliary epithelium and muscles of other molluscs referred to by Zhirmunsky (1960a) and Dzhamusova (1960a, 1960b,

1960c, 1961, 1963, 1965a). Essentially similar observations have also been reported for the cells of other groups of poikilothermal animals: in crawfish (B. Ushakov, 1956a), in leech muscles (B. Ushakov, 1956b), and in the sea urchin spermatozoa (Andronikov, 1968).

One can discover in the literature a number of findings indicating similar thermostability of animal tissue cells recorded in cold and warm seasons. As early as 1919, Thörner observed an arrest of nerve conductivity in winter frogs at 34.8° C, when he heated a nerve-muscle preparation isolated from a frog at the rate of 1° C per 30 sec. In summer frogs no response could be recorded even at a lower temperature (33.4° C). A comparison has been made of the heat-resistance of the ciliary epithelium from the frog *Rana temporaria*, and from the molluscs *Anadonta cygnea* and *Unio pictorum* in winter and summer. Both the animals themselves and tissues isolated from them have been kept in winter at zero degrees prior to testing for thermostability. Despite this experimental design, the level of thermostability of the cells was found to be the same in winter and in summer. No seasonal variations in thermostability were detected by Mikhalchenko (1958, 1959) in the ciliary epithelium of frogs. Arronet (1959) confirmed that there was no difference in the heat-resistance of the ciliary epithelium of *Rana temporaria* and *Unio crassus* in winter and in summer. He found, however, that thermostability of the animals themselves was appreciably lower in winter than in summer.

Observations reported in other experimental studies are in agreement with those referred to above. Licht (1967) kept the lizards *Uma notata* for three weeks at 36, 18, and 10° C. This pretreatment did not affect the resistance of muscles to heat. Mikhalchenko (1956, 1958), exposed adult frogs to 26, 17, and 3° C for a month or longer, and failed to detect any shift in the heat stability of the ciliary epithelium cells. Zhirmunsky and Shlyakhter (1963) exposed frogs in winter time to temperatures of 2 and 18° C. The resistance of the organisms to heat changed proportionally; however, no difference in heat stability of the ciliary epithelium cells appeared. Similarly, thermostability of spermatozoa taken from frogs kept for a month at 2 and 23° C was the same in both instances (Svinkin, 1959). This was also the case when thermostability of *Unio crassus* ciliary epithelium was compared in the molluscs that had been exposed to 1–2° C and 17–18° C for four months (Arronet, 1959). Rumyantsev (1960) cultured frog myocardium cells at 26, 18, and 11° C: the thermostability of the cells was unchanged.

Experiments performed with other groups of poikilothermal animals gave similar results. Ushakov and Kusakina (1960) kept the leeches *Hirudo medicinalis* during an autumn–winter period at 3, 19, and 32° C. The heat stability of the "warm" leeches increased appreciably, whereas that of the muscles isolated from the organisms showed no dependence upon the cultivation temperature. Actinia were able to double their thermoresistance during a 7–10-h exposure at 15–20° C as compared to those left at 5–7° C. Stability of their ciliary epithelium against heat, however, remained at the original level (Zhirmunsky, 1959). In this context, the results obtained by Amosova (1963) are very convincing: the meat fly was allowed to grow for many months at 17 and 24° C. Seven generations appeared during this period, but the heat stability of muscles was not modified. The same results have also been reported for the muscles of the fly larvae that have developed at 14 and 34° C. The larvae themselves that developed at an elevated temperature exhibited higher resistance to heat as compared to their "cold" counterparts. And, finally, a reference should be made to the experiments reported by Konev and Burtseva (1970) who transferred the mollusc *Littorina mandzhurica* from 20 to 28° C. Thermostability of these organisms was found to be higher, whereas that of the muscles isolated therefrom remained unaltered.

The capability of organisms to adapt their thermoresistance to changes in the temperature has been reported for various animal types (Shkorbatov and Salo, 1959; Brattstrom and Lawrence, 1962; Bowler, 1963; Precht, 1964a; Sergeeva, 1969; Furch, 1972; and others). Interestingly, despite this fact one could frequently observe, either in different seasons or under experimental conditions, no influence of the cultivation temperature on the temperature preferred by the animals, although the preferred temperature in different animal species corresponded to their temperature ecologies (Graevsky, 1946; Licht, 1967; Gromysz-Kolkowska, 1967). (The preferable temperature is estimated in the Herter apparatus, in which a temperature gradient is produced and the temperature in the zone of a maximal accumulation of organisms is recorded.)

It can be concluded unequivocally from the data presented that tissue cells of multicellular organisms, in contrast to those of *Protozoa* and algae, are incapable of temperature adjustment, i.e. of changing their primary thermoresistance if exposed to wide-range temperature fluctuations in the environment which frequently exceed the limits of the tolerant zone. Multicellular animal organisms themselves, as opposed to their constituent cells, but similar to *Protozoa* cells, readily modify their resistance to heat during variations in the temperature environment. The entire picture, however consistent it may seem, becomes distorted by a series of apparently contradictory findings, which deserve consideration in order to evaluate the degree of their interference with the validity of the conclusion stated above.

Ushakov and Sleptsova (1968) repeated their experiments with leeches in spring, and, in contrast to the evidence obtained previously (B. Ushakov and Kusakina, 1960) were able to detect higher thermostability of muscle tissue in those animals which had been kept for 15–40 days at 30° C, compared with those maintained at 3° C. This difference, however, could be revealed only if heat exposures of mild intensity were employed, during which the loss of muscle excitability developed for several hours. With respect to short-term intensive heatings effective in suppressing the excitability within half an hour, no difference in the heat stability of muscles isolated from the "cold" and "warm" leeches could be found. Similarly, Ushakov and Zander (1961) compared the excitability of muscles isolated from *Rana ridibunda* living in normal and hot waters. The temperatures at which the loss of excitability produced by intensive heat treatments occurred proved to be the same in both instances. Nevertheless, muscles from the warm frogs were found to be more resistant than those from the "cold" ones against mild heat exposures, which while not completely inhibiting excitability of the muscles, interfered with this process (increased the rheobase, shifted the curves for the potential vs. time dependence).

These data suggest that during changes of ambient temperature within the tolerant zone the primary thermoresistance of tissue cells is not modified. The general thermostability of cells may display, under these conditions, an adaptive shift, which is apparently determined by the effects of the cultivation temperature not on the stability of protoplasmic components, but on the reparatory capacity of the cells. The evidence gained in plant cells will be presented in favor of this thesis (see p. 36).

Consequently, the cells, when tested for their ability for temperature adaptation, display different behavior depending upon whether an intensive or mild heat treatment is being used for assessing their sensitivity to heat. Therefore, extreme caution should be exercised in considering the positive data when evaluating the capacity of tissue cells for temperature adjustment of their primary thermostability. Schlieper (1960) and Vernberg et al. (1963) kept littoral and sublittoral molluscs at 10 and 25° C. They did not observe any changes in the heat-resistance of the gill cells of the ciliary epithelium during acclimation of sublittoral molluscs, but they reported changes occurring during acclimation of the littoral species. It is possible that an enhancement of the reparatory capacity of cells of the "warm" molluscs was responsible for the changes detected. In the experiments on thermal adaptation of muscles in fish exposed to different temperatures reported by Precht (1964a, 1964b) and Berkholz (1966), an experimental design was used that should not be employed for evaluating the level of the primary thermoresistance of a tissue.

The evidence mentioned above shows the absence of any differences in the heat stability of tissue cells isolated from animals during warm and cold seasons. On the other hand, detailed investigations have confirmed the existence of regular seasonal variations in the thermostability of animal cells. An analysis of these observations revealed, however, that seasonal fluctuations in thermostability are, in fact, the reactions of animal cells to hormonal shifts occurring in the organisms, and do not represent an immediate response of the cells to changes in ambient temperature. This conclusion is supported by a wealth of experimental evidence.

Stier and Taylor (1939) reported that cessation of the frog heart beat ensued at a lower temperature in winter than in autumn and spring. When pituitary bodies were implanted into the abdominal cavity of winter frogs, the heat-resistance of the heart increased up to the level characteristic of the autumn–spring animals. These authors consider the effect observed to be due to stimulation of the thyroid gland, whereas seasonal variations of thermostability of the heart they regard as a response to a disbalance of the hypophysis/thyroid gland interrelationships. As a confirmation of this conclusion they refer to a publication by Carter (1933), who observed a return of the temperature dependence of the heart-beat rate from the winter level to that typical for summer frogs, if isolated hearts, taken from winter frogs, were perfused with Ringer solution plus thyroxin. Hajdu (1951), working with *musculus sartorius* isolated from the frog *Rana pipiens* found that the temperature optimum for isometric tension, produced by an electric stimulation, was 20° C in summer and 15° C in winter. If the summer frogs were kept for 6 weeks at 6° C, the plot of temperature dependence of the tension resembled that of the winter animals.

Later, in the laboratory of B. Ushakov, seasonal variations in thermostability of various tissue cells were studied in various animal species, and a correlation between these fluctuations of thermostability and changes in the functioning of the endocrine system was analyzed. In the first publication of this series,

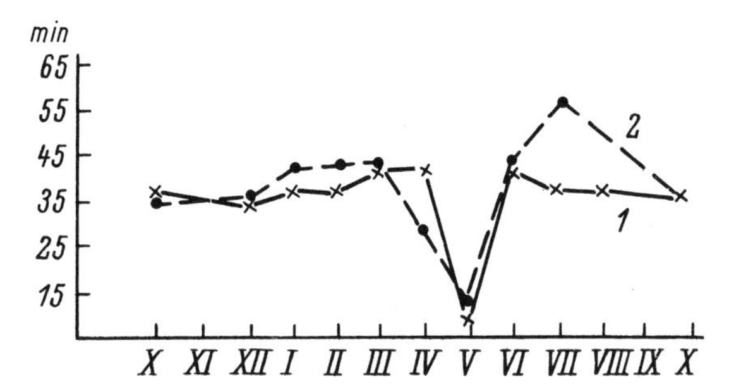

Fig. 14. Seasonal changes in thermostability of frog muscle tissue in animals in nature *(1)* and in those exposed for two weeks to an elevated temperature *(2)* (Pashkova, 1963b). *Abscissa:* months; *ordinate:* persistence of muscle excitability in Ringer solution at 36° C (*Roman numbers:* months of year)

Shlyakhter (1961) demonstrated an enhancement of the heat-resistance of the *Rana temporaria* muscles in spring and summer as well as a decrease of thermostability below the winter level after a period of reproduction. Injections of hypophyseal extractions into females in winter induced spawning and subsequent reduction in the heat stability of muscles. If the frogs collected late in January were maintained at room temperature, by the end of February they displayed an increase in thermostability of their muscles, which reached the level typical for the frogs taken from nature in March. The endocrine nature of the changes in the level of thermostability of frog cells has been confirmed in a series of observations reported by Pashkova (1962a, 1962b, 1963a, 1963b, 1965), by Dzhamusova and Pashkova (1967) in muscles, as well as by Vainman (1966) in spermatozoa.

Figure 14 depicts a curve of seasonal variations of thermostability of *R. temporaria* muscle tissues that differs somewhat from the curve reported by Shlyachter. It should be noted that the level of thermostability is the same in winter and summer, decreasing during the reproductive period. Administration of pituitary hormones to the winter frogs diminished the heat-resistance of muscles, whereas feeding frogs with thyroidin and injections of the thyrotropic hormone tended to elevate the muscle thermostability. In immature frogs, spring shifts of thermostability of tissues do not occur, and their thermostability cannot be modified by administration of an hypophyseal extract. The dashed line in Figure 14 shows the effect of exposing frogs for two weeks to elevated temperatures (to 16–18° C in winter and 26–28° C in summer) upon the heat stability of muscles in various seasons. It can easily be seen that, for most of the year, such a 2–week exposure of the animals to a temperature exceeding the initial one by 10–12° C does not affect the cellular thermostability level. The shifts observed in April (a decrease) and in July (an increase) are readily explained as due to changes in the activity of the thyroid gland. In thyroidectomized animals, variations of ambient temperature do not affect thermostability of their tissues (Pashkova, 1965).

A somewhat dissonant note to this uniform complex of experimental evidence is given by the work of Dzhamusova (1971) performed in the same laboratory. In her experiments, thyroxin injections given to frogs resulted in enhancing the heat stability of one group of muscles, at the same time reducing it in the other group.

Zhirmunsky and Shlyachter (1963) found that exposure of the winter frogs to 2 and 18° C for 20–25 days produced no difference in thermostability of the ciliary epithelium and muscles. If, however, the summer frogs, which exhibited an elevated thermostability of their tissues, were used in the experiments, a difference could be observed. Thermostability of muscle tissues from the summer bellied toads *Bombina bombina* was reduced when the animals were transferred to cold (for 15 days at 5–8° C) (Ushakov and Glushankova, 1970). Juvenile frogs display a pattern of seasonal thermostability of muscles differing from that of adult frogs. In addition, it does not correspond to seasonal variations in ambient temperature (Chernokozheva, 1967b).

An illustration of the endocrine–related nature of the level of cellular thermostability is provided by the results of the experiments reported by Pashkova (1962b). She succeeded in obtaining a decrease in thermostability of frog muscles when the animals had been frightened by a grass-snake. The author explains this phenomenon by a massive release of adrenalin into the blood. The significance of the endocrine system for temperature adaptations of *Rana temporaria* has been demonstrated by Jankowsky (1960). The frogs kept in the cold (7° C) exhibited a compensatorily intensified respiration activity as compared to that in the animals maintained at 25° C. The rate of respiration has been estimated in both groups of animals at the same temperature. After thyroidectomy, respiration in the "cold" frogs fell to the level reported for their "warm" counterparts. Moreover, in the "cold" frogs, an adaptive enhancement of glycine incorporation into muscle proteins was found to be inhibited after extirpation of the anterior lobe of hypophysis.

Changes of the cellular thermostability level determined by the sex-cycle-dependent events occur also in other animals: in lizards (B. Ushakov, 1963d), fish (Altukhov, 1963), molluscs (Dregolskaya, 1963b; Friedrich, 1967), sea urchins (Zhilenko and Vasilyeva, 1969), actinia (Dregolskaya, 1962). In addition to this group of findings one may, presumably, also add the data communicated by Gorodilov (1961) and Ivleva (1964) for sea worms.

In the warm-blooded Siberian marmot and field mouse, Skholl (1965, 1970) detected a regular decrease in the heat resistance of muscles in winter and an increase in summertime when kept at a constant temperature.

Ample evidence is available in the literature on adaptation of the level of thermostability of animal tissue cells to fluctuations in environmental temperature (Mikhalchenko, 1956; Shkorbatov and Kudriavtseva, 1964; Ushakov et al., 1971, 1972; Ushakov and Pashkova, 1972; and others). Considering the experimental evidence presented above, doubt still remains as to whether the described adaptive modifications of thermostability might be viewed as constituting a direct cellular response to a temperature shift. It is highly likely that the investigators referred to in the foregoing discussion were dealing with reaction of cells to shifts in the hormonal balance. Furthermore, one cannot be sure that it was the primary thermostability of the cells that was assessed. In this respect, the numerous negative data prove to be much more convincing when they substantiate the concept of the constancy of the level of the cellular primary thermostability, even against the background of prolonged and appreciable deviations of ambient temperature to the lower and upper extremes. Such negative data convince us that animal cells have lost the capacity for temperature adjustment characteristic of unicellular organisms and algae. Different responses of animal tissue cells and *Protozoa* to temperature fluctuations are vividly demonstrated by Sukhanova (1962a, 1962b, 1963) in her experiments with *Opalina* and infusoria, which parasitize in a frog's intestine. When, for instance, winter frogs are brought to room temperature, the heat resistance of the host tissues cells remains unaffected, whereas that of the parasitizing *Protozoa* markedly increases. All this taken together forces one to agree that the conclusion, suggested by Kamshilov (1960) that "tissues, like integral organisms, are capable of the temperature adaptation" (p. 232) is not consistent.

1.2.1.4 Plant Tissue Cells

In the literature, evidence concerning variations of the thermostability level of plant tissue cells provoked by changes in ambient temperature and occurring within the limits of the tolerant zone, is not as vast, but no less diverse. There are data that show plants of the same species, growing under different temperature conditions, differ in the heat stability of their tissues. Lange and Lange (1962) observed in a number of species that thermostability of leaves, taken from plants growing at high altitudes was lower than that of the leaves from the same plant species found in the plains. In some lichen species, Lange (1953) was able to demonstrate that, within one and the same species, a correlation exists between the thermostability of a lichen and habitat temperatures. In other species, however, despite essentially different environmental temperatures for different populations, individual organisms in those populations exhibited practically no difference in their thermolability. Lange (1955) also reported similar findings for the mosses: in eleven species the difference in the heat stability of individual plants of a given species, grown under contrasting temperature conditions, was found to be "surprisingly small". One species—*Hypnum cupressiforme*—was an exception: in this case thermostability of the plant depended on ambient temperature. Mooney and Billings (1961) found that the *Oxyria digyna* leaves taken from plants of a southern population exhibited greater resistance to a 40° C heat treatment than did the leaves in a northern population.

Positive evidence of this kind, regarded as supporting the idea of phenotypic lability of the cellular thermostability level, must be considered with a certain degree of caution. It cannot be ruled out that various biotopes may be inhabited with genetically different ecological races. Experimental findings are more convincing in this regard. Here again, certain complications in their interpretation may arise, as will be shown later in this book.

Contradictory findings have been obtained already in the experiments made by Sapper (1935). She cultivated *Elodea* and *Vallisneria* at 9–15° C and 25–30° C. The "warm" *Elodea* proved to be more heat resistant than the "cold" one. *Vallisneria*, on the other hand, yielded opposite results: the "warm" plants displayed somewhat lower stability against heat. Kappen (1964) maintained the fern *Blechnum spicant* for 3 weeks at 25, 16, and 4° C. In the first two groups of plants, thermostability was the same, whereas in those exposed to cold it turned out to be even higher. Lange (1962a) reported different results. He grew *Commelina africana, Phoenix dactilifera* and *Veronica persica* for 5–10 weeks at 28, 18, and 8° C.

Table 2. Thermostability of epidermal cells of the higher plants taken from different habitats

Species	Habitat	Minimal temperature of a 5-min heating which stops protoplasmic streaming	Reference
1	2	3	4
Allium altissium Rgl.	Leningrad	43.7	Experiments of Kamentseva
	Ashkhabad	43.5	Experiments of Kamentseva
A. giganteum Rgl.	Leningrad	43.1	Experiments of Kamentseva
	Ashkhabad	42.9	Experiments of Kamentseva
A. paradoxum Don.	Leningrad	43.3	Experiments of Kamentseva
	Ashkhabad	43.0	Experiments of Kamentseva
A. schoenoprosum L.	The Taimyr Peninsula	47.0	Experiments of Alexandrov and Shukhtina
	Leningrad	46.6	Experiments of Alexandrov and Shukhtina
Catalpa speciosa Warder.	Leningrad	48.3	Shukhtina, 1965
	Ashkhabad	48.3	Shukhtina, 1965
Chrysosplenium alternifolium L.	The Taimyr Peninsula	42.2	Experiments of Alexandrov and Shukhtina
	Leningrad	41.7	Experiments of Alexandrov and Shukhtina

In his experiments the leaves of those plants which had been exposed to an elevated temperature exhibited higher thermostability.

The data accumulated in our laboratory brought us to a conclusion that in higher plants the cells which have completed their growth do not change their primary thermostability, provided variations of ambient temperature are within the wide limits of the biokinetic zone. This contention has been substantiated primarily by the evidence concerning similar cellular thermostability in plants of one species, but growing in different sites of an area and exposed to contrasting temperature environments (Table 2). This table, however, does not show observations on the cellular thermostability of the Middle Asian plants at periods of intensive heat, when the temperature maximum exceeds the limits characteristic of a given species. Further details of this phenomenon will be given later. Antropova (1971) evaluated the resistance of the moss *Mnium affine* cells to a 5-min heating by the capacity of the cells for plasmolysis. During the August–September period, the air temperature varied considerably. Antropova subdivided the data

Table 2 (continued)

Species	Habitat	Minimal temperature of a 5-min heating which stops protoplasmic streaming	Reference
1	2	3	4
Dactylis glomerata L.	The Kola Peninsula	44.9	Alexandrov and Feldman, 1958
	Leningrad	44.7	Alexandrov and Feldman, 1958
	Dushanbe	44.7	Alexandrov and Feldman, 1958
Gagea lutea (L.) KerGawl.	Leningrad	42.3	Experiments of Feldman and Lutova
	Tbilisi	42.4	Experiments of Feldman and Lutova
G. minima (L.) KerGawl.	Reservation "Les na Vorskle"	42.0	Experiments of Feldman and Lutova
	Tbilisi	42.0	Experiments of Feldman and Lutova
Morus alba L.	Leningrad	47.0	Shukhtina, 1965
	Ashkhabad	47.1	Shukhtina, 1965
Sternbergia lutea L.	Leningrad	44.7	Experiments of Feldman and Lutova
	Tbilisi	44.6	Experiments of Feldman and Lutova
Swertia marginata Schrenk.	The Pamirs, 4800 m	45.0	Experiments of Alexandrov and Denko
	The Pamirs, 3860 m	44.7	Experiments of Alexandrov and Denko
Zostera marina L.	The White Sea	36.8	Feldman and Lutova, 1962
	The Tartar Strait	36.8	Feldman and Lutova, 1962

she obtained into three groups. The first comprised findings on plants taken from nature, when the air temperature was 2–9° C; the second embodied observations on plants taken at an air temperature of 10–17° C; and the last group covered the plants collected at 18–22° C. Thermostability in all three of the groups was found to be the same. A somewhat different result was reported by Dircksen (1964) for two species of water mosses—*Fontinalis antipyretica* and *Platyhypnidium riparioides*, which grow in springs where the water temperature (average annual temperature) is 9° C ±1°. These mosses had been removed from a spring and placed for three weeks at 16–19° C and 2–4° C (in the latter case they terminated their growth but appeared unaffected). As a result, both in the "warm" and "cold" mosses, thermostability proved to be 1.5–2.0° C higher than that of naturally growing mosses.

When discussing the problem of thermostability of animal tissue cells as related to environmental temperature, we make extensive use of the evidence accumulated in studies concerning seasonal variations in cellular thermostability.

Table 3. Comparison of the cellular thermostability of plants exposed for various time periods to 20° (W) and 10° (C). (From Lutova and Zavadskaya, 1966)

Species	Time elapsed from the beginning of the experiment (days)	Temperature of a 5-min heating which stops protoplasmic streaming, °C	
		W	C
Triticum aestivum	8	43.2	43.4
	18	44.2	44.1
	26	—	45.1
Pisum sativum	14	42.3	42.2
Tradescantia fluminensis	15	45.6	45.9
	35	45.0	45.7[a]
	55	44.7	44.9
	84	44.2	45.1[a]

[a] Values in C which statistically significantly differ from corresponding values in W.

With respect to plants, this approach is fraught with complications since, as winter comes closer, many a plant tends to enter a specific state that is associated with cold-hardening and the rest period. These states may be accompanied by general stabilization of cells, and more specifically, with enhancement of the cells' resistance to heat. The problem of seasonal variations of thermoresistance of plant cells will be dealt with elsewhere in this book (see Sect. 1.2.2.3).

The constancy of the cellular thermostability level observed in plants exposed to appreciably differing temperature environments, but within the limits of the biokinetic range, has been confirmed in a number of experimental studies (Alexandrov, 1956, 1964a; Alexandrov and Feldman, 1958; Denko, 1964; Lutova and Zavadskaya, 1966; Lomagin and Antropova, 1968; Antropova, 1971). These observations made in many plant species contradict the data cited above referred to by Lange (1962a). The reliability of the results reported by Lange is certainly beyond any doubt, therefore it is suspected that the cause of this contradiction lies in the differences in the objects studied and/or in the experimental approaches employed. In the work of Lange, thermostability was assessed macroscopically by the appearance of necrotic lesions in leaf plates during the 6–12 days following a 30-min heat shock. By using such an experimental design, the result of heating not only becomes determined by the primary thermoresistance of cells, but also by the destructive after-action and the cells' reparatory capacity. On the other hand, in our experiments with pulse-heat shocks, followed by estimation of the effects by the criterion of the protoplasmic streaming or plasmolysis, it was the primary thermostability that was assayed. In order for these discrepancies between the data of Lange and that mentioned earlier to be resolved, Lutova and Zavadskaya (1966) carried out a specially designed research project. Wheat and pea plants were grown for 18 and 14 days at 10 and 20° C, respectively. In the "cold" and "warm" wheat, a 5-min heating stopped the streaming of protoplasm in epidermal cells at 44.1 and 44.2° C; in the pea plant cells at 42.2 and 42.3° C (Table 3). No shift in the heat stability was detected. The next step was to cultivate *Tradescantia fluminensis* at 28, 20, and 10° C. After varied exposures of the plants to different

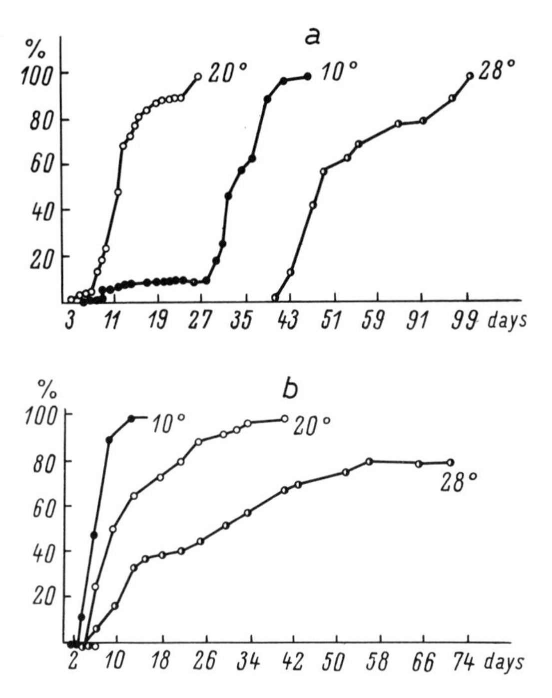

Fig. 15 a and b. Changes in extent of *Tradescantia* leaf injury following exposure for 30 min to 46° C as function of duration of plant growth at different temperatures and as function of time period elapsed from end of heating (Lutova and Zavadskaya, 1966). *Abscissa:* time period elapsed from end of heating; *ordinate:* area of leaf plate with coagulated protoplasm. Length of cultivation period: (a) 15 days, (b) 57 days

temperatures (up to 84 days), thermostability of the cells was evaluated by two methods: (1) by that used in the author's laboratory, i.e. immediate assessment of the effect after the end of a short-term (5–40-min) heating—primary thermoresistance, and (2) by the method of Lange—general thermostability. In the latter case, an increase in heat-stability could be observed with prolongation of exposure of the plants to elevated temperatures: fully conforming with the data reported by Lange (Figs. 15a and b). No increase could be detected when the former method was used for estimating the primary thermostability of cells, in the "warm" plants (Fig. 16). On the contrary, the heat-resistance of the "cold" wheat cells was found to be enhanced, whereas that of the "warm" *Tradescantia* was reduced (Table 3; Fig. 16).

From the data obtained by Lutova and Zavadskaya it can be concluded that the primary thermostability of cells, as shown earlier, is, in fact, a rather constant parameter, not modified by changes in ambient temperature occurring within the biokinetic range. On the other hand, the experiments of Lutova and Zavadskaya described above made us suspect that processes of repair of thermal injury become activated during cultivation of plants at an elevated temperature.

Assessment of the capacity of cells to repair thermal injury can be made using two approaches: first, the rate of repair of a function can be estimated when it has

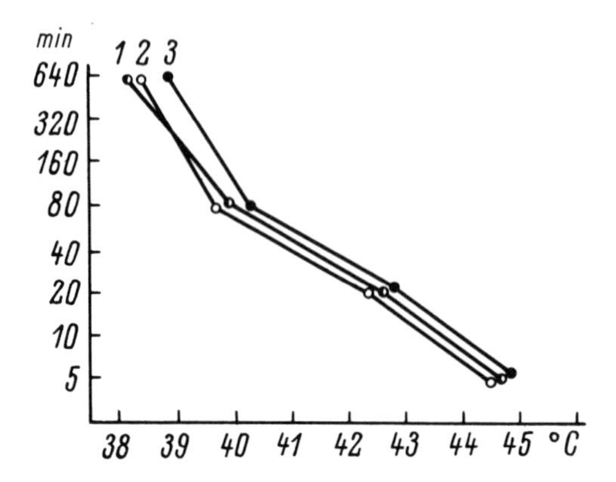

Fig. 16. Effect of cultivation temperatures on thermostability of *Tradescantia* leaf epidermal cells (Lutova and Zavadskaya, 1966). *Abscissa:* temperature of heating; *ordinate:* time period during which protoplasmic streaming persisted (log scale). Cultivation temperatures: *1:* 28° C, *2:* 20° C, *3:* 10° C. Plants were grown for 55 days

been suppressed to a certain degree; second, the *temperature zone of repair* can be determined, i.e. that portion of the temperature scale within which an inhibition of a given function, produced by a heating of specific duration, proves to be still reversible (Alexandrov, 1964b). In the experiments of Lutova and Zavadskaya, a 5-min heating at 45.1° C completely suppressed the streaming of protoplasm in *Tradescantia* leaf cells after the plants had been grown for 2–3 months at 10° C; the maximal temperature of heating, after which the streaming could be resumed was 49.7° C. Consequently, the reparatory zone for these plants amounted to 4.6° C. For the plants grown at 20° C this value was 1.4° C higher, and in those cultivated at 28° C the reparatory zone was larger by 2.7° C, namely as great as 7.3° C. The data obtained by Alexandrov and Barabalchuk (1972) agree with these findings: after a 3-h treatment of *Tradescantia* leaves at 38° C, the rate of repair of the protoplasmic streaming, inhibited by heat shocks, accelerates. Gorban (1974) found that a preliminary heat treatment of *Zebrina* and *Podophyllum* leaves carried out for 20 h at 30° C did not change the cellular primary thermoresistance. This pre-treatment, however, enhanced the rate of repair in the cells after a subsequent intensive heat shock and widened the reparatory zone.

In summary, it can be said that despite unchanged primary thermoresistance of plant cells, a significant alteration of the efficiency of the cells' reparatory mechanism may occur. This modification is dependent on the cultivation temperature or it may follow a preliminary short-term heat treatment. These changes will not affect the results of an experiment analyzing the outcome of a heat treatment, if the method employed allows for estimation of the level of the primary thermoresistance; and the changes can be revealed with prolongation of either a heat treatment or an interval between the end of heating and beginning of an estimation of the effect produced. Ushakov and Zander (1961) succeeded in acclimating "warm" frog muscles to moderate heat treatment, but failed to attain this with intensive heatings. That was also the case with leeches (Ushakov and Sleptsova, 1968). These findings indicate that also in animals the reparatory apparatus can adapt to environmental temperature despite constant primary thermostability of their cells. Effects of the intensity of an agent being used on the results of an evaluation of the sensitivity to that agent have been noted already by Child (1915, 1928). In organisms exhibiting a gradient of sensitivity along the body axis, the same agent, applied in high or mild doses, produced sensitivity gradients, both "direct" and "reverse" gradients oriented in opposite directions.

It seems appropriate at this point to express some hypothetical speculations pertinent to the evidence referred to above. It is known that the cell disposes of a specific enzymatic apparatus which functions as a "repair shop" to eliminate errors in DNA strands (see: Zhestyanikov, 1968; Setlow and Setlow, 1972). It might be visualized that there are in the cell some enzyme systems designed for renaturation of proteins. The enzymes that accelerate renaturation of denatured proteins have been described repeatedly (Venetianer and Straub, 1963a; 1963b; Givol et al., 1964; Yutani et al., 1967). Such enzyme systems are expected to use denatured protein macromolecules as substrates. Admitting that the concentration of altered macromolecules increases with elevation of temperature, it might be expected that a temperature increase would somehow enhance the efficiency of this hypothetical system responsible for renaturation of cellular proteins.

It has been shown that an analysis of the discrepancies between the data of Lange and that of the author enabled some regularities of interest to be revealed. At the same time, this analysis demonstrated once again the complicated nature of the concept of "cellular resistance" and a necessity for carrying out studies on the components of which it consists. This is emphasized also by the work of Kappen and Lange (1968) that will be dealt with later (p. 65).

Summary. Despite the seemingly controversial nature of the experimental evidence presented in this section it, nevertheless, warrants a definite conclusion that the capacity of cells to adjust the level of their primary thermoresistance to ambient temperature, when the latter fluctuates within the tolerant zone, characteristic of *Protozoa* and algae, has been lost in animal and plant tissue cells. The fact that this loss has occurred in the course of evolution of both animals and plants testifies to its profound biological significance, the full implications of which still remain unclear. If primary thermoresistance is regarded as an indicator of the resistance of cellular proteins to the denaturing action of heat, then one should agree that in evolution of animals and plants a cardinal modification of the relation of cellular proteins to the temperature of environment has occurred. Polyansky and Sukhanova (1964) wrote concerning this problem: "In the course of establishment of the multicellular organization, the heat stability of cells and of their proteins has acquired a relative stabilization and the function of the individual adaptation has been transferred to other levels of the organization of the living being (to that of an organ, entire organism, to the biocenotic level)" (p. 141). One may, moreover, admit that more sophisticated regulatory mechanisms which do not require the primary thermostability to be altered, have appeared in the cells of multicellular organisms designed to maintain normal functioning of cells during variations of ambient temperature within the zone of optimal temperatures.

In working with primitive multicellular organisms, it would be of interest to elucidate at what particular stage in phylogenesis cells were deprived of their ability for temperature adjustment. The dearth of pertinent evidence so far precludes any attempts at meaningful answers. The findings obtained by Dregolskaya (1963a) in *Hydra* seem to testify to the capacity of the cells of this organism for temperature adjustment. Additional research is, however, needed in order that a conclusive opinion might be formed in that case. Turning to plants, one may note that temperature adjustment does not exist in mosses (Antropova, 1971, 1974a) or in the myxomycete *Physarum polycephalum* (Lomagin and Antropova, 1968).

1.2.2 Effects of Habitat Temperature Variations Outside the Tolerant Zone Upon the Primary Thermostability of Cells (Temperature Hardenings)

1.2.2.1 Heat Hardening of Plant Cells

The preceding chapter concluded that higher plant cells which have terminated their growth display a rather constant level of primary thermostability, maintained even during considerable fluctuations of ambient temperature. This holds true, however, only in the case when temperature variations do not extend into the region of supraoptimal heatings. Plant cells react to supraoptimal heat by elevating their thermostability; this phenomenon has been called heat hardening.

a) Experimental Objects. Loginova (1945) was probably the first to enhance the heat stability of plant cells by a short-term heat exposure. She heated a suspension of yeasts for 30 min consecutively at 34, 36 and 38° C which resulted in an increase in the resistance of the yeasts to elevated temperature[2]. Later, Laude and Chaugule (1953) obtained an increase in heat resistance of plants after a preliminary thermal treatment. Seedlings of three brome-grass species were treated over a period of many days at various time intervals with 30-min heat exposures to temperatures of 43–54° C followed by a test heating at 54° C for several hours. Seven days after the end of the experimental heat treatment, the extent of the injury was evaluated by measuring the necrotized leaf area on a percentage basis. In preliminary heated plants the percentage of necrotized areas proved to be appreciably lower than in control plants.

Further in-depth studies on heat hardening have been carried out in the author's laboratory at the cellular and molecular levels (Alexandrov et al., 1970). The method for testing injury enabled an estimate to be made of the effects of heat hardening on the primary thermostability of cells. Subsequently, this phenomenon has attracted attention in a number of laboratories both in the Soviet Union and other countries. Up to the present time heat hardening has been obtained under experimental conditions and in nature in almost 60 species of higher plants from 27 families. Four of these families are monocotyledons (22 species), 21 dicotyledons (34 species) and two gymnosperms (Alexandrov, 1956, 1963; Coff-

[2] A communication by Yarwood (1967), that cites the adaptation of grass seeds presumably achieved as early as 1877 by Yust, is somewhat misleading. Yust reported in his publication only a gradual desiccation of seeds at various temperatures, which procedure resulted in enhancing their thermostability.

man, 1957; Alexandrov and Feldman, 1958; Bukharin, 1958; Lutova, 1958, 1962, 1963a, 1963b; Kiknadze, 1960; Alexandrov and Yazkulyev, 1961; Lange, 1961; Lomagin, 1961; Yarwood, 1961, 1962, 1964a, 1967; Kislyuk, 1962; Oleynikova and Uglov, 1962; Feldman and Lutova, 1962; Altergot, 1963, 1964; Lomagin et al., 1963, 1970; Zavadskaya, 1963a, 1964; Schroeder, 1963, 1967; Oleynikova, 1964a, b; Sevrova, 1964; Shkolnikova and Shterman, 1964; Shukhtina, 1964, 1965; Yazkulyev, 1964a; Engelbrecht and Mothes, 1964; Wagenbreth, 1965; Feldman, 1966, 1969; Denko, 1967; Kinbacher and Sullivan, 1967; Shukhtina and Yazkulyev, 1968; Barabalchuk, 1969; Shcherbakova, 1969; Bauer, 1970; Zavadskaya and Shukhtina, 1971; Barabalchuk and Chernyavskaya, 1974).

In addition, heat hardening has been obtained in liverworts (3 species) and leafy mosses (4 species) (Antropova, 1971, 1974a), in the myxomycete plasmodium (Lomagin and Antropova, 1968), in various species of the rust, mildew and plant viruses (Yarwood, 1961, 1963, 1967; Yarwood and Holm, 1962; Joshi and Holmes, 1968), in some algae (Lutova et al., 1968), and in bacteria and their spores (Alderton et al., 1964; Bausum and Matney, 1965). This list will be certainly incomplete by the time this book appears in print.

Heat hardening is a reaction realized at the cellular level. Under experimental conditions it has been achieved both in intact plants and in isolated shoots, leaves, leaf pieces, and roots, as well as in tissue culture of fruits.

b) Enhancement of Thermostability of Various Cellular Functions. Heat hardening was found to be capable of elevating heat resistance of all the functions studied so far. Hardened cells require the application of higher temperatures for the following functions to be completely inhibited: the streaming of protoplasm (Fig. 2; Alexandrov, 1956; and many others), phototaxis of chloroplasts (Lomagin et al., 1966; Barabalchuk, 1969), movement of stomata (Barabalchuk and Chernyavskaya, 1974) and photosynthesis (Lutova, 1958, 1962; Feldman, 1968; Barabalchuk, 1969; Bauer, 1970, 1972). Ageeva and Lutova (1971) hardened pea leaves for 18 h at 35–38° C. After heat hardening the heat resistance of photosynthesis increased as well as its partial reactions, i.e. cyclic and non-cyclic phosphorylation, the Hill reaction with 2,5 dichlorophenolindophenol, photoreduction of nicotinamide adenine dinucleotide phosphate (NADP). These reactions were assayed in isolated chloroplasts derived from hardened and non-hardened leaves following a 5-min test heating of the leaves. An elevated heat resistance of phosphorylation has also been recorded when chloroplasts, isolated from hardened leaves, were tested for their thermoresistance (as opposed to a suspension of chloroplasts isolated from control leaves). Heat hardening of the spiderwort *Tradescantia fluminensis* and the dandelion *Taraxacum officinale* leaves resulted in an enhancement of the thermostability of the association of chlorophyll with the lipoprotein complex of the chloroplast (Lutova, 1963a). The selective permeability of the protoplast also increased; a more intensive heating was needed for suppressing the ability for plasmolysis (Alexandrov and Feldman, 1958; Antropova, 1971; and others), for inducing leakage of antocyan from the vacuoles into the medium (Lomagin, 1961), and causing leakage of electrolytes (Oleynikova, 1964a, b; Kinbacher and Sullivan, 1967; Sullivan and Kinbacher, 1967). After heat hardening the heat stability of respiration (Lutova, 1962; Kinbacher and Sullivan, 1967;

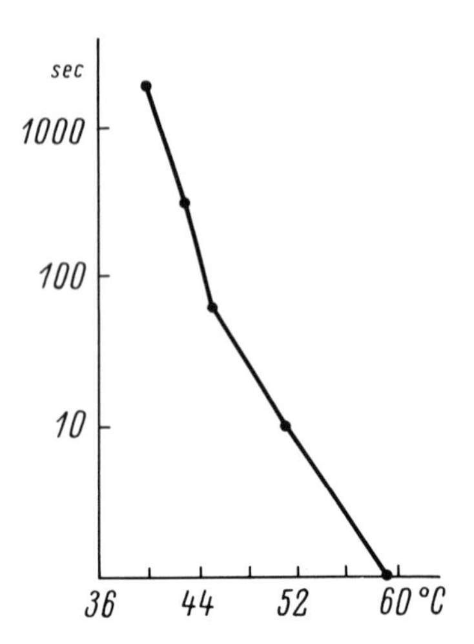

Fig. 17. Correlation between temperature of heat hardening and duration of heat hardening that is effective in producing a maximal increase in heat stability of the *Campanula persicifolia* leaf epidermal cells (Lomagin, 1961). *Abscissa:* temperature of hardening; *ordinate:* optimum time of exposure to a hardening temperature. Thermostability was assayed by temperature of a 5 min heating that stops protoplasmic streaming

Barabalchuk, 1969) and of growth (Kislyuk, 1962; Schroeder, 1963, 1967; Lomagin et al., 1970) also increased. Engelbrecht and Mothes (1964) succeeded in increasing the heat stability of protein synthesis and degradation by heat hardening.

Taking into account the data referred to above and the fact that in all the cases studied in the authors laboratory heat hardening inevitably led to enhancement of the stability of the function under study against heat, it may be concluded that supraoptimal heating, if properly administered, increases the level of thermoresistance of all principle cellular functions.

c) The Dependence of the Effect of Heat Hardening on Temperature and Duration of Hardening. A result of hardening is quantitatively estimated as the difference between the temperatures characteristic of the heat stability of the object concerned prior to and after hardening. This parameter depends on the heat dosage. Duration and temperature of hardening are found to be in reciprocal interrelationships: the higher the temperature of heat hardening, the shorter must be the duration of a heat treatment to attain a maximal effect (Fig. 17). Heat hardening the *Tradescantia* leaves at 37° C produced the maximal effect in 18 h. At 50–56° C, however, an increase in the heat resistance of *Tradescantia* leaves can be obtained as quickly as in 10 sec. Heat hardening effects achieved with exposures in the order of seconds have been reported by Lomagin (1961) and Zavadskaya (1963a) in *Tradescantia fluminensis, Dactylis glomerata, Campanula persicifolia* and *Leucanthemum vulgaris* leaves, and also by Yarwood (1961, 1962) in several other plants.

Given a certain duration of hardening, the effect depends on the hardening temperature. The curve describing a hardening effect as a function of the tempera-

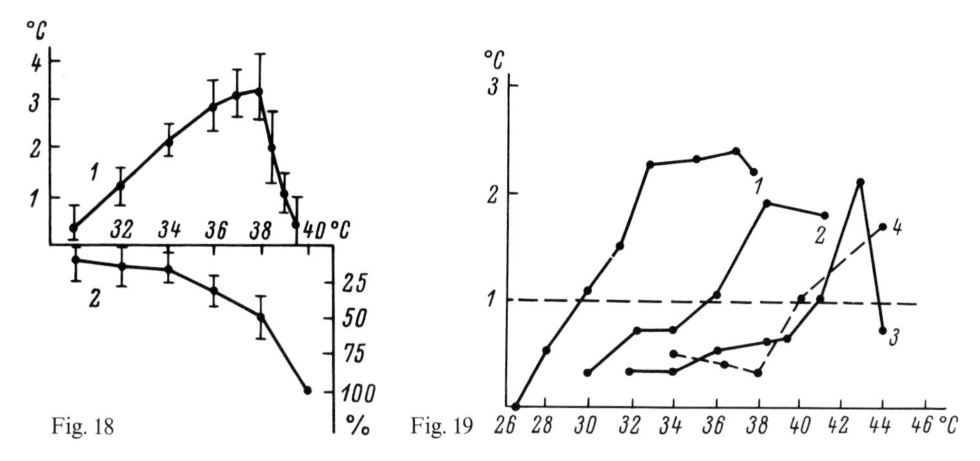

Fig. 18. Changes in thermostability *(1)* and protoplasm flow rate *(2)* in *Tradescantia* leaf epidermal cells after 3-h heat hardenings at different temperatures (Barabalchuk, 1969). *Abscissa:* temperature of hardening; *ordinate:* left: difference between minimal temperature of a 5-min heating that stops protoplasmic streaming in hardened and control cells (effect of hardening), right: reduction of protoplasmic flow rate in hardened cells

Fig. 19. Effect of temperature of 18-h heat hardening on heat stability of epidermal cells in leaf sheath of grasses which differ in their thermostability (Alexandrov and Feldman, 1958). *Abscissa:* temperature of hardening; *ordinate:* difference between temperature of a 5-min heating that stops protoplasmic streaming in hardened and control cells (effect of hardening). *1: Dactylis glomerata, 2: Phragmites communis, 3: Panicum miliaceum, 4: Eleusina indica*

ture gradually reaches a maximum as the temperature is elevated. Further increase in the hardening temperature results in progressive reduction of the hardening effect (Fig. 18, curve 1). This usually occurs at hardening temperatures that produce appreciable suppression of the function concerned during the process of hardening (Fig. 18, curve 2).

At a given duration of hardening, the temperature effective in producing the hardening effect must be raised is proportion with the initial thermostability of the study object (Alexandrov, 1956; Alexandrov and Feldman, 1958). Figure 19 illustrates the dependence of an 18 h hardening effect upon the hardening temperature in four grass species differing in cell thermostability. To arrest the protoplasmic streaming in the epidermal leaf sheath cells by a 5-min heating one must expose *Dactylis glomerata* to 44° C, *Phragmites communis* to 46° C, *Panicum miliaceum* to 48.5° C and *Eleusina indica* to 49° C. To elevate the cellular thermostability of these plants by one degree centigrade, the hardening temperature must be 30, 36, 40, and 40° C, respectively. The temperature of a 5-min heating, which arrests the protoplasmic streaming in the leaf epidermal cells of the grass *Aristida karelinii* is 3° higher than that for analogous cells in the grass *Arundo donax* (49 and 46° C, respectively). Accordingly, a 3-h heat hardening produces a maximal effect in the former at 44.5° C, whereas in the latter at 41° C (Shukhtina and Yazkulyev, 1968).

There are investigators who have obtained an increase in the heat stability of plant objects with the aid of repeated heat exposures, alternating heating the plants to superoptimal temperatures (constant or elevated at each subsequent exposure) with exposing them to optimal temperatures (Laude and Chaugule, 1953; Yarwood, 1961; Altergot, 1963; Engelbrecht and Mothes, 1964; Novoselova et al., 1966, 1971; Sevrova, 1964; Kinbacher et al., 1967; Sullivan and Kinbacher, 1967). Yarwood (1967) reported that for certain plant species, the greatest enhancement of their thermostability can be observed after a short-term intensive heat treatment followed by a prolonged heating at moderately high temperature. In his earlier communication Yarwood (1961) concluded, however, that after a series of 15 consecutive brief heat shocks, performed over a period of four days, the kidney bean and cowpea leaves only slightly increased their thermoresistance from the level after a single treatment. Shukhtina (1964) succeeded in demonstrating that, provided an optimum dose of the first heat treatment has been properly chosen, subsequent exposures to the same temperature do not affect thermostability in the least. If, however, the temperature of a subsequent heat treatment is higher than that of a previous one, the effect of the subsequent heat hardening, if any, is negligible, as opposed to that achieved in the first hardening, and sometimes thermostability grows less. Consequently the cells, by responding to the first, optimally chosen, hardening heat treatment, apparently attain a maximal enhancement of their thermostability (see also Shukhtina and Yazkulyev, 1968).

According to Yarwood (1963, 1964a, 1964b, 1967), under the influence of a preliminary heat treatment, there may sometimes occur a decrease in the heat stability of undetached kidney bean leaves and that of kidney bean rust. This phenomenon, designated by Yarwood as "sensitization" to heat, is promoted by a number of factors, i.e. by increasing the interval between a preliminary hardening and subsequent test heating; by lowering the dosage of the preliminary heating (compared with the doses effective in producing heat hardening), diminishing the dose of the test heating, and by excluding light for 48 h prior to the test heating. Nonetheless, when dealing with the primary thermostability of plant cells, and despite extensive variations of temperature and duration of the test heating, as well as of the interval between the hardening and test heat treatments, sensitization to heat was not observed (Alexandrov, 1956; Alexandrov and Feldman, 1958; Lomagin, 1961; Barabalchuk, 1969; and others). It is possible, however, that sensitization is related to changes in the general thermostability of cells. It would be appropriate, in this context, to take into account the data reported by Mothes and Engelbrecht (Mothes, 1960; Engelbrecht and Mothes, 1964) who showed that heating *Nicotiana rustica* leaves at moderately high temperatures, which fail to produce immediate, noticeable injury, eventually results in rapid yellowing, and finally, in the death of the leaf concerned. This can be explained, according to Mothes and Engelbrecht, by the ability of neighboring, untreated plant organs to attract amino acids from the heated leaves, and, in this way, weaken the latter. It should be anticipated that the general thermoresistance of cells in such leaves is lowered. The "sensitization" reported by Yarwood might be conceivably accounted for by the same phenomenon.

The experiments which follow support the arguments that heat hardening can simultaneously elevate the primary thermoresistance of cells and lower their general thermostability (Alexandrov and Barabalchuk, 1972). Earlier (p. 36), reference was made to a 3-h heating of *Tradescantia* leaves at 38° C that enhanced the rate of repair of thermal injury. That enhancement was accompanied with elevation of the cellular primary thermoresistance. A 3-h hardening of leaves at 42° C increased the primary thermostability of respiration in the cells and that of selective permeability of the protoplast. Following this treatment, however, the reparatory zone for the protoplasmic streaming grew less, the rate of repair lowered, and the duration of the survival of detached leaves was reduced to several days. At the same time, control *Tradescantia* leaves detached from a plant survived in water or under humid chamber conditions for over two months. This difference may be explained by the development of the destructive after-action that ensues even after heating leaves at 42° C. It is clearly evident that, depending on the choice of a method for testing thermostability of leaves hardened for 3 h at 42° C, the results may be controversial. They will largely depend on whether it is the primary or the general thermoresistance that is being evaluated.

It is not surprising that an intensive heat hardening can increase the primary thermostability and simultaneously suppress the reparatory capacity of the cells, because, as will be shown in the next section, various cellular functions are suppressed by heat hardening. Taking into consideration this phenomenon, the results reported by Fries (1963) on vegetative cells of the fungi *Ophiostoma multiannulatum* and *Rhodotorula glutinis* are easily understood. Usually these fungi grow at 30 and 28° C, respectively. If they are exposed for several minutes to a sublethal heating, their ability to grow at temperatures above 25° C is lost. The author called this phenomenon "induced thermosensitivity". It is, presumably, related to a decrease in the reparatory capacity of the cell that is required for growth at temperatures close to the supraoptimal ones. In all probability, the cells of these fungi can also exhibit, after a thermal treatment, an increase of their primary thermostability. As noted earlier during prolonged cultivation of plants at an elevated but not yet hardening temperature, the primary thermoresistance undergoes no changes, whereas the reparatory capacities of the cells grow (see Sect. 1.2.1.4).

d) Suppression of Cellular Functions by Heat Hardening. As will be shown, in order to obtain the effect of hardening, it is necessary to act on a tissue with sufficiently high doses of heat. The higher the initial thermostability of the cells, the more intensive the heat must be. This observation leads us to suggest that heat hardening is a reaction of cells to a certain degree of thermal injury. In fact, as the dose of a hardening heat treatment is augmented and thermostability of a function grows higher, it can be noticed that the function is being suppressed. Curve 1 in Figure 18 depicts an enhancement of thermostability of the streaming of protoplasm in *Tradescantia* leaf cells as the temperature of a 3-h hardening is elevated. The lower curve 2 shows that, coincident with an increase in thermostability, the rate of streaming diminishes. The same situation is seen in Figure 20 for chloroplast phototaxis (see also: Lomagin et al., 1966), and in Figure 21 showing photosynthesis in *Tradescantia* leaves. A decreased rate of the protoplasmic flow (by 30%) under optimum conditions of hardening of *Campanula persicifolia* leaves has been reported by Shkolnikova and Shterman (1964). Kislyuk (1962) observed an inhibition of growth by 15–25% caused by hardening in rye, barley and wheat seedlings. It should be noted, however, that intensity of the most heat-resistant

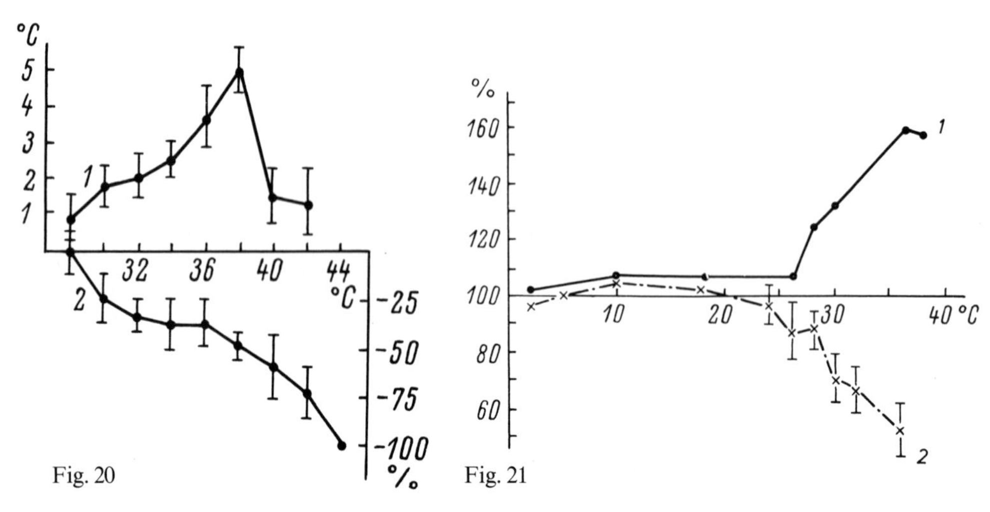

Fig. 20. Changes in thermostability *(1)* and intensity of chloroplast phototaxis *(2)* in *Tradescantia* leaf cells following 3-h heat hardening at different temperatures (Barabalchuk, 1969). *Abscissa:* temperature of hardening; *ordinate left:* a difference between temperature of a 5-min heating that inhibits chloroplast phototaxis in hardened and control cells; *ordinate right:* percent of reduction of intensity of chloroplast phototaxis in hardened cells

Fig. 21. Changes in thermostability *(dots)* and intensity of photosynthesis in *Tradescantia* leaves *(crosses)* after 18-h heat hardening at different temperatures. (Lutova, 1962). *Abscissa:* temperature of hardening; *ordinate, dots (1):* intensity of photosynthesis in hardened leaves following 5-min heating at 45° C (as percent of intensity of photosynthesis of heated non-hardened leaves); *ordinate crosses (2):* intensity of photosynthesis in leaves after hardening (percent of control)

cellular functions may, under optimum conditions of heat hardening, show no alteration whatsoever. In some plants, e.g. in *Campanula persicifolia* and *Tradescantia fluminensis*, an elevation of the heat resistance of respiration after heat hardening is not accompanied by a reduction in the respiration intensity (Lutova, 1962).

e) Reversibility of Heat Hardening. A hardened state is reversible. Enhanced thermostability of hardened cells gradually diminishes, and in higher plants, eventually reaches the control level in 5–7 days after a hardening treatment (Alexandrov and Feldman, 1958; Lomagin, 1961; Lutova, 1962; Schroeder, 1963; Yarwood, 1964b). In *Phaseolus vulgaris* rust, heat hardening can not be detected as early as 48 h after a preliminary heat treatment (Yarwood, and Holm, 1962), whereas in the myxomycete *Physarum polycephalum* plasmodia it is retained for no longer than 24 h at room temperature (Lomagin and Antropova, 1968). The rate of loss of hardened state is a function of ambient temperature. A decrease in thermostability of hardened leaves of *Dactylis glomerata* at 0° C proceeds at a significantly lower rate than at 20° C (Fig. 22). The same retardation of the loss of thermostability with transfer to cold has been observed in hardened plasmodia of *Physarum polycephalum* (Lomagin and Antropova, 1968).

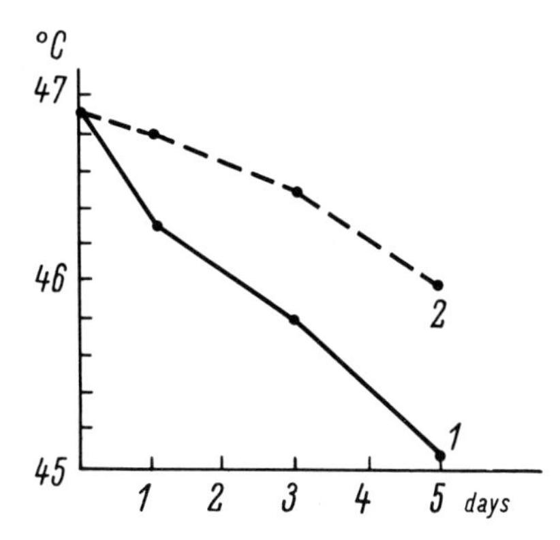

Fig. 22. Decrease in heat stability of epidermal cells of *Dactylis glomerata* leaf sheath after 18-h hardening at 37° C as a function of the temperature at which hardened leaf pieces were kept thereafter (Alexandrov et al., 1970). *Abscissa:* the time period during which plant tissue pieces were kept after hardening; *ordinate:* temperature of 5-min heating effective in stopping protoplasmic streaming. *1:* leaf pieces kept at 20° C, *2:* at 0.3–2.0° C

f) Heat Hardening in Nature. In the foregoing, the hardening obtained in a laboratory experiment was covered. It is appropriate to ask whether this phenomenon can be found in nature. Do plants utilize their ability to elevate the level of thermostability of their cells while adapting to overheating? Non-systematic tests performed over a period of several years in the Leningrad climate have never revealed any noticeable changes in the primary thermostability of the various plant species tested. This may be accounted for by the fact that the maximal air temperature in this area has never reached the threshold required to provoke a hardened state. During the hot summer of 1972, however, heat hardening could be detected in four *Allium* species (Feldman and Kamentseva, 1974). Still earlier, Lange (1961) during a hot dry summer, was able to observe an elevation of thermostability in some plants growing in the vicinity of Göttingen. Next year, when the summer was cooler, no elevation of thermostability could be found. In that same region, according to Kappen (1964), thermostability of some ferns did not increase, whereas in others the heat resistance grew higher. In the Khibini, to the north of the Polar Belt, Shukhtina (1962) was lucky to observe natural heat hardening of *Dactylis glomerata* during the unusually hot summer days of 1957 and 1960.

Pronounced natural heat hardening can be clearly seen in various plants (*Aristida karelinii, Arundo donax, Spartium junceum, Catalpa speciosa, Morus alba*, and others) in the hot climatic conditions in Turkmenia (Alexandrov and Yazkulyev, 1961; Yazkulyev, 1964a, 1970; Shukhtina, 1965; Shukhtina and Yazkulyev, 1968). In Tadzhikistan, heat hardening has been observed in maize (Popova, 1964), it has also been reported in plants growing on the Southern Crimean Coast (Falkova and Galushko, 1974; Falkova, 1975). Heat hardening in nature features the same regularities that are typical of an artificial heat hardening. Research on heat hardening, conducted in Turkmenia and the Crimea, showed that in spring when the air temperature fluctuations never reach the range of hardening temperatures, the level of primary thermoresistance of those cells

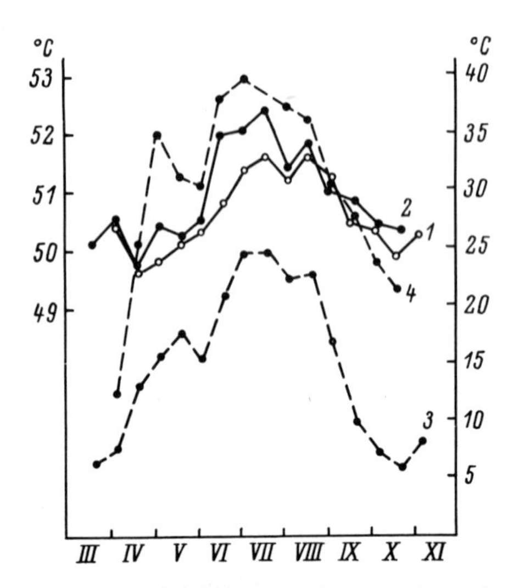

Fig. 23. Seasonal changes in thermostability of epidermal cells of the grass *Aristida karelinii* leaves and air temperature (Yazkulyev, 1964a, b). *Abscissa:* months; *ordinate left:* temperature of 5-min heating that stops protoplasmic streaming; *ordinate right:* air temperature. *1:* thermostability of cells in the morning, *2:* daytime, *3:* minimal, *4:* maximal air temperature on days when thermostability was evaluated

which have terminated their growth, is almost constant. At that time the cellular thermostability, estimated early in the morning, is the same as that recorded in the afternoon. The situation changes when the maximal diurnal temperatures become effective enough to produce heat hardening. In summer on hot days the cellular heat resistance in the afternoon is, as a rule, higher than that in springtime. This elevation is reversible, however, and in the morning, after a cool night, thermostability decreases by approximately 1° C. An overall rise of heat resistance is in the order of 1.5–3.5° C, whereas diurnal variations of thermostability persist over the whole hot period. Toward autumn, after the end of the hot period, thermostability decreases to the spring level, whereas diurnal fluctuations either disappear or become less pronounced (Figs. 23, 24).

The presence of a hardened state may be frequently confirmed by taking advantage of its reversibility. Thus, if leaves cut from a plant that had been heat-hardened in its natural environment are placed into a humid chamber at moderate temperatures, a decrease in the cellular primary thermostability can be observed within the next one to two days. Non-hardened leaves, if kept under similar conditions, usually exhibit a certain elevation of their thermostability.

For a heat hardening to appear in nature, different temperatures are needed for different plant species. In the grass *Aristida karelinii*, the state of heat hardening can be detected on those days when the air temperature rises to 37° C and higher. In a less thermoresistant grass, such as *Arundo donax*, heat hardening develops at 30° C. In *Catalpa*, heat hardening occurs at 35° C, and in the mulberry plants the heat-hardening state only occurs at temperatures above 41° C (experiments by Yazkulyev).

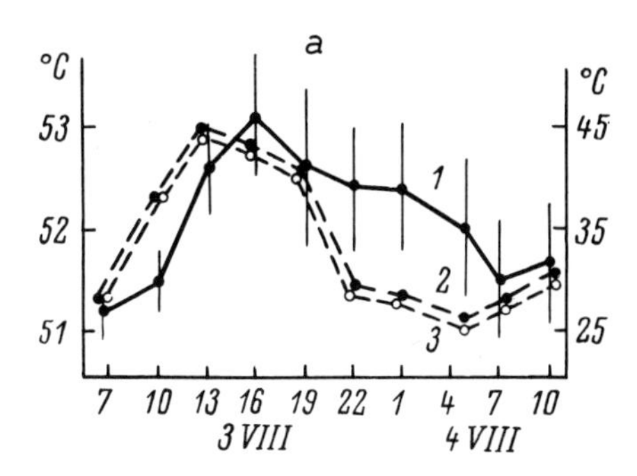

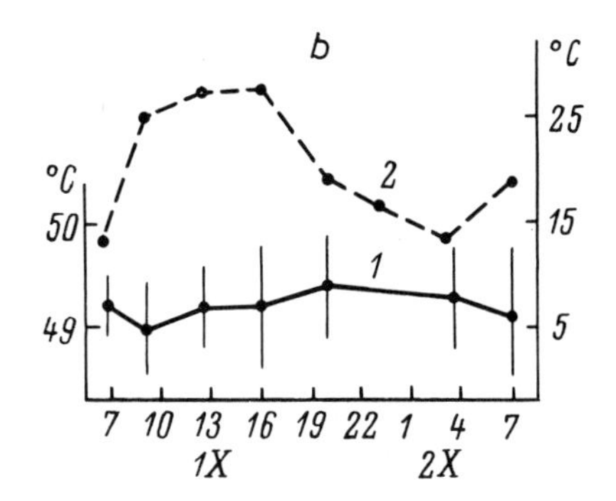

Fig. 24 a and b. Diurnal variations of thermostability of epidermal cells of grass *Aristida karelinii* leaves on days with hardening (a) and non-hardening (b) temperatures (Yazkulyev, 1970). *Abscissa:* time of day; *ordinate left:* temperature of 5-min heating that stops protoplasmic streaming; *ordinate right:* temperature of air or leaf temperature. *1:* thermostability of cells, *2:* air temperature, *3:* temperature of sheath part of leaf

There is evidence in the literature of diurnal variations of plant cell thermostability, which are dependent not on heat hardening, but on changes in illuminance. According to Laude (1939), the maximal thermostability in many plants can be observed at midday; the minimal in the morning. Conversely, Schwemmle and Lange (1959) and Lange (1964) reported that the heat resistance of *Kolanchoë* reached its maximum at night, and its lowest point in the daytime. Experiments by Alexandrov and Shukhtina (unpublished) included day-long observations on the primary thermostability of leaf epidermis of plants growing in different geographical localities: in Leningrad, working with *Cardamine pratensis, Myosotis silvatica, Tussilago farfara, Petasites officinalis* and *Carex acuta;* on the Taimyr (73°17′ North): *Cardamine pratensis, Myosotis asiatica, Nardosmia gmelinii, Carex stans;* on Franz-Joseph Land (80°37′ North): *Cardamine billidifolia, Poa abbriviata, Phippsia algida, Cerastium hyperboreum.* Every four hours the temperature of a 5-min heating that stopped protoplasmic streaming was determined. In the overwhelming majority of cases the heat resistance during the whole day remained constant. In those experiments when variations did occur, they could hardly be related to alterations in period of illumination or to temperature variations.

Summary. Tissue cells of fungi, mosses, gymnosperms and flowering plants, as opposed to algal cells, maintain a constant level of primary thermostability despite temperature fluctuations within the wide range of the tolerant zone. They respond to the action of supraoptimal temperatures, both under experimental conditions and in nature, by reversibly increasing their thermostability, by heat hardening. The latter is effective in enhancing the primary thermoresistance of all functions so far studied in this regard. Since primary thermoresistance reflects the thermal sensitivity of protoplasmic proteins, it should be deduced that the stabilization produced by heat hardening extensively embraces the protein constituents of the cell. Heat hardening has also been found in those objects capable of temperature adjustment, among then multicellular algae. In the experience of the author, heat hardening could not be elicited in the unicellular alga *Chlamydo-*

monas eugametos (Luknitskaya, 1967). Whether this inability to respond with heat hardening is specific for only this particular study object or for a wider group of unicellular algae is difficult to say with certainty. At any rate, apparently almost all plant cells possess the ability to react to a supraoptimal heating by increasing their primary thermoresistance. Such an omnipresent capability of plant cells for heat hardening justifies attempts to gain insight into the so far unrealized fundamental biological significance of this cellular reaction. This problem will be taken up once again elsewhere in this book (p. 260) in order to approach an understanding of its significance.

1.2.2.2 Heat Hardening of Animal Cells

Much less effort has been put into research on this phenomenon in animals. Thörner (1919, 1922) reported an inhibition of conductivity of the frog ischiadic nerve that can be produced by a heat treatment. Cooling restored the conductivity. After several repeated thermal treatments, the resistance of the nerve to heat grew higher, so that a maximal effect attained was an increase in the heat stability by 4° C. This process of "accommodation" of the nerve to heat was accompanied by a decrease in the rate of impulse conductivity, prolongation of a latent period of excitation, and by diminished utilization of oxygen. The state of hardening is reversible. The findings reported by Thörner have been confirmed on the same study object by Zhukov (1935). Similar effects were obtained by Yamada (1924) with non-medullated frog nerves. A preliminary sublethal heat treatment elevated thermostability of HeLa cells in tissue culture (Salawry and McCormick, 1956).

Working with the *Rana temporaria* muscles, Shlyakhter (1959) and Chernokozheva and Shlyakhter (1963) also succeeded in obtaining an elevation of thermostability after a short-term preliminary exposure of a muscle preparation to supraoptimal temperatures. A maximal effect—an increase of 30–40% in the time required for the onset of non-excitability, assayed by a test heating at 38° C—was achieved with preliminary 15 min heatings at 33–34° C. Further experiments were performed by Chernokozheva (1965) on *Rana ridibunda* muscles, the latter being approximately 4° C more thermostable than the corresponding muscles of *R. temporaria*. Interestingly, in order for the maximal effect of heat hardening to be attained with *R. ridibunda*, a more intensive heat treatment was needed. The effect of hardening ensued after a 10-min exposure to 39° C. Heating muscles to temperatures below 37° C never enhanced their thermostability. Consequently, similar to the situation with plant cells, the more thermostable an object is, the higher must be the dose of hardening heat treatment. Chernokozheva failed to produce heat hardening in the ciliary epithelium of the frog palate.

Friedrich (1967) placed mussels in water heated to 25–35° C for a period of 15 min and subsequently estimated the time interval during which the ciliary beating persisted at 36° C. A preliminary heating of the mussels at 25–29° C gave no results, but after exposing them to 32–34° C a significant increase in the heat stability of the gill ciliary cells could be noted. The effect could be obtained at a lower temperature by prolonging the exposure: thus, at 27° C the largest increment of thermostability was recorded with a 20-min exposure. Exposures of 30-

and 40-min to this temperature were less effective. Given an optimum duration and temperature of treatment, an increase of thermostability was accompanied by a reduction of oxygen uptake by the cells. This elevation of thermostability was reversible. Similar results were obtained not only when a whole animal had been heated, but with detached pieces of gills as well.

Contrasting findings in this organism have been communicated by Skholl (1971). Heating pieces of mussel gills to 29 and 33° C for 1, 2, and 3 h enhanced the heat resistance of the ciliary cells only when the original level of their thermostability had been low, and diminished their resistance to heat if the original level had been high. In other words, a preliminary heat treatment reduced individual differences in thermostability by bringing them closer to an intermediate level. Essentially the same results have been gained by Ushakov and Amosova (1972) in frog muscles.

Moderate heating at sublethal temperatures for 1–2 h was effective in elevating thermostability of the silk-worm nursery by 36.4% on average (Astaurov et al., 1962; Astaurov, 1964). These authors account for the effect produced by heat hardening and compare it with heat hardening of plants.

Since no heat hardening could ever be elicited in the unicellular alga *Chlamydomonas eugametos*, it would be of interest to see whether unicellular animals were capable of heat hardening. Answers to these questions can be found in the works of Jacobs (1919) and Polyansky and Irlina (1967). In the experiments of Jacobs, the *Paramecium caudatum* cells transferred from room temperature directly to 41° C died in 2 min 20 sec. If, however, elevation of the temperature from room temperature to 41° C was a gradual one, extending over a period of 2 min, after which the infusoria were kept at that temperature for another 2 min, a quarter of the cells survived the transfer. Prolongation of the temperature shift to 12 min allowed half of the cells to survive. Polyansky and Irlina succeeded in elevating the primary thermoresistance of *Paramecium caudatum* after preliminary heat exposures of the cells to 40° C for only 30 sec. Heating had to be prolonged if the hardening temperature had been lowered. Heat hardening of the infusoria proved to be reversible, and in 6–7 h the difference in the heat stability of the hardened and control cells disappeared. Precht and his collaborators (1966) could not increase thermostability of the colonial infusorium *Zoothamnium hiketes* when preliminary heat treatments were conducted at supraoptimal temperatures.

Summary. Despite the dearth of data concerning heat hardening of animal cells, one cannot refrain from noticing an indubitable similarity of this phenomenon to that observed in plant cells. Animal cells also respond to heat shocks if the latter are applied in doses comparable or equal to the injurious ones, by increasing their primary thermostability. Again the more thermoresistant cells require higher doses of the hardening heating. As with plants, the effect may be obtained with pulse exposures as brief as scores of seconds. According to Thörner (1919, 1922) and Friedrich (1967), animal cells found in a hardened state, just as plant cells, demonstrate an inhibition of a number of their vital functions. In both cases, the heat hardening is a reversible state. Further studies will show to what extent animal and plant cells are similar in this respect.

1.2.2.3 Modification of Thermostability of Plant Cells by Cold Hardening

This subsection describes a change in the level of the primary thermostability of plant cells produced by cold hardening. Other aspects of cold hardening of plants, despite the practical and theoretical significance they may have, are beyond the scope of the present study. They have been dealt with, however, in numerous good reviews, both new and old (Maksimov, 1913; Tumanov, 1940; Levitt, 1956a, 1958, 1966, 1972; Levitt and Dear, 1970; Biebl, 1962c; Santarius, 1971; Alden and Hermann, 1971). Scientists working in thermobiology return repeatedly must to an excellent paper by Sapper (1935), a pioneer in describing a whole series of phenomena concerning the response of plant cells to environmental temperature. Except for a short report by Illert (1924) on an increase in the heat stability of *Oxalis acetosella* toward autumn, it must be acknowledged that Sapper was the first to observe the paradoxical phenomenon, namely, in seedlings of some plants *(Hedera helix, Linaria cymbalaria, Prunus laurocerasus)* which displayed high thermostability immediately after the end of winter, but a lower thermostability after a long warm and rainy period. Again, *Eranthis hiemalis* plants were taken from nature early in March, a portion being left at 10–15° C, and another at −4° C. The "cold" plants were found to be more thermoresistant than their "warm" counterparts. Referring to data of this kind, Levitt (1951, 1956a, 1956b, 1972) repeatedly advocated a common mechanism as responsible for the cold, drought and heat resistance of plants.

The pattern of seasonal variations of primary thermoresistance and its relation to cold hardening has been studied in detail in the author's laboratory both in nature and under experimental conditions. The resistance of the protoplasmic streaming in epidermal cells to heat and cold has been assessed in some plants over the period of a year. As a measure of cold resistance the temperature of a 5-min cooling was chosen, which was effective in depriving the cell of its ability to resume streaming of the protoplasm upon return of the tissue under study to room temperature. Leaf pieces were immersed for 5 min in a silicone oil precooled in a semiconductor microrefrigerator. This method yields evidence primarily concerning the readiness with which an intracellular ice formation occurs. It is known that cold hardening interferes with the process of ice formation inside the cells (Levitt, 1956a; Salcheva and Samygin, 1963; and others). As with any method used for assessing cold resistance, the approach employed was certainly a relative one and valid only for a comparative evaluation to be carried out under identical experimental conditions. Prior to immersion in silicone oil, all study objects were kept at room temperature for 1–2 h. After cooling the leaf pieces were immediately placed in silicone oil at room temperature and afterward examined with the light microscope. In these experiments, a loss of the protoplasm's ability for streaming practically coincided with death of the cells.

Preliminary thawing of leaf pieces prior to test cooling was necessary, since if the thawing was omitted the results were essentially dependent on the temperature of the plant at the moment of collection in the field (Alexandrov et al., 1964b). The difficulty is that there occurs a redistribution of water between the cell and the intercellular ice, a process depending upon the temperature of the plant. A relative water content in the cell can be evaluated by measuring the thickness of

Table 4. Thickness of the wintergreen plant cells as a function of temperature. (From Alexandrov and Shukhtina, 1964)

Object and tissue	Date	Air temperature (° C)	Thickness of cells (nm)
Campanula persicifolia, leaf epidermis	20.11.63	+ 4.8	31 ± 0.8
	28.11.63	− 7.2	14 ± 1.3
	29.11.63	− 1.4	36 ± 2.1
Sedum spurium, leaf epidermis	12.11.63	+ 0.7	59 ± 3.2
	28.11.63	− 7.7	16 ± 0.9
	26.11.63	− 1.5	43 ± 0.9
Ribes nigrum, epidermis of bud scales	12.11.63	+ 0.8	12 ± 0.9
	21.12.63	−10.5	7 ± 0.3
Same pieces after transfer to room temperature		+22.0	14 ± 3.6

the cells. By using a specially designed tool for measuring thickness of cells out-of-doors and after transfer of a plant indoors, (Alexandrov and Shukhtina, 1964) variations in the thickness of some plant cells were detected and these variations were found to be directly related to the temperature of the plant (see Table 4). The free water content in the cell determines the temperature of an intracellular ice formation, i.e. one of the variables of the frost resistance of plants. If plant leaves taken from under snow are immersed in silicone oil precooled to a temperature corresponding to that of the original temperature of the plant, if then the temperature of a 5-min cooling that stops the protoplasmic streaming is recorded, this temperature will be directly determined by the temperature of the plant at the moment of its being taken for an experiment. If, on the other hand, the cold resistance is evaluated by this method, but the plants were preliminarily exposed to room temperature for a short while, the results would be independent of the temperature of the plant prior to an assay (Table 5). This indicator of cold resistance reflects the state of the cell, which is independent of the water-content variations caused by the extent of intercellular ice formation on a given day.

The latter version was employed to assess the cellular cold resistance, and a conclusion was reached that in certain plants the resistance both to cold and heat varies concordantly over a year, i.e. the minimal resistance is observed in the spring–summer season; towards autumn the resistance increases, and in spring it declines. This has been found to hold true for the following plants: *Dactylis glomerata, Elymus arenarius, E. angustus* (Alexandrov and Feldman, 1958; Alexandrov et al., 1959), *Carex rostrata* and *C. rotundata* (Shukhtina, 1962), *Geum rivale, Calluna vulgaris, Leucanthemum vulgare, Catabrosa aquatica* (Alexandrov et al., 1964b) (Figs. 25, 26). An average level of thermostability estimated by the cessation of protoplasmic streaming after a 5-min heat shock in winter (December–February) exceeded that in summer (July–August) by 1.5–3.0° C.

Similar findings have been reported by Lange (1961) and his colleague Kappen (1964). Lange detected an elevation of thermostability of *Erica tatralix, Asarum europeum* and *Taxus baccata* leaves in winter, whereas Kappen found the same to be true for some ferns. A conspicuously high winter maximum of thermo-

Table 5. Cold resistance of epidermal leaf cells estimated with and without preliminary thawing. (From Alexandrov et al., 1964)

Plant species	Date	Temperature (° C)			Temperature of a 5-min cooling which leads to a loss of the protoplasm's ability to streaming		Difference between I and II
		air	plant under snow	oil prior to test	without preliminary thawing (I)	after thawing (II)	
Dactylis	21.12.62	−18	−10	− 8.5	−23.8	−13.4	10.4
glomerata	27.12.62	− 4	− 4	− 4	−14.7	−13.9	0.8
Hepatica	21.12.62	−18	− 8	− 8.5	−25.3	−12.4	12.9
nobilis	26.12.62	0	0	0	−12.0	−12.4	− 0.4
Saxifraga	28.12.62	− 7	—	− 7	−12.4	− 9.2	3.2
cuneifolia	4. 1.63	−14	− 8	−14	−16.4	− 9.4	7.0
	12. 1.63	−21	− 5	− 7	−16.9	− 9.4	7.5
Geum rivale	20.12.62	−18	−10	−10	−23.3	−12.3	11.0
	27.12.62	− 4	− 4	− 4	−16.4	−12.3	4.1
Sedum	19.12.62	−20	− 9	−11	−23.7	−10.4	13.3
spurium	24.12.62	− 2	− 2	− 2	−13.1	−11.0	3.1
Chelidonium	20.12.62	−18	− 8	−18	−22.4	−12.8	9.6
majus	26.12.62	0	0	0	−11.6	−12.8	− 1.2
Vinca	24.12.62	− 2	− 2	− 2	−16.2	−15.0	1.2
minor	11. 1.63	−18	− 2.5	−16	−22.4	−14.7	7.7
Campanula persicifolia	12. 1.63	−21	− 5	− 7	−17.5	−12.7	4.8

stability was recorded in *Polypodium vulgare*, *Blechnum spicant*, *Dryopteris spinulosa* and *D. filix-mas*. Dircksen (1964) reported that thermostability of some mosses was higher in winter than it was in summer. Again, Biebl and Maier (1969) found the heat resistance of *Chelidonium majus* and *Senecio vulgaris* to be at its highest point in the cold season, whereas *Saxifraga aizoon* showed less increase in thermostability in winter. According to Schwarz (1969, 1970), the resistance of cedar and rhododendron was concomittantly modified both to cold and to heat. Bauer (Bauer, 1970; Bauer et al., 1971) described a rise in the heat stability of various organs of fir and maple during the autumn–winter season. Parker (1971) has reported similar findings for various woody plants. Bannister (1970) indicated that the three species of heather and blackberry found in the vicinity of Glasgow display minimal thermostability of their tissues in summer, and maximal thermostability in winter. After a frosty night in Turkmenia during the autumn–winter season, one could observe an augmentation of the heat stability of *Poa bulbosa* and *Hordeum leporium* epidermal cells (Yazkulyev, 1964b). In the Crimea, Falkova (1973) detected a winter elevation of thermostability in honeysuckle and *Daphne* cells.

An increase of plant cell thermoresistance can be achieved experimentally by cold hardening even in summer plants. *Dactylis glomerata* plants were cooled for two nights to −2.5° C and, as a result, thermostability of the protoplasmic

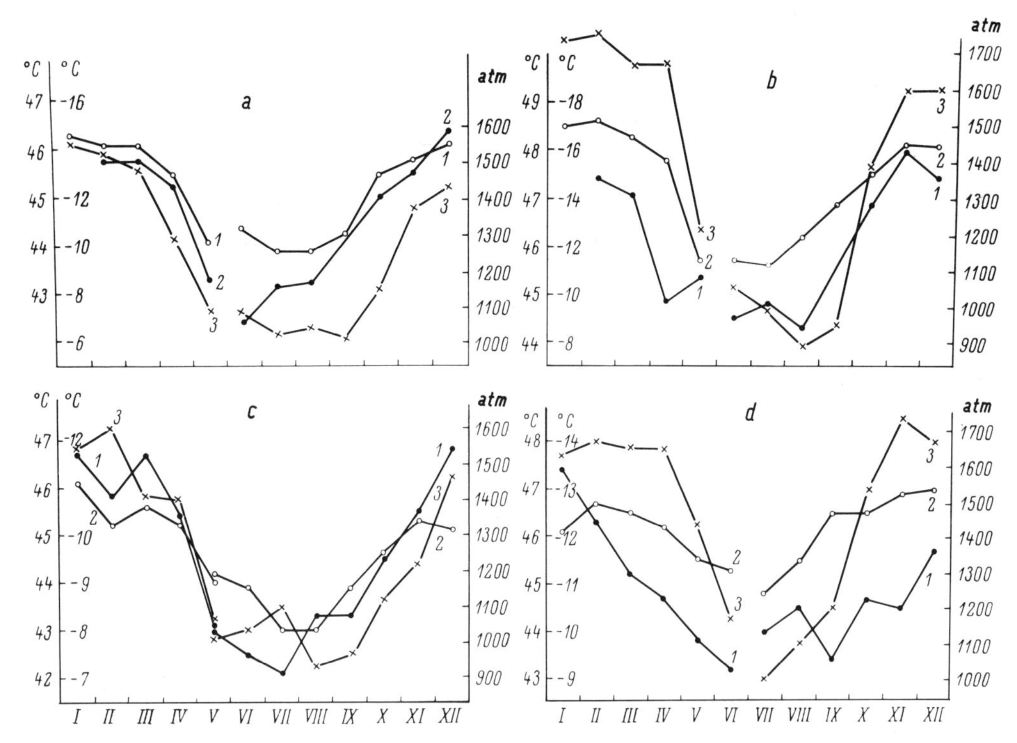

Fig. 25 a–d. Seasonal variations of cellular resistance to cold *(1)*, heat *(2)* and high hydrostatic pressure *(3)* in various plants (Alexandrov et al., 1964). *Abscissa:* months; *ordinate left:* maximal (left) and minimal (right) temperatures of 5-min heating after which protoplasmic streaming still persisted; *ordinate right:* maximal hydrostatic pressure that does not stop protoplasmic streaming. (a) sheath epidermis of grass *Dactylis glomerata*, (b) same in grass *Elymus arenarius*, (c) leaf epidermis of avens *Geum rivale*, (d) same in heather *Calluna vulgaris*

streaming was elevated by 1° C (Alexandrov et al., 1959). Shcherbakova (1969) hardened winter wheat seedlings by exposing them to 2.5–5.0° C for five days in the light, and then for an additional four days at 0.5–1.0° C in the dark. This led to an increase in the heat resistance of the protoplasmic streaming by 1.6° C, whereas thermostability of the tetrazolium reduction test was elevated by 2.0° C. According to Denko (1964), exposure of the water plant *Cabomba aquatica* to 20° C for 50 days does not modify thermostability of the protoplasmic streaming in this plant; however, in plants maintained at 5 and 10° C, both the cold resistance and heat resistance grew higher.

Cold hardening is known to be a reversible state; in accord with this the winter plants transferred to room temperature reduce their cold resistance in several days. Such a decline in the resistance to heat was observed when *Dactylis glomerata* (Fig. 27) and sedge plants, or leaf pieces taken from under snow, were brought indoors (Alexandrov et al., 1959; Shukhtina, 1962).

One might ask whether the autumn–winter enhancement of thermostability is an immediate response of cells to cold, or whether it depends on other factors. An

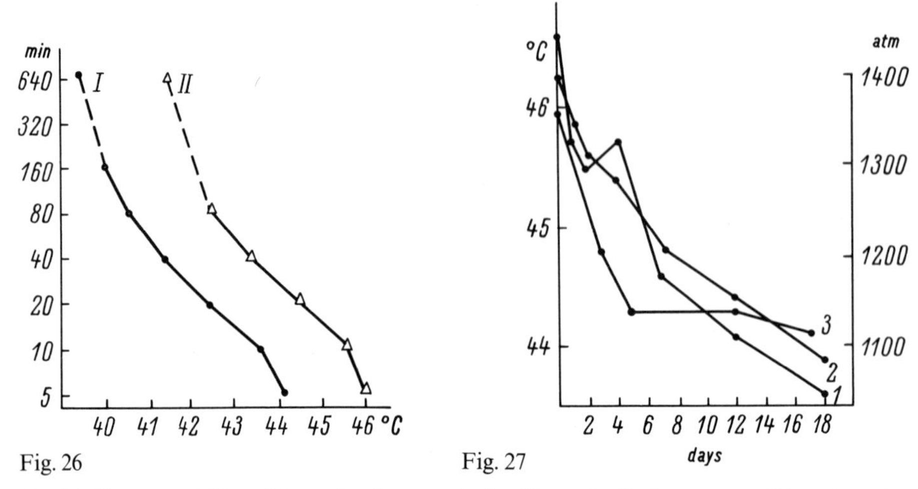

Fig. 26 Fig. 27

Fig. 26. Thermostability of *Dactylis glomerata* leaf epidermal cells in summer *(1)* and winter *(2)*. (Alexandrov et al., 1959). *Abscissa:* temperature of heating; *ordinate:* time period during which protoplasmic streaming persisted (log scale)

Fig. 27. Changes in resistance to heat *(1)* and high hydrostatic pressure *(2)* of epidermal cells of *Dactylis glomerata* leaves after transfer of an intact hibernating plant to room temperature and changes in heat stability of same cells after exposure of leaf pieces to same temperature *(3)*. (Alexandrov et al., 1959). *Abscissa:* days elapsed after transfer to room temperature; *ordinate left:* temperature of 5-min heating that stops protoplasmic streaming; *ordinate right:* value of high hydrostatic pressure effective in stopping protoplasmic streaming in 5 min

unequivocal conclusion valid for all plants apparently cannot yet be reached. In experiments with *Dactylis glomerata*, the plants that had been growing out-of-doors were divided into two groups. Plants in the first group were left in the open air, and the remaining plants were transferred to a greenhouse in which the temperature was maintained at about 8–10° C all winter. The plants taken to the greenhouse exhibited only a slight elevation of their thermostability as compared with those plants that were left out-of-doors. Hence it might be concluded that the autumn–winter increase in the heat resistance of this plant is mostly, although not entirely, due to lowered temperature. Essentially contrasting findings have been reported by Schwarz (1968, 1970) for cedar and rhododendron. In these plants transferred to a temperature-controlled greenhouse, the annual pattern of variations of their cold and heat resistance remained the same as in the plants grown out of doors: the lowest resistance in summer and the highest in the autumn–winter season. These seasonal fluctuations were reduced if the photoperiod was maintained at a constant level throughout the year. It has been concluded that the variations observed were largely determined by the length of day. It takes two months for cedar to respond to a change in the length of day. Biebl arrived at the same conclusion (1967a, 1968; Biebl and Maier, 1969) with regard to various plants of the cold and temperate zone. An artificially produced reduction in the length of day elicited in the Arctic plants an enhancement of cold and heat resistance of their leaves as early as the 10th day of the experiment (Biebl, 1967c). Maier (1971) studied many plants of the temperate zone both in nature and under experimental laboratory conditions. He concluded that an elevation of the heat stability was accounted for by a shortening of the light period and enhancement of the intensity of illumination in daytime.

Induction of a cold-hardened state accompanied by an increase in thermoresistance is, presumably, achieved through the concordant action of low temperatures, shortened daylight periods, and involvement of the inner rhythm, the relative contribution of these factors varying from plant to plant.

Considering the experimental material referred to above, one might deduce that enhanced thermostability of cells always accompanies cold hardening. Such a view seems to be realistic, because a correlation between elevated cold and heat resistance in the autumn–winter period has been reported for representatives of extremely diverse plant groups. However, it has already been shown in our experiments that many plants behave otherwise.

To begin with during the author's test not all the winter plants studied revealed a distinct increase in their cold resistance in winter. No change was detected in the resistance of *Hepatica nobilis* either to cold or to heat from July until April, and it was only in May that the resistance of young leaves to both extremes was appreciably decreased (Alexandrov et al., 1959). This was the case also for the rockfoils *Saxifraga cuneifolia* and *S. umbrosa*. In the periwinkle *Vinca minor* which shows a pronounced winter maximum of cold resistance, only a minor enhancement of its thermostability could be found. Antropova (1971, 1974b) studied seasonal variations of resistance to cold and heat in two moss species. Again, both species exhibited significantly higher cold resistance in winter than in summer. Moreover, the heat resistance of the *Mnium affine* cells was also higher in winter. In a greenhouse, where the autumn–winter temperature depression was eliminated, the winter rise of the cold resistance persisted in this moss, whereas only a slight increase in its thermostability could be detected. In another species *(Atrichum undulatum)* an elevated cold resistance in winter was not accompanied by any changes of its thermostability even outdoors.

Biebl and Maier (1969) distinguish three groups of plants with respect to seasonal variations of the plants' thermostability: (1) plants in which the highest thermostability is observed in summer; (2) plants displaying a thermostability maximum in the cold season; and, (3) those plants, which show an enhanced thermostability both ih winter and in summer. Bauer (1970) recognizes five types: (1) thermoresistance in plants remains the same all year round, being slightly lower in spring; (2) maximal thermoresistance is observed in winter, it is lower in spring and summer; (3) the highest increase in thermostability occurring in winter, whereas in summer an elevation of thermostability is somewhat lower; the lowest resistance to heat is observed in spring and autumn; (4) an equal enhancement of thermostability recorded in summer and in winter; and, finally, (5) maximal thermostability occurs only in summer, otherwise a summer maximum by far exceeds that observed in winter (a summer maximum, as we have seen, may be accounted for by heat hardening).

A source of this variation in the seasonal pattern of cold and heat resistance of plant cells in various species may, presumably, be found primarily in the vast diversity of mechanisms responsible for cold hardening, as well as in the various phenomena accompanying this tate in different groups of plants. Elucidation of the biochemical bases underlying an increase of the resistance of cells to low temperature produced by cold hardening is of primary importance for both theoretical biology and agriculture. Accordingly, a vast amount of literature exists on the mechanisms of cold hardening (see: Levitt, 1972). No satisfactory concept, however, has been developed so far.

Investigations along these lines have been centered on elucidation of the relationship of variations of cold resistance, observed both in nature and under experimental conditions, to shifts in cellular metabolism such as changes in the content of carbohydrates, soluble proteins, nucleic acids, ascorbic acid; to changes in the content and composition of lipids and fatty acids; to shifts in the ratio of S—S vs. SH groups; to variations in the viscosity and permeability of the protoplast, osmotic pressure of the cell sap, etc. Frequently, a close parallelism can be observed between seasonal variations of a plant's cold resistance and this or that biochemical and/or physiological indicator. Difficulties occur, however, when investigations are carried out on other study objects, or when studies on artificial hardening or dehardening of plants are undertaken. Then, it seems, the inseparable links between cold resistance and the character connected with it tend to disappear: either a dissonance interferes with the concordant variations of the character concerned usually accompanying changes in the cold resistance, or the character in question undergoes modification which does not affect the level of cold resistance. In such situations, one is forced to abandon an attempt to establish a causative relationship between cold resistance and the character normally associated with this state. One finds oneself in the situation explicitly depicted in the words of an owner of a zoological panopticum, who declares, showing a hyena: "Hyena—a beast of prey, hunts *only* in the moonlight, and if there is none—he hunts just *as good*".

This situation is clearly demonstrated in an interesting work of Sakai and Yoshida (1968), among others. In a number of plants, these authors succeeded in dissociating variations in cold resistance from accompanying changes in the concentrations of soluble proteins and carbohydrates, and were led to suggest an essentially hypothetical idea of a very broad nature, stating that the primary factor controlling seasonal variations of the cold resistance is conformational changes of cellular membranes. At any rate, it is beyond doubt that the complex changes which accompany cold hardening in various plants, differ by a wide margin. It has been rightly pointed out by Levitt (1972) that many of the variations involved may be related only to maintaining metabolism during the winter rest and to providing for the spring outburst of metabolism, associated with awakening of plant growth. They do not contribute, however, to the establishment of the resistance of plants to cold.

Another aspect of this problem, which complicates and diversifies the seasonal pattern of thermostability variations, is the recognition of other factors besides the low and high temperatures, length of day and inner rhythms that come into the picture and affect thermostability of cells during a season. Among these factors one finds primarily the growth processes and age of tissues, and, in many localities, the water deficiency. As a result, situations may arise, in which the resistance to both cold and heat will be modified in opposite directions (p. 261).

A few words are appropriate here concerning the effects of low temperatures on the heat stability of animal cells. Friedrich (1967) exposed mussels for 60 min to −3° C or for 20 min to −10° C and recorded a simultaneous enhancement of the resistance of ciliary epithelium cells of the mollusc gills to both cold and heat. Kiro (1967) observed an increase in the heat resistance of the *Rotifera* corona ciliary cells when the organisms were cultivated at both elevated (23° C) and lowered (7° C) temperatures as compared with thermostability recorded in a control culture maintained at 17–18° C. As regards seasonal variations of the primary thermostability of animal cells, they are, as has been seen, determined largely by the cyclic shifts of the hormonal balance.

1.3 Changes in the Cellular Primary Thermostability Produced by Non-Temperature Factors

1.3.1 Variations in Cellular Thermostability During Growth and Development of Plants

References made to age-dependent variations of thermostability of leaves can be found as early as in the writings of Sachs (1864). According to his findings, in *Tropaeolum majus* and *Phaseolus vulgaris*, the middle-aged leaves were the most sensitive to heat; in some other plants—*Papaver somniferum*, *Tanacetum vulgare*, *Cannabis sativa* and *Solanum tuberosum*—it was the young leaves that proved to be the most labile. Later numerous writers touched on this problem, but, as a rule, in passing and only in the context of other investigations. Some of them, however, provided confirmatory evidence of the findings of Sachs, and also reported higher sensitivity of the middle-aged leaves as opposed to that exhibited by the young and old ones (in moss, Scheibmair, 1938; in *Kolanchoë*, Schwemmle and Lange, 1959). The majority of investigators, however, demonstrated thermostability of young cells to be low, whereas thermostability was enhanced as leaves approached termination of their growth: in algae such data have been obtained by Ewart (1903); in various gymnosperms—Shirley (1936), Lange (1961); and in angiosperms—Illert (1924), Konis (1949), Badanova (1957), Alexandrov et al. (1959), Lange (1961), Lange and Lange (1963), Karimov and Popova (1966), Popova (1968). According to Klyachko and Kulaeva (1969), protein synthesis in young, growing *Nicotiana* leaves is appreciably more sensitive to heat than that in leaves which had terminated their growth. Bauer (1970) believes the effects of length of day upon thermostability of the fir and maple organs to be accounted for by the changes produced in growth processes. Activation of growth diminishes, whereas termination of growth enhances the resistance of tissues to heat. Conversely, young cells of *Rheo discolor* display higher thermostability than mature ones (Bogen, 1948). Lange (1964) concluded that "no general rule can be established as to the relationship between the resistance of plant organs and their age" (p. 93).

In greater detail an analysis of the relationship between growth and the primary thermostability of cells has been undertaken by Gorban (1962, 1963, 1964,

1968). In various plants *(Echeveria secunda, Zebrina pendula, Tradescantia fluminensis, Sempervivum grandiflorum, Kolanchoë blossfeldiana, Dactylis glomerata)* the epidermal cells of growing leaves in the upper layers of plants were found to be more heat sensitive, as revealed by 5-min heat treatments, than were those in the leaves which had terminated their growth. For all of the objects studied, a criterion of their thermoresistance was the temperature of a heat treatment which arrested the streaming of protoplasm. Additional indicators were also employed: for *Zebrina pendula* and *Echeveria secunda* the temperature of heating effective in suppressing the cells' capacity for plasmolysis was estimated; tests on *Dactylis glomerata* were supplemented by measuring the temperature which altered the pattern of an intrinsic chlorophyll fluorescence in the cells. It was found that epidermal cells located at the base of a leaf plate were more sensitive to the action of high temperatures in growing leaves than were the more differentiated cells at the leaf apex. With termination of the growth process, the difference in the heat resistance of the cells at the base of a leaf plate and of those in the leaf apex disappears, although differences still can be found between the size and shape of the cells.

The findings presented above clearly show the relationship between growth processes and thermostability of cells. Experiments on *Catalpa speciosa* leaves (Shukhtina, 1965) taken from plants growing in the garden of the Botanical Institute in Leningrad also confirm this relationship. In the middle of summer, at the time of intensive growth of leaves in the upper layer of the plant, the heat resistance of small apical leaves was significantly lower than was that of large leaves which had arrested their growth. Thus, in growing leaves measuring 25 × 15 mm, the protoplasmic streaming in epidermal cells was halted by a 5-min heating at 46° C, whereas in those grown to dimensions reaching 260 × 200 mm at 48.2° C. In October, when growth processes stop, the leaves in the upper layers, despite their miniature size, increase their thermostability up to the level typical of large mature leaves.

Feldman and Kamentseva (1963) also indicated that a relationship exists between growth and the primary heat resistance of cells. They studied changes in thermostability of *Gagea lutea* epidermal cells in the course of the plants' development both in nature and under experimental conditions. In the former case, thermostability of seedlings, estimated by the temperature of a 5-min heating that stopped the streaming of protoplasm, was low (Table 6). As the growth declined, the heat resistance grew higher until the florescence stage, it then levelled off and remained unchanged until completion of vegetation. The fact that thermostability of a plant is associated with its growth, but not with its developmental stages, is shown in the following experiments.

Gagea lutea bulbs were forced in winter in a greenhouse. For several months prior to forcing, the bulbs were kept at 0 and 5° C. In the greenhouse the temperature was maintained within 8–12° C, with daylight illumination. The plants that had been forced from bulbs kept at 0° C displayed subdued growth, but eventually entered the florescence stage. In these plants, similar to those grown out of doors, leaf cells initially exhibited low thermostability (Table 7). As the florescence stage approached, the heat resistance grew higher. The growth of the plants forced from bulbs kept at 5° C was markedly inhibited; however, onset of developmental

Table 6. Thermostability of epidermal cells of *G. lutea* leaves during development in nature. (From Feldman and Kamentseva, 1963)

Phase of development	Length of leaves (cm)	Temperature of a 5-min heating which stops protoplasmic streaming (° C)
Appearance of seedlings	1.5–2.0	40.5±0.1
Flow bud formation	14.0–16.0	42.1±0.1
Flowering	20.0–30.0	42.9±0.1
Fruiting	20.0–30.0	42.9±0.2

Table 7. Thermostability of epidermal cells of *G. lutea* leaves during forcing in a greenhouse. (From Feldman and Kamentseva, 1963)

Phase of development	Temperature at which the bulbs (° C)	Length of leaves (cm)	Temperature of a 5-min heating which stops (° C)
Appearance of seedlings	0	1.0– 4.0	40.7±0.2
Flowering	0	7.0–17.0	42.2±0.1
Appearance of seedlings	5	1.2– 1.8	43.4±0.1
Flowering	5	2.0– 8.0	43.7±0.1

stages was not retarded. Accordingly, the level of cellular thermostability was already high in the seedlings and it showed no elevation with development of this dwarf plant. The absence of stage-related effects on thermostability of mature cells has also been reported by Alexandrov (1956), Yazkulyev (1964b) and Falkova and Galushko (1974).

Gorban (1962) found that the cells of *Kolanchoë* growing leaves (1st and 2nd layers from the top) were less resistant to heat than were those of non-growing leaves in lower layers. If, however, a plant was treated with maleic hydrazide, this difference disappeared after termination of leaf growth due to an increase in thermostability of the cells of leaves in the upper layers. Thermostability of the leaves was assayed three weeks following treatment of the plant with maleic hydrazide. Subsequent studies on wheat coleoptiles (Gorban, 1968) showed that, as usually happens, the relationship between the process of cellular growth and the level of cells' thermostability is more complicated than it might appear at first glance. The primary thermostability of cells in coleoptiles undetached from plants, as well as in those which had been separated from a plant and grown in an artificial medium, increased as the growth declined. Supplementing growth medium with indolylacetic acid and sucrose promoted elongation of the cells, and, accordingly, their thermostability decreased. In the absence of sucrose, elongation of the cells was stimulated by indolylacetic acid, whereas their resistance was not reduced. Again, maleic hydrazide, although effective in preventing elongation of detached coleoptiles, did not affect thermostability of the cells. Interestingly, the treated coleoptiles whose growth had been inhibited, eventually elevated their thermostability simultaneously with retardation of growth of the control coleoptiles.

At any rate, in summarizing the data referred to in the foregoing, it may be concluded that, as a rule, growth of plant cells normally proceeds against the background of a reduced level of primary thermostability. Decline of growth is accompanied by an increase in cellular thermoresistance. Under experimental conditions, however, it is sometimes possible to uncouple changes in the growth process and those of resistance of cells.

Simultaneous modifications of these cellular parameters in norm are likely to be accounted for by a dependence of both growth and thermostability of cells upon a common factor, or upon different, but synchronized factors. The problem of the relationship of cellular thermoresistance to ageing and developmental stages is a much more debatable issue.

A controversy found in the relevant literature may well be explained as due not only to differences in the objects being studied, but also to the diversity in approach to assessing resistance of cells. In the latter case, evaluation of the primary and general thermostability of cells may yield dissimilar results. Gorban (1963) showed in *Tradescantia* and *Zebrina* that reduced primary thermostability of growing leaf cells is accompanied by the ability of these cells to repair thermal injury at a higher rate than that typical of mature cells. It is this very phenomenon that accounts for the dependence of the results on a particular experimental design employed for evaluating thermostability of growing and mature cells. To illustrate this point, if a temperature of a 5-min heating that stops the streaming of protoplasm in *Zebrina* cells is estimated immediately after the end of a heat treatment of leaves, then the growing cells may be considered more thermolabile. The protoplasmic streaming in these cells is arrested after exposure of the leaves to 44.8° C, whereas mature cells must be heated to 46.7° C in order to achieve the same result. If, however, a result of heating is checked 12 h after the heat treatment, during which time the cells will have partially repaired the injury, it may be concluded that there is no difference between the heat resistance of young and mature cells. With this experimental design it is seen that in order to stop protoplasmic streaming in the former case the cells must be heated to 47.2° C, and practically the same heating—47.4° C—is sufficient in the latter case.

The cellular thermostability changes that are related to the dynamics of growth contribute to the pattern of seasonal variation of heat resistance of plant cells. An elevation of thermostability occurring in spring, which, it seems, frequently accompanies a rise in air temperature, may, in fact, reflect a forthcoming retardation of growth processes, but not an adjustment of thermostability of tissue cells to ambient temperature (Alexandrov et al., 1964 b).

1.3.2 Changes of Thermostability of Animal Cells in Ontogenesis

Resistance of cells to heat does not remain constant during the cell cycle. This can be seen also in the work of Polyansky and his collaborators (1967) on *Amoeba proteus*. They have found the stability of cells to be reduced soon after division of the cells, so that toward the next division it grows higher. Irlina (1972) studied changes of thermostability of the infusorian *Tetrahymena pyriformis* occurring over the cell cycle and detected there were two periods during which resistance of the cells was diminished: immediately following a division, and at initial stages of the oral morphogenesis. Another pattern of response to pulse heat shocks has been reported by Westra and Dewey (1971) in Chinese hamster cells in a synchronous culture. The highest sensitivity these cells exhibited was during mitosis and S-period, whereas in G_1 the cells displayed the highest resistance.

Evidence has accumulated concerning changes of the level of thermoresistance of cells in the course of development of a multicellular organism. Svinkin (1962b) showed that the process of fertilization of the *Rana temporaria* egg cell did not affect thermostability of the egg, nor could changes be found at the first stages of cleavage; however, starting from the sixth generation of blastomere, thermostability began to increase and at the stage of 128 cells a striking elevation of thermostability could be detected. Andronikov (1963) found no difference in the heat stability of the egg cells and zygotes of the sea urchin *Strongylocentrotus nudus*. Changes in the heat resistance of developing salmon embryos, reported by Gorodilov (1969), will be discussed elsewhere in this book (p. 129). Tissue cells have been studied in this context by Chernokozheva (1967a). She noticed an appreciable enhancement of thermostability of muscles during the development of immature frogs. Following the onset of maturity, the age-dependent variations in thermostability of the frog muscles diminish significantly.

Impressive manifestations of thermostability variations have been reported to occur during the diapause of insects. A diapause is a temporary suspension of development, which, in different species, occurs at different stages of ontogenesis of the insects. Usually, the diapause coincides with a period that is unfavorable for development of an organism (e.g. cold winter or hot summer). Astaurov and his colleagues (Astaurov et al., 1962; Astaurov, 1964) showed that when a developing silk-worm egg enters the summer–autumn diapause, it becomes, at the same time, more resistant to both heat and cold. With transition from the diapause to active development there ensues a decrease in the resistance to heat as well as to cold. Astaurov believed that there was a similarity between a non-specific increase in the cellular thermostability during a diapause and a concordant elevation of the resistance to both cold and heat observed in many plant cells during the state of cold hardening.

Ilyinskaya (1966) compared the stability of intersegmental muscles of the *Carpocapsa pomonella* pronymphs in an active phase and during the autumn-winter diapause. The muscles of the pronyphs during the diapause were found to possess an elevated resistance to high temperature. Further studies are needed to disclose to what extent an analogy can be drawn between the events occurring in plant and animal cells during periods of their relative rest.

1.3.3 Changes in Thermostability of Plant Cells Caused by Water Deficiency

The evidence presented above demonstrates that the level of primary thermostability of plant tissue cells is affected by both cold and heat, and also by non-temperature factors, such as the length of day and growth processes. Water deficiency is also among these factors.

As for the plants capable of withstanding an air dry status, the data are unequivocal: desiccated tissues possess high thermostability. Lange (1955), for instance, has studied some 73 moss species. The maximum temperature of a 30-min heating that did not adversely affect vitality of the mosses varied in the range of 70 to 110° C in different species. The temperature of a heat exposure that the mosses survive in a desiccated state may exceed that effective in killing water saturated mosses by as much as 50–60° C. This was the case also with the lichens. Their thermostability in a swelled state was 40–50° C lower than that of desiccated lichens (Lange, 1953). According to Biebl (1970), air dry littoral algae display significantly higher thermostability than the plants with normal water content.

Partial dehydration of cells causes the relationship between the water content and heat stability of the cells to become more complicated. Ample evidence is available in the literature concerning an enhancement of thermostability in plants produced by water deficit. Thus, Sapper (1935) subdivided the plants she has dealt with into two groups. Plants of the first group were generously supplied with water, whereas those in the second group were maintained in a next-to-wilting state. Despite the fact that the plants in the first group developed much better, their thermostability proved to be lowered. A most pronounced difference was found in those plants which usually endure water shortage rather well. Julander (1945) reported similar experiments made on four grass species and arrived at similar conclusions. Desiccating the fronds of two fern species and the *Ramonda myconi* leaves, Kappen (1966) succeeded in elevating their thermostability. Heat resistance of *R. myconi* increased by 1° C at a water deficit reaching 45%, whereas at a 90% deficit this increase was as great as 7° C. Hammouda and Lange (1962) desiccated *Commelina africana*, *Hedera helix* and *Phoenix reclinata* leaves, and obtained a water deficit of up to 22–27% that elevated the heat stability of the leaves by about 3° C. Similarly, Bannister (1970) reported an appreciable enhancement of the heat stability of leaves in plants belonging to four species of a heather community, when water deficit produced under experimental conditions reached 20%.

All these observations seem to testify unambiguously in favor of an increase in heat resistance produced by reduction of the water content in the cell. In all the works referred to above, an assessment of the resistance was made macroscopically many days following a test heating. Essentially different results have been gained in evaluating the effects of water deficiency on the primary thermoresistance of cells. Zavadskaya (1963b) produced dehydration of detached *T. fluminensis* leaves experimentally and evaluated changes in the water content in leaves, viscosity of the protoplasm, and the primary heat stability of the leaves' cells, assayed by the temperature of a 5-min heating that arrested the streaming of protoplasm. These parameters were related to the time elapsed from the beginning of dehydration. Despite the loss of almost half the water content (by 12–14 days) and a considerable rise in viscosity of the protoplasm, no elevation of the primary thermostability could be detected. Later, Zavadskaya and Denko (1966) working with 21 flowering plant species belonging to different families, desiccated detached leaves followed by assessment of the primary thermostability of epidermal cells by the criterion just mentioned. In 9, mostly xerophyte species, the

Table 8. Thermostability of the cells in detached leaves during water deficit. (From Zavadskaya and Denko, 1966)

Species	Water loss	Temperature of a 5-min heating which halts protoplasmic streaming in leaf cells (° C)		Difference in the thermostability of the cells in desiccated and control leaves
	%	Control	Desiccated	(° C)
First group				
Arabis alpina	33	42.5	43.4	0.9
A. caucasica	25.5	42.3	43.0	0.7
Elymus arenarius	40	44.0	44.8	0.8
Festuca sulcata	15	44.3	45.2	0.9
Hordeum sativum	22	44.7	45.9	1.2
Ruscus aculeatus	20	46.1	46.8	0.7
R. hypoglossum	30	45.8	47.2	1.4
Stipa capillata	15	44.4	45.4	1.0
Veronica prostrata	30	46.1	46.7	0.6
Second group				
Agropyron repens	22	45.0	44.1	−0.9
Beta vulgaris	28	45.0	45.0	0
Bromus inermis	27.5	44.9	45.3	0.4[a]
Campanula carpatica	18	44.5	43.3	−1.2
C. persicifolia	24	44.4	43.3	−1.1
Dactylis glomerata	23	42.8	43.1	0.3[a]
Festuca ovina	35	45.1	45.0	−0.1[a]
Hedera helix	26	42.6	42.6	0
Nymphaea rubra	26	46.3	45.3	−1.0
Panicum miliaceum s. Saratovskoe	25	46.9	47.1	0.2[a]
P. miliaceum, s. Minskoe	24.5	47.3	47.2	−0.1[a]
Tradescantia fluminensis	27	46.3	45.8	−0.5

[a] An unreliable difference. In other cases, the difference is statistically trustworthy ($P<0.05$).

deprivation of the plants of 15–40% of their water content resulted in an increase in the cellular primary thermostability. In the other 12, more mesophytic species, thermostability was either not modified or was somewhat decreased (Table 8).

Still later, Zavadskaya and Denko (1968) studied the primary thermoresistance in 13 plant species of the Eastern Pamirs and four species of the Western Pamirs. The temperature of a 5-min heating that suppresses the ability of epidermal cells to produce plasmolysis was taken as a criterion. All plants, except the mesophyte *Ranunculus pseudohirculus*, grow on dry desert soils. Many of the plants were found to exhibit a considerable water deficit in their tissues. The cellular thermostability of the plants growing in naturally arid localities was compared with that of those found in watered places. In all cases, when water deficit in plants of the first group reached a certain limit, the heat stability of their cells proved to be higher than that in plants found in watered places. By way of control, *Eurotia ceratoides* and *Stipa glareosa* leaves had been detached from the plants and the leaves were then air dried at room temperature. As soon as a desired degree of desiccation was attained the cellular thermostability was estimated. At certain water deficit values, thermostability of the cells reached the same level as in the plants growing in nature under similar water-deficient conditions. Desiccation of the mesophyte *Ranunculus pseudohirculus*, which lives along river banks and irrigation ditches, never increased the primary thermostability of cells.

Diurnal variations of the water content in cells can be observed in the Pamirs xerophytes, which are related to variations in air humidity in arid localities. These variations, however, never generate diurnal fluctuations of the cellular thermostability. This is accounted for by the fact that an increase in thermostability produced by water deficit is notably slow to reverse.

The dramatic enhancement of thermostability of mosses and lichens caused by exhaustive desiccation can be readily explained by the well-known difference in the response of solubilized and dehydrated proteins to thermal denaturation. A heat treatment in the order of 120–150° C is needed to denature dehydrated proteins, whereas these same proteins in a water solution are denatured at temperatures around 60–70° C.

The mechanisms responsible for enhancement of cellular thermostability in partially desiccated tissues are, however, more complicated. It should be noted in this context that the experiments performed by Zavadskaya and Denko (1966) have shown that a partial water loss in numerous plants caused no elevation of the cellular thermostability at all. When, however, a shift in the heat resistance occurs, it can hardly be regarded as being due to the direct effects of a reduction in the water content in the cell. These authors subjected a drought-resistant variety of barley (var. Krasnodarsky 2929) to a soil drought lasting four days, during which time the leaves lost about 36% of their water. This treatment resulted in an increase in the viscosity of the protoplasm, the streaming of protoplasm slowed down and the primary thermostability of cells increased. The plants were then profusely watered. Six hours after watering the leaves were found to regain their full water content. Viscosity and the rate of streaming of the protoplasm returned to normal values. This gave ground to conclude that water balance in the protoplasm had been fully restored. Despite this fact, however, an elevated level of thermoresistance persisted for another two days after commencement of the watering and only after three days did it decrease to the original level. These observations suggest that, in this case, it was not the reduction of the water content itself that was responsible for the enhancement of the thermostability, but some other secondary processes that had developed as a consequence of a partial desiccation of tissues. The authors compared this phenomenon with an increase of the cellular thermostability that can be observed in cold and heat hardening.

The findings reported by Zavadskaya and Denko have found support in the experiments of Falkova (1973, 1975) with a number of Crimean plants. Again, she found that water deficit caused thermostability to grow higher in those plants which were resistant to water loss, and it never resulted in elevation of the heat resistance in the mesophytes.

These findings were at variance with those of the majority of other authors who reported that heat resistance increases in all cases of water deprivation. It was only natural to suspect, therefore, that the reason for this controversy might be found in the differences in the methods used for assessing the heat resistance. Zavadskaya and Denko assayed the primary thermoresistance, whereas other investigators determined the general heat stability. This became apparent after the work of Kappen and Lange (1968) on the mesophyte *Commelina africana*, in which they evaluated the effects of desiccation of detached leaves upon thermostability using two approaches: (1) by determining the temperature of a 30-min heating that arrested the streaming of protoplasm (the leaves were examined microscopically immediately after the end of a heat treatment), and (2) the leaves were exposed to various temperatures for 30 min and their resistance was estimated 10–14 days after the heating by measuring an area of leaf injury. When the first approach was used, these writers could not observe an elevation of thermostability in desiccated leaves, whereas the second method allowed an increase in thermostability to be detected. These experiments demonstrate once again the complicated nature of the concept of "resistance" and the necessity for discriminating between primary and general thermoresistance which reflect different properties of the cell. Kappen and Lange agree with the importance of studies concerning the primary thermoresistance for analyzing cytophysiological problems; for ecological investigations, however, they consider an evaluation of heat resistance by delayed effects to be of greater value. Later it will be shown that the level of primary thermostability too is of principal importance for ecological investigations.

In analyzing the findings referred to in this section, it becomes evident that seasonal variations of water deficit may, in a number of plant species, provoke corresponding fluctuations in their cellular thermostability. If a marked water deficit occurs in summer, it may bring about an increase of thermostability even when a maximal air temperature is not high enough to produce heat hardening. It is this phenomenon that Hellmuth (1971) believes to be the source of summer elevations of thermostability of photosynthesizing organs of the plants growing in the desert and semi-desert regions of Western Australia. Bannister (1970), on the other hand, explains the occurrence of a winter maximum of thermostability in plants of a hearth community in England due by a developing water deficit in winter.

1.3.4 Effects of Salinity on Thermoresistance of Cells

In order to understand the ecology of marine animals and plants found in the littoral, or in other periodically desalted regions, it is essential to know in what way changes in salinity affect the level of cellular thermostability. In the laboratory of Schlieper several investigations of this kind have been performed on the ciliary cells of the mollusc gills (Schlieper and Kowalski, 1956a, 1956b; Reshöft, 1961). An elevation of cellular thermostability has been shown to occur by increasing the salinity of water. Reshöft, among others, was able to demonstrate this phenomenon by adapting *Mytilus edulis* and *Congeria cochleata* to various salinities (6, 15, 30 per thousand). Dregolskaya (1961) reported similar findings for the *Actinia* ciliary cells taken from organisms living in the Black Sea at a salinity of 17.5 per thousand and from those adapted to 35 per thousand. In the latter case, the cells proved to be 68% more thermostable. For five days Ivleva (1962) adapted *Polychaeta* of three Mediterranean species to the Black Sea water, in which the salinity is almost twice as low as in the Mediterranean Sea. As a result, thermostability of muscle fibers decreased. At a later date, Ivleva (1964) adapted *Polychaeta* from the Black Sea to different salinities ranging from 5 to 35 per thousand and found that as the salinity was elevated the heat stability of the muscle tissue increased too. Similarly, Dregolskaya (1963 b) found that the ciliary cells of the mussel *Mytilus galloprovincialis* from the Adriatic Sea (salinity 36 per thousand) were more thermoresistant than those of *M. galloprovincialis* living in the Black Sea (salinity 16–18 per thousand). In the experiments described, lowering of the salinity decreased thermostability of the cells in the first case, whereas increasing the salinity, in the second case, also resulted initially in a decrease of the cellular thermostability, but by the 5th day the heat resistance was found to be elevated. According to Vogel (1966), the heat resistance of the marine infusorian *Zoothamnium hiketes* was the highest in the salinity values to which this organism had been adapted in nature, namely 20 per thousand. Schlieper and Kowalski indicated that changes in the concentrations of ions are the most important factors contributing to the effects of variation of the general salinity. Thus, a decrease in the heat stability of the mussel gill ciliary cells, occurring during shifts in the salinity from 30 to 15 per thousand, can be prevented by doubling the concentrations of Ca and Mg ions.

The experiments reported by Biebl (1972) are in accord with these findings. The freshwater alga *Ulva pertusa* was kept in diluted and concentrated solutions of sea water (0.3 to 4-fold concentration). The higher the concentration, the higher the heat resistance of the algal cells proved to be.

Consequently, as a rule, enhancement of the general salinity provokes an elevation of the cellular heat resistance.

1.3.5 Effects of Wound Injury on Thermostability of Plant Cells

In studying thermostability of leaf epidermal cells, leaf pieces cut out with a razor blade from a leaf plate were used in the author's laboratory rather than sections of leaves. It has been noticed in some objects that following a heating that arrested the streaming of protoplasm in a middle portion of a leaf piece, the protoplasmic streaming can still be seen in the cells adjacent to the periphery of the piece neighboring on the line of incision. This phenomenon has been studied in detail by Feldman (1960). The cells located along the edge of a leaf piece cut out from the *Gagea lutea* plants, found at the forcing stage or in the course of the flower bud formation, exhibited 1–2° C higher heat resistance than the cells situated farther from the mechanically injured tissues. At later stages of plant development this difference disappeared. Similar enhancement of the primary thermostability has also been described for *Campanula persicifolia* leaves. In many a plant this phenomenon cannot be observed.

Summary. The evidence concerning modificational changes of the level of primary cellular thermostability may be summarized as follows. In *Protozoa* and algae, the primary heat resistance is adequately and reversibly shifted during fluctuations of ambient temperature within the whole range of the biokinetic scale. This property, designated as the temperature adjustment, has been lost in tissue cells of multicellular animals as well as in the cells of more highly organized plants. Tissue cells of plants are capable of responding to a short-term superoptimal heating with a reversible increase of their thermostability—heat hardening. Since tissue cells of higher plants can react to exposure at superoptimal temperatures and show no response to temperature fluctuations within the biokinetic zone, a conclusion may be suggested that heat hardening and temperature adjustment are, in essence, different phenomena. However, due to lack of detailed evidence concerning these phenomena in algae, which are able to respond to shifts within both moderate and superoptimal temperature ranges, it is as yet too early to acknowledge the existence in algae of two different mechanisms of response.

Cells of a number of plants reactively alter their thermoresistance in response to low temperatures, water deficit and variations of the length of day. As a rule, thermostability of growing cells increases as the growth declines. The factors listed are essentially those which determine the pattern of seasonal variation of plant cell thermoresistance.

Although under experimental conditions one can obtain an increase in the heat stability of animal cells when they are exposed to extreme temperatures, this elevation can hardly play a significant role in nature during an active period in the life of animals. A cardinal factor affecting the level of thermostability of animal cells is the activity of the endocrine glands. Seasonal variations of the heat stability of animal cells are determined primarily by cyclic shifts in the hormonal balance. The effects of ambient temperature may, in many cases, be explained as an indirect influence mediated through modification of the activity of the endocrine system.

The reasons for the numerous controversies found in the literature, and, in particular, in the contrasting views on the ability or inability of tissue cells to modify their thermal stability with changes in the temperature within the tolerant zone, can be traced down to the fact that authors ignore the complicated nature of the notion of "resistance". Discrepancies arise when the primary and general thermostability of cells are not distinguished from each other. Even if the primary thermoresistance remains unaltered, the cells are able to modify their reparatory capacity adaptively and also the extent of the destructive after-action. Given this complexity, an immediate recording of the result of an intensive pulse heating, on the one hand, and a delayed evaluation of the extent of the thermal injury, produced in 1–2 weeks after the injury, provide an investigator not only with essentially different, but even opposite answers to the questions posed.

The biological significance of variations in the level of cellular thermostability will become clearer after discussing the molecular basis of this phenomenon.

Chapter 2
Genotypic Changes of the Primary Thermoresistance of Cells

The preceding chapter discussed modificational changes in the level of cellular primary thermostability that are caused by various factors of the external and internal milieu of organisms. In some cases, such changes are definitely related to temperature adaptation of organisms. A question may be asked as to whether a change occurred in the primary thermoresistance of cells in the course of the divergent evolution which generated species adapted to life in different temperature habitats. To obtain an answer to this question, one must compare the heat stability of analogous cells in closely related species, differing in their temperature ecology during an active period in the life of the organisms. Research along these lines should take into account the fact that a modificational variability of this parameter exists. Taking aside the modificational effects of extreme temperature, the level of thermostability of animal and plant tissue cells, in terms of its relation to environmental temperature, should be regarded as a conservative property. It is important to eliminate other influences capable of altering cellular thermostability during the life of an individual. In other words, when comparing the cellular thermostability in various plant species, one should use for the studies only mature cells which have terminated their growth: those which have not been affected by either heat or cold hardening, have been spared water deficit, etc. Likewise, for studies carried out in animal cells, only those organisms in the same phase of their sex cycle should be taken for experiments.

Comparative evaluation of the habitat temperature conditions, specific for taxonomically distant species, may be a very difficult task to perform, and it is frequently, no easier with related species. It is the microclimate, and not the macroclimate, that should be considered in the first place (Biebl, 1967b), and only then the active periods in the life of an organism. In this regard, the investigations on gas exchange of lichens performed by Lange and his co-workers in the Negev desert are very instructive (Lange, 1969; Lange et al., 1970a, 1970b). These authors have unexpectedly found that in spite of the hot desert climate, the lichens displayed a surprisingly low temperature optimum for photosynthesis. It was demonstrated that active gas exchange in the lichens occurred only in the first hours after sunrise, when they had not yet expended the water accumulated in dew or adsorbed from the humid air. During a hot period in the day, the lichens entered an inactive air-dry state. The findings reported by Lange and his colleagues demonstrate that, in some cases, when analyzing the thermal ecology of a species, one should take into consideration the time of an active life not only in terms of seasons, but also with regard to the time of day.

Levitt (1966) recognizes the adaptation of living systems to the action of an injurious factor as being achieved by either avoidance or tolerance. A most effective adaptation for avoiding extreme temperatures has proved to be the homeothermal nature of animals. For motile animals, the term "avoidance" may also have a literal connotation. Based on vast evidence, Herter (1943) has shown that a correspondence exists between the perferable temperature and the conditions and way of life of the animals, and also between the temperature preferred by the animals and the specific features of their anatomy, etc. Animal behavior may be of decisive significance for temperature ecology of the animals in one and the same biotope. Suffice it to mention the animals found in hot deserts, both those active is the daytime, and those preferring the night lifestyle. Aside from behavioral acts, the action of extremal temperatures may be subdued in poikilothermal animals through a number of adaptive devices operating at different levels of the organization of a living being. Some of them were mentioned in the introduction to this monograph. A detailed analysis of these devices is outside the scope of the present study, however a cytologist, working in the field of zoology, should not neglect these devices when comparing thermophily of different species.

Although the homeothermal devices of plants as well as their behavioral acts to avoid the action of ambient temperature are severely restricted (Huber, 1935), they, nevertheless, utilize highly efficient means for avoiding extreme temperatures in their biotopes. The selection of a season for a start of vegetation may be regarded among the first means employed by plants to avoid adverse effects of extreme temperatures. Of considerable importance to many plants exposed to overheating, is their capacity for cooling through transpiration (Gates, 1968). In this context, Lange (1959, 1962a, b) distinguishes „Untertemperaturarten", the species which are able to maintain their leaf temperature at a level appreciably lower than the air temperature in the shade. Such plants achieve cooling by vigorously transpiring if exposed to unshielded sunlight at a hot time of day. Sometimes this difference was found to be as large as 15° C. The plants which are incapable of cooling their leaves were called by Lange „Übertemperaturen". These and a host of other devices used by plants and animals to avoid extreme temperatures necessitate cytologist's exercising the utmost caution in estimating the degree of a species' thermophily. These precautions being mentioned, we may now turn to a discussion of the evidence concerning a correlation between primary heat resistance of cells and the environmental conditions of a species' life.

2.1 Animals

Abundant evidence has accumulated in the literature concerning the correspondence between thermoresistance of entire organisms and habitat temperatures (see Fry, 1958). Data of this sort are not directly related to the problem in question and here only the relevant information at the cellular level will be covered. Pertinent findings are very numerous and cover representatives of different types and numerous classes, but are, nevertheless, rather unequivocal. Vernon (1899) was apparently the first to observe that the thermal contracture of the muscles isolated from the southern, lake frog *Rana ridibunda* occurs at a higher temperature than that in the muscles from the northern, grass frog *Rana temporaria*. Still later, Thörner (1919) working with these frog species reported a difference between thermostability of their nerves and muscles in the order of 6° C. Such observations were largely a byproduct of that author's principal interest, and were not considered from an ecological point of view. For the first time since Purkinje and Valentin, the problem of a correspondence between thermostability and thermophily was formulated as such by Battle (1926), who studied tissue preparations from two ray species—*Raja erinacea* living in warmer waters and *R. radiata*. Thermostability of the somatic muscles, gastric and intestinal smooth muscles, nerves and the myonerval junction was found to be higher in *R. erinacea* by 2.2–3.3° C. At a still later date, Patzel (1933) confirmed the data reported by Vernon on the thermal contraction of the *Rana esculenta* and *Rana temporaria* muscles. In order for a contraction to appear in the muscles isolated from the warm-loving *R. esculenta*, an exposure of the preparation for 5 min had to be made at a temperature of 4° C higher than that needed for the same experiment with the *Rana temporaria* muscles. In her experiments, contraction of the muscles taken from two southern lizards occurred at 3–11° C higher than the temperatures which produced contraction of the muscles from two northern lizards. Essentially similar data have been obtained by Adensamer (1934) when comparing the heat resistance of the ischiadic nerve from two northern and two southern lizard species.

After a prolonged spell of silence in relevant research, this problem was approached once again, first in the Soviet Union and later in other countries. In 1952, the present author wrote a paper dealing with a comparison of the resistance to a 5-min heating of the ciliary epithelium of four mollusc species; namely two species from the Barents Sea (one littoral species, *Mytilus edulis*, another, the sublittoral species *Pecten islandicus*) and two freshwater species taken from water bodies in the vicinity of Leningrad (*Anadonta cygnea* and *Unio pictorum*). A difference in the temperature of heatings, which stopped the ciliary beat in the molluscs was found to correspond closely to the difference in the respective temperature habitats; thus, the ciliary beat was arrested at 39.5, 30.0, 43.0, and 44.0° C respectively. The palate ciliary epithelium of the southern frog *Rana ridibunda* was 4° C more heat-resistant than that isolated from *R. temporaria*. Furthermore, a correlation has been found between body temperature and thermostability of the ciliary epithelium in homeothermal organisms, a discussion of which follows.

Findings of this kind have been obtained in abundance by B. Ushakov and his associates. The results of their studies, commenced as early as 1955, have been summarized in reviews by Ushakov (1959b, 1964a, b) and Zhirmunsky (1971). Several hundred species, belonging to almost all types of the animal kingdom have been studied: *Coelenterata* (Zhirmunsky, 1959), *Vermes* (B. Ushakov, 1956b), *Arthropoda* (Crustacea—B. Ushakov, 1956a; Zhirmunsky, 1960a; Makhlin, 1961), *Insecta* (Amosova, 1962), *Mollusca* (B. Ushakov, 1956b; Dzhamusova,

1960b, 1960c, 1963; Zhirmunsky, 1960b, 1964, 1969a, 1973; Zhirmunsky and Tsu Li-tsun, 1960, 1964; Svinkin, 1961; Andronikov, 1968; Vasilyeva et al., 1969), *Echinodermata* (B. Ushakov, 1959a; Andronikov, 1963, 1968; Vasilyeva et al., 1969; Zhirmunsky, 1973), *Chordata* (Ascidia—Zhirmunsky, 1963), *Pisces* (Kusakina, 1960, 1962a, 1962b; B. Ushakov et al., 1962; Zhirmunsky, 1973), *Amphibia* (B. Ushakov, 1955, 1960b; Svinkin, 1959; Shlyakhter, 1960; Krolenko and Nikolsky, 1963; Dzhamusova, 1965b; Glushankova et al., 1967; Pravdina, 1967), *Reptilia* (B. Ushakov, 1958, 1960a, 1962, 1963a; Ushakov and Darevsky, 1959).

In the majority of these studies, the resistance of cells to intensive short-term heatings has been estimated in taxonomically related species, so that the response of a tissue under study has been recorded immediately after the end of heating. Therefore, it was the primary thermostability of cells that was mainly studied in the papers listed. The investigations carried out by Ushakov and his co-workers, in almost all the cases reported, have also revealed a correlation between thermophily of species and thermostability of their cells. Here are several examples to illustrate this point.

Figure 28 depicts curves of the heat resistance of the ciliary cells of marine molluscs living at different depths and found at different geographical latitudes. A difference in the heat resistances of extreme representatives spans over 16° C. There is no doubt that a close correlation exists between the cellular thermostability and the temperature conditions of a species' habitat. In greater detail this can be seen in the work of Zhirmunsky and Tsu Li-tsun (1960). These authors evaluated thermostability of the gill ciliary epithelium in the tropical molluscs belonging to the genus *Nerita*, which live at different littoral horizons of the South Chinese Sea. Figure 29 schematically presents the habitats typical of three species of this genus and the curves of the heat resistance of their ciliary cells are given in Figure 30. It can be seen that the higher and warmer the horizon of a species' habitat, the higher is the level of its cellular thermostability. Similar data have been reported by Kusakina (1963a) for bullheads living in the Baikal Lake (Table 9).

Interesting examples of a parallelism between temperature ecology and cellular thermostability can be found in the work of B. Ushakov (1960a). He studied nine species of desert lizards and two species of turtles living in the Kara-Kum desert. Differences in the ecologies of these species are largely determined not by their habitat, but rather by their ways of life, which fact is reflected in the different heat resistance of their cells. Thus, in the three species of lizards which are active late in the afternoon or at night, the muscle tissues lose their excitability at less intensive heatings than do those isolated from the six species usually active in the afternoon. Muscle fibers, cartilaginous cells and the sympatic and spinal ganglia cells were found to be more thermostable in the warm-loving steppe turtle than were those isolated from the bog turtle.

Reference will not be made further to the countless findings accumulated by B. Ushakov and his collaborators, only to several observations reported by other authors who have found a correlation between the temperature ecology of a species and the heat resistance of the cells. Among these a mention should be made of the data obtained in the laboratory of Schlieper concerning the marine molluscs from the littoral and sublittoral zones (Schlieper, 1960, 1964; Schlieper

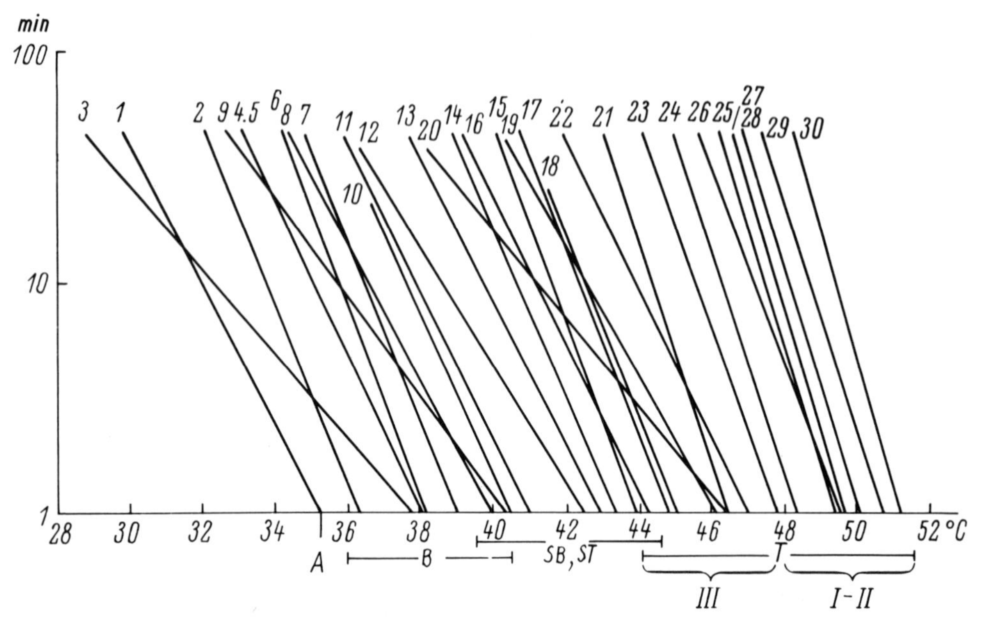

Fig. 28. Curves of thermostability of ciliary epithelium cells of 30 littoral mollusc species (Zhirmunsky, 1964). *Abscissa:* temperature of heating; *ordinate:* time length of ciliary beat (log scale). *A:* an Arctic species, *B:* boreal species, *SB:* south boreal species, *ST:* subtropical species, *T:* tropical species. *I* and *II:* upper and medium littoral horizons, *III:* lower horizons. *1: Saxicava arctica* (L), *Bivalvia; 2: Protolaca staminea* (Conrad.), *Bivalvia; 3: Mya arenarya* L., *Bivalvia; 4: Acmaea scutum* Esch., *Gastropoda; 5: Modiolus modiolus* (L.), *Bivalvia; 6: Natica janthostoma* Deshayes., *Gastropoda; 7: Acmaea cassis* Esch., *Gastropoda; 8: Nucella lima* (Martyn.), *Gastropoda; 9: Buccinum percrassum* Dall., *Gastropoda; 10: Mytilus edulis* L., *Bivalvia; 11: Nucella elongata* Golikov et Kusakin., *Gastropoda; 12: Neptunea artritica* (Bernardi), *Gastropoda; 13: Mya japonica* Jay., *Bivalvia; 14: Venerupis variegatus* (Sowerby), *Bivalvia; 15: V. philippinarum* (Adams et Reeve), *Bivalvia; 16: Tridacna crocea* L., *Bivalvia; 17: Modiolus medcalfei* Hanley., *Bivalvia; 18: M. philippinarum* Hanley., *Bivalvia; 19: Placenta sella* Gmelin., *Bivalvia; 20: Lingula shantungensis* Hata, *Brachipoda; 21: Donax semigranosus* Dunker *tropicus* Scarlato, *Bivalvia; 22: Placenta placuna* L., *Bivalvia; 23: Donax dysoni* Deshayes, *Bivalvia; 24: D. cuneatus* L., *Bivalvia; 25: Modiolus atrata* (Lischke), *Bivalvia; 26: Donax faba* Gmelin., *Bivalvia; 27: Nerita albicilla* L., *Gastropoda; 28: N. plantospira* Anton., *Gastropoda; 29: N. scabriocosta* Lam., *Gastropoda; 30: N. plicata* L., *Gastropoda*

et al., 1960; Reshöft, 1961; Vernberg et al., 1963). Likewise, Licht (1964a, b) showed a correlation between the thermophily and thermostability of muscles in five lizard species. Similar data have been communicated by Tabidze (1972) for two worm species of the genus *Eisenia*; again, a correspondence has been noted between the upper temperature limit for development of five frog species and an area of their distribution (Moore, 1939); the same was the case with five toad species (Volpe, 1953, 1957) and a number of other *Amphibia* (Bachmann, 1969).

In the majority of the species compared, the correlation between their thermal stability and the temperature ecologies has been estimated in one or two tissues. Most frequently, thermostability has been evaluated in muscle fibers or the ciliary epithelium cells. In working with these tissues, the easiest approach is to deter-

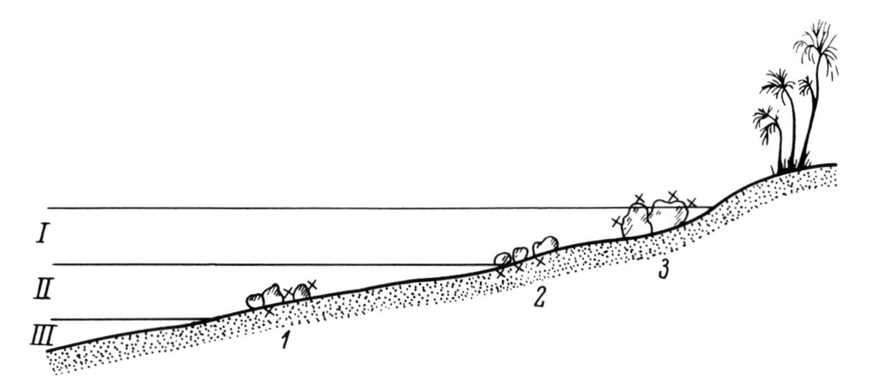

Fig. 29. Distribution of three mollusc species of *Nerita* genus along the littoral (Zhirmunsky and Chu Li-chun, 1960). *1*: *N. albicilla*, *2*: *N. scabriocosta*, *3*: *N. plicata*. The littoral horizons: *I*: upper, *II*: medium, *III*: lower

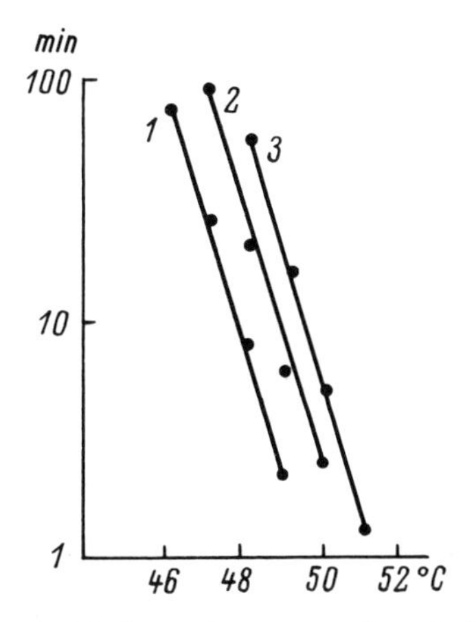

Fig. 30. Effect of heat on ciliary epithelium of *Nerita* molluscs' gills (Zhirmunsky and Chu Li-chun, 1960). *Abscissa*: temperature of heating; *ordinate*: time length of ciliary beat (log scale)

mine the temperature at which their intrinsic motile functions are suppressed. There is every reason to believe, however, that the differences in the cellular thermostability are not limited to individual types of cells, but are found in all cells of the organism, irrespective of their morphophysiological type. This can readily be seen, for example, in two frog types—the northern grass frog *(Rana temporaria)* and the southern lake frog *(Rana ridibunda)*, which have been studied in great detail in this regard (Tables 10 and 11).

The difference in the cellular heat resistance between *R. ridibunda* and *R. temporaria* persists even following a prolonged cultivation of their tissues in vitro. According to Rumyantsev (1960), the *R. ridibunda* myocardium explants retained their higher thermostability, as opposed to that of *R. temporaria* explants, for 84 days of cultivation at 16–19° C. Changing the cultivation temperature to 26° C had no effect on the frogs' tissue thermostability.

Table 9. Thermostability of organism, muscles and muscle cholinesterase from the *Cottoidea* family fishes of the Baikal Lake. (From Kusakina, 1963a)

Species	Zone of spawning (according to Taliev, 1955)	Temperature of thermo-anesthesia of fish organisms (° C)	Temperature at which excitability of muscles is lost in 30 min (° C)	Temperature of 50% inactivation of cholin-esterase (° C)
Cottocomephorus grewinki (Dybowski)	Littoral (0–5 m)	20.1	27.6	42.2
Batrachocottus baicalensis (Dybowski)	Sublittoral (5–100 m)	20.6	26.2	40.8
Procottus jeitelesi (Dybowski)		19.0	25.3	38.2
Asprocottus megalops (Dybowski)	Profundal (100–300 m)	17.5	24.4	36.8
Abyssocottus godlewskii (Dybowski)		17.0	23.2	35.6
Comephorus dybowskij (Korotheff)	Cannot be allocated to any particular bathymetric zone	9.8	—	32.4
Comephorus baicalensis (Pallas)	Are found from sublittoral down to abyssal (deeper than 500 m)	8.7	—	31.2

Table 10. Thermophily of two species—the grass frog (*Rana temporaria* L.) and the lake frog (*R. ridibunda* Pall.)

Characteristics	*R. temporaria*	*R. ridibunda*	Reference
Area boundary			
northern	70° Noth	60° North	Mertens and Müller (1928)
southern	43° North	40° North	Mertens and Müller (1928)
upper	3048 m	2438 m	Terentyev (1950)
Duration of hibernation in Moscow region	180–200 days	210–230 days	Terentyev (1950)
Body temperature recorded in nature	6–26° C	11–29.5° C	Ryumin (1939); Bannikov (1943)
Preferred temperature	13–26° C	18–28° C	Bannikov (1943)

Table 11. Thermostability of organisms, cells and proteins of two frog species, *Rana temporaria* L. and *R. ridibunda* Pall.

Object and indicator of thermostability	Temperature (° C)		Duration of heating (min)	Reference
	R. temporaria	*R. ridibunda*		
Loss of reflex excitability of organism	30–32	35–36	30	Ushakov (1956)
Spinal ganglia neurons, alteration of vital staining pattern	39–40	40–42	30	Ushakov (1960b)
Somatic muscles, complete loss of excitability	36.3	40.4	30	Ushakov (1955)
Spermatozoa, loss of motility	39.2	41.4	30	Svinkin (1959)
Egg cells, loss of ability for development	36.6	38.5	10	Andronikov (1965)
Palate ciliary epithelium, cessation of beating	41.7	44.4–44.8	5	Alexandrov (1952)
Cornea epithelium, alteration of vital staining pattern	39–41	44–46	30	Ushakov (1960b)
Cartilaginous cells, alteration of vital staining pattern	41–42	45–46	30	Ushakov (1960b)
Erythrocytes, alteration of the pattern of vital straining, brightness in the dark field	44	48	15	Braun and Fizhenko (1963)
Myocardia explants, complete loss of excitability	42	44.5	30	Rumyantsev (1960)
Dehydrogenase activity of cornea tissue, 50% inactivation	38	44.5	30	Makhlin (1963)
Glycerol model of plate ciliary epithelium, cessation of beating	40	44	5	Alexandrov and Arronet (1956)
Succinate dehydrogenase activity in homogenates of muscles, 50% inactivation	42	46	30	Vinogradova (1961)
Acetylcholinesterase activity of brain tissue homogenates, 50% inactivation	48.1	48.3	15	Konstantinova and Grigoryeva (1969)
Cholinesterase activity in homogenates of liver, 50% inactivation	41–42	44–45	30	Kusakina (1961)

Table 11 (continued)

Object and indicator of thermostability	Temperature (° C)		Duration of heating (min)	Reference
	R. temporaria	*R. ridibunda*		
ATP-ase activity of homogenates of muscles, complete inactivation	40–42	46–50	5	Komkova and Ushakov (1955)
ATP-ase activity of erythrocytes, 50% inactivation	46	49.6	15	Braun and Fizhenko (1963)
Hemoglobin, 50% denaturing	60	64	15	Braun and Fizhenko (1963)
Actomyosin, 50% inactivation of ATP-ase activity	40.5	44.5	15	Braun et al. (1959)
Myosin, 50% inactivation of ATP-ase activity	35.5	37	30	Konstantinova (1974)
Actin, 50% inactivation	52	56.4	30	Glushankova (1967b)
Adenylate kinase, 50% inactivation	62.5	73.5	30	Konstantinova (1972)
Alkaline phosphatase, 50% inactivation	48.6	50	30	Glushankova (1967a)
Water-soluble esterase, 50% inactivation	44	49	30	Pravdina (1970)
Aldolase in the myogenic fraction of muscle proteins, 37% inactivation	30	35	30	Pravdina (1965)
Aldolase, cathode fraction	No difference in thermostability			Pravdina (1967)
Glutamate dehydrogenase in liver	No difference in thermostability			Yakovleva et al. (1972)
Tendon collagen, 50% contraction (T_c)	51.5	57.6	—	Alexandrov and Andreyeva (1967)
Skin tropocollagen, 50% denaturing (T_D)	25.5	32	—	Alexandrov and Andreyeva (1967)
Serum albumin, beginning of coagulation	68	75	30	Alexandrov and Vitvitsky (1970)

Numerous illustrative examples were referred to in Section 1.2.1.3, in which it was shown that thermostability of cells in various subspecies and populations may be the same within one species, in spite of appreciable differences in their temperature ecologies. This fact, as well as the occurrence of distinct interspecific differences in the resistance of cells to heat, made B. Ushakov (1959a, 1959b) regard the level of thermostability of tissues as a species-related characteristic of poikilothermal animals. This criterion has been repeatedly brought forward in analyses of certain problems of taxonomy and genesis of species (B. Ushakov,

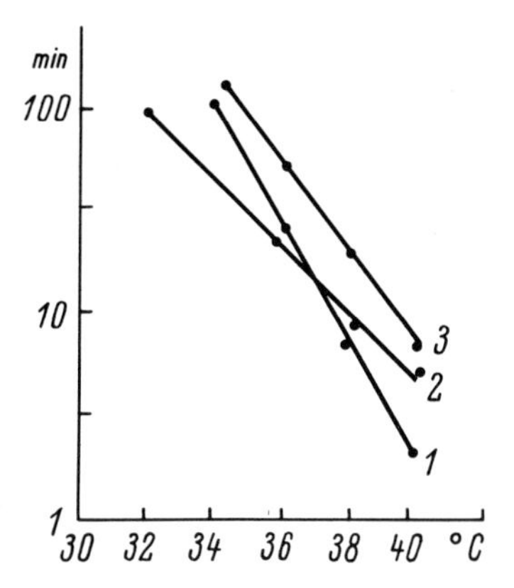

Fig. 31. Effect of temperature on the excitability of muscles of *Littorina* genus molluscs (Dzhamusova, 1960a). *Abscissa:* temperature; *ordinate:* time length, during which excitability persisted (log scale). *1: L. littorea, 2: L. obtusata, 3: L. rudis*

1959a, 1959b; Dzhamusova, 1963; B. Ushakov, 1964; Zhirmunsky, 1964; Altukhov et al., 1968; Meier et al., 1969, 1972; Payusova and Koreshkova, 1972; and others). More detailed investigations revealed, however, as might have been expected, that thermostability, like any other characteristic, is labile in terms of evolution. This accounts for the fact that in some species an intraspecific differentiation of cellular thermostability can be detected. Intraspecific differences in thermostability have been observed in the Baikal Lake fishes: omul and grayling (B. Ushakov et al., 1962), in marine bass (Altukhov et al., 1967), in lizards (B. Ushakov, 1962), in toads (Volpe, 1953) and in *Rana ridibunda* (B. Ushakov, 1963b). The contention is justified that in these cases it is the genotypic, not modificational, variations that are being revealed; the former denoting the onset of a divergence of species. In cytecological investigations of this kind, manifestations of the evolutionary conservatism of cellular thermostability become apparent. Dzhamusova and Shapiro (1960) found, for instance, that the endemic mollusc species which live in the Baikal Lake, display lower cellular heat resistance than the species which have arrived there at later times. Accordingly, the endemic species occupy the cooler habitats. Numerous marine poikilothermal species, characteristic of a wide ecological area, tend to settle deeper in the sea the farther they move to the south, thereby remaining within the limits of the same isotherm. Similarly, the terrestrial species, found in the southern regions of an ecological area, achieve the same goal by moving to higher places in the mountains.

In some cases, no correlation can be detected in related species between thermophily and thermostability of the tissues (Makhlin, 1961). Sometimes, the curves of the cellular thermostability of related species cross each other (Fig. 31). When this happens, the question of which of the species possesses more resistant cells is certainly left open.

The findings described above have been obtained in poikilothermal organisms. Additional evidence is available concerning warm-blooded animals. Comparisons have been made of the temperatures of 5-min heatings that arrested the ciliary beat in the trachea of the European hedgehog (44.6° C), albino mice

(45.1° C), albino rats (46.3° C), guinea-pigs (46.9° C) and rabbits (47.7° C). The differences in thermostability of the cells corresponded to those in body temperatures of the animals studied. In birds (pigeon, cock), conspicuous for their high body temperatures, the cessation of the ciliary cell beating occurred at temperatures 3–4° C higher (49.2° C) than those reported for mammals (Alexandrov, 1952). These data have been corroborated by the findings of Vasyanin (1960) concerning the heat resistance of the muscles from several finch species exhibiting different body temperatures. Skholl (1963a) has studied thermostability of the internal and underskin muscles from the Siberian marmot, albino rat and the bat. She found that thermal stability of the deeper located muscles, which function at higher temperatures, was higher, as opposed to that of the relatively superficial muscles. Intriguing results have been obtained by Skholl (1967) in her studies on thermostability of the muscles in 16 rodent species. She detected slight interspecific differences, but no correlation between the level of the muscle heat resistance and body temperatures could be established. At the same time, Skholl indicated that there was a "distinct correlationship between this characteristic and the temperature conditions of a species development and life" (p. 167).

The difficulties arising in attempts to correctly evaluate the temperature ecology of species, and the acknowledged interference of the so far unknown factors which affect cellular thermostability have been stated above. However, it may be argued, based on the large body of positive findings, that in the process of the evolution of animals, the level of primary thermostability of tissue cells has been brought into correspondence with an average temperature of a species' existence in an active period of the life of organisms.

It was shown in Section 1.2.1.1 that both under experimental conditions and in nature, the level of thermostability of *Protozoa* can readily be modified by transferring the organisms to a different temperature environment within the tolerant zone. When cultivating *Protozoa* at one single temperature, it can be detected, however, that not only interspecific, but also interclonal, genetically determined differences exist in the heat resistance of the cells (Osipov, 1966a, 1966b). This justifies a question regarding those organisms capable of temperature adjustment, whether a correlation exists between their thermoresistance and the temperature ecologies of species. Positive evidence has been provided in a number of investigations. Irlina (1960), for instance, reported the heat stability of *Spirostomum minus* (Rome) to be higher than that of *S. ambiguum* (Leningrad region). Lozina-Lozinsky (1961) showed that the difference in the heat resistance of *Paramecium aurelia* and *P. multimicronucleata* clones, collected in springs with different water temperatures, persisted during the cultivation of the organisms at a constant temperature of 22.5° C for three to four months. Impressive results have been gained by Sukhanova (1959, 1961, 1962a, 1963) in her studies on the heat resistance of the *Protozoa* parasitizing in taxonomically related hosts which differ in their thermophily. A close correlation has been established in these experiments between thermostability of a parasite, thermophily of the host, and the heat resistance of tissue cells of the latter. Both *Balantidium duodeni* and *O. ranarum* may be placed in the following sequence of their decreasing thermoresistance in reference to the heat resistance of the host frog: *Rana ridibunda* > *Rana terrestris* > *Rana temporaria*. Thermostability of the muscle fibers of these frogs

(B. Ushakov, 1961) and that of the ciliary epithelium cells (Alexandrov and Andreyeva, 1974) may be arranged in a similar order. In *R. ridibunda*, the parasitic *Protozoa Balantidium entozoon*, *B. elongatum* and *Nyctotherus cordiformis* also proved to display higher thermostability compared to that of the *Protozoa* dwelling in *R. temporaria*. Similar observations were made by Sukhanova, when comparing the parasitic *Protozoa* taken from toads and worms which differ in their thermophily.

Unfortunately, the experimental approach employed in those experiments does not rule out the possibility that the difference in the heat resistance of parasitic *Protozoa* in various hosts is due, to a greater or lesser extent, to modificational shifts, i.e. to temperature adjustment. In other experiments, however, Sukhanova (1961) revealed decisively the genotypic nature of these differences. She showed that *Balantidium duodeni* and *Nyctotherus cordiformis*, taken from the *R. ridibunda* intestine, retained their heat resistance, which was higher than that of the *Infusoria* of the same species derived from the *R. temporaria* intestine in spite of 8–10-week cultivation at 15° C. Consequently, it may be concluded that *Protozoa* combine the capacity for modificationally adapting the level of their primary thermostability to ambient temperature through hardening and adjustment on the one hand, and with the genotypic adaptation of the primary thermoresistance, on the other hand. The latter is responsible for the establishment of both interspecific and interpopulational differences in thermoresistance.

2.2 Plants

Two remarks are appropriate to precede a description of the findings bearing on a correlation between the cellular and tissue thermostability and the temperature ecology of plants. First, in numerous investigations, especially in the most interesting and vast researches reported by Biebl, Lange and their associates, information is usually provided relating to the general, but not the primary, heat resistance of cells. Biebl preferred prolonged 12-h heat treatments and believed that such exposures were more appropriate in ecological studies. Lange usually employed 30-min exposures to heat, but assessment of the results of thermal injury is carried out macroscopically in one or several weeks following heating of mosses or flowering plants. In the studies that have been made in the author's laboratory concerning a comparative evaluation of the cells in related species exhibiting different thermophily, it was the primary thermoresistance that was estimated. More often than not plots were constructed of the dependence of the duration of heating required to produce arrest of the streaming of protoplasm as a function of heating temperature. Occasionally, when comparing thermostability of cells in various species, our studies have been limited to measuring the temperature of a 5-min heating effective in halting the protoplasmic streaming. In other studies, aside from the protoplasmic streaming, other functions have been employed as indicators. In spite of the differences in methodology, the regularities revealed by various researchers are similar in many respects.

Second, it is no simple problem to give a correct comparative evaluation of species thermophily, when working with plants, much the same as when dealing with animals. Plants, being unable to translocate themselves actively, manage, nevertheless, by resorting to diverse contrivances, to avoid overheating in hot climates, and, conversely, to accumulate warmth where it is scarce. Among these means employed by plants is the choice of an ecological microniche and of the seasons for vegetating, not to mention the whole armamentarium of physiological and morphological adaptations (Huber, 1935; Lange, 1959; Tranquillini, 1964; Billings and Mooney, 1968; and others).

In her universal researches, Sapper (1935) touched on this problem too. She has worked with many plant species belonging to various taxonomic groups and established a correlation between thermostability and the temperature habitats. Sapper recognized six ecological groups. Most sensitive were found to be the freshwater plants *(Elodea, Vallisneria)*, whereas the succulents exhibited the greatest resistance. According to the criteria chosen by that author, the difference between the extreme representatives exceeded 15° C (38.5 and 54° C). Resistance of the plants in the remaining four groups could be distributed within this range according to their respective temperature ecologies.

The evidence offered by Lange and his co-workers has been obtained in a large number of lichen, moss, fern and flowering plant species. In his first relevant paper, Lange (1953) reported findings concerning ca. 100 lichen species taken from 23 different biotopes ranging from the northern Swedish Laplandia to the vicinity of Marseille. The temperature of a 30 min heat treatment that suppressed respiration intensity in the lichens by 50% was chosen as a criterion for assessment of the thermal injury. The writer concluded that a definite correleation existed between the plant cell thermostability and the temperature conditions of the habitats. This regularity held for both swelled and desiccated lichens, the latter being 40–50° more heat resistant than the former ones. Lange remarked that, in contrast to a significant interspecific difference in the sensitivity to heat, the difference within a species was surprisingly small, if present at all, even in the plants growing in different climatic conditions. For some species (e.g. *Cladonia rangiferina*) the intraspecific difference was found to be trustworthy.

In their later researches, Lange and his associates (Lange, 1965, 1969; Lange et al., 1970a, 1970b) have studied in great detail thermostability of the gas exchange in desert lichens and found that these, in particular *Ramalina maciformis*, exhibit low thermostability in a watered state; the heat resistance showed no correspondence to the macroclimate of the hot desert. As mentioned earlier in the book, this discrepancy could be explained by the fact that these lichens are only active early in the cool morning. They are rescued from the heat by entering an air-dry state in which thermostability of the plants appreciably increased.

Later, Lange (1955) studied more then 50 moss species and found a correlation between their thermostability in a desiccated state and the habitat temperatures. Again, the author indicated the absence of any differences in the heat stabilities of the plants of the same species taken from contrasting climatic conditions. But, as in the case of the lichens, there has been detected one moss species, *Hypnum cupressiforme*, a plastic species in terms of its morphology, which displayed differences in its thermostability depending on the habitat. Conceivably, in that case the author was dealing with ecological races or with the state of heat hardening. A temperature maximum for the resistance of mosses is close to a maximal environmental temperature, and, according to Lange, may be a limiting factor in their distribution.

A major study carried out by Lange (1959) was dedicated to the analysis of 45 species of flowering plants found in the Mauritania desert, in the savannas and in the tropical forests of the Ivory Coast. A comparison of taxonomically distant plant species revealed no relationship between their thermoresistance and the temperature habitats. Such a correlation can be seen, however, when comparing closely related species (four species of the genus *Acacia*). At the same time, in unrelated species, a distinct correlation could be noted between thermostability of the leaves and the intensity of their transpiration. In those plants which are capable of appreciably reducing their leaf temperature through transpiration (*Citrullus colocynthis, Cucumis melo, C. prophetarum, Abutilon muticum, Chrozophora senegalensis* and others) the heat stability of their leaves was lower than in species exhibiting low transpirational cooling (*Phoenix dactylifera, Aristida pungens, Boscia senegalensis* and others). The temperature of the date palm *Phoenix dactylifera* leaves may reach above 53° C. It should be noted that as early as 1932, Khlebnikova ascribed the lower thermostability of the watermelon leaves, compared to that of pumpkin, vegetable marrow and melon leaves, to a more intensive cooling by transpiration achieved by the watermelon leaves.

A similar correspondence between the transpirational cooling and the leaf thermostability was observed by Lange and Lange (1963) in 39 Mediterranean plant species: the scleromorphous, poorly transpiring leaves proved to display a higher resistance than did the mesomorphous, vigorously cooling leaves. A difference in the heat stability recorded in extreme representatives was as large as 12° C.

The majority of the studies performed by Biebl have been carried out on algae. They will be discussed elsewhere in this book, and at this point reference will be made to his findings in the plants which are incapable of temperature adjustment. Regarding the ferns and flowering plants in Puerto-Rico, Biebl (1964) concluded, after comparing his data with those reported by Sapper and other writers, that thermostability of the Puerto Rican plants shows no dissimilarity to that of the plants found in central and southern Europe. An explicit relationship can, however, be revealed between thermostability of those plants and the temperature conditions in their respective habitats. The temperature limit for the resistance of plants in the tropical virgin forests (16 species) lies somewhere within 43–48° C, that of the mangrove trees (4 species)—around 50° C, whereas that of the trees, shrubs and grasses in the parks (9 species) is between 50 and 56° C. These measurements have been assessed by the method of Lange, i.e. a 30-min heating followed by assessment of the heat injury several days later.

Likewise, Biebl (1967e) has studied five moss species in the Ceylon forests of Sri Lanka, and, having compared his findings with the data provided by Dircksen (1964), found no significant difference in thermostability of these mosses and the moss species in central Europe. Similar results have been obtained in 10 moss and 23 flowering plant species in western Greenland, and Biebl (1968) concluded that thermostability of Greenlandic mosses was the same as that of the mosses found in temperate zones and in tropical forests. This was also the case with flowering plants: their thermostability was similar to that of the plants in central Europe,

Table 12. Thermostability of adaxial epidermal cells in leaf sheath of grasses

Species	Tribe	Temperature of 5-min heating which stops the protoplasmic streaming (° C)	Locality
Elymus nutans Griseb.	Triticeae	45.8	The Pamirs, Chechekty
E. arenarius L.	Triticeae	45.6	Leningrad
E. angustus Trin.	Triticeae	45.7	Leningrad
Aegilops triuncialis L.	Triticeae	45.9	Ashkhabad
Triticum vulgare Host.	Triticeae	45.3	Leningrad
Secale cereale L.	Triticeae	45.1	Leningrad
S. silvestre Host.	Triticeae	45.9	Ashkhabad
Eremopyrum orientale (L.) Jaub. et Spach.	Triticeae	45.7	Ashkhabad
Hordeum sativum Jessen.	Triticeae	45.6	Leningrad
H. bulbosum L.	Triticeae	44.9	Dushanbe
H. vulgare L.	Triticeae	47.2	Ashkhabad
H. leporinum Link.	Triticeae	45.8	Ashkhabad
Bromus inermis Leyss.	Bromeae	44.0	Leningrad
Zerna pumpeliana (Scribn.) Tzvel.	Bromeae	43.4	Wrangel Island
Deschampsia caespitosa (L.) Beauv.	Aveneae	44.1	Khibin Mntns
D. flexuosa (L.) Trin.	Aveneae	45.2	Khibin Mntns
D. sukatschevii (Popl.) Roshev.	Aveneae	45.3	Taimyr
Anthoxanthum odoratum L.	Phleeae	43.8	Leningrad
A. alpinum Love et Love	Phleeae	44.5	Khibin Mts.
Phleum alpinum L.	Phleeae	45.1	Khibin Mts.
Alopecurus alpinus L.	Phleeae	45.0	Taimyr
A. pratensis L.	Phleeae	43.8	Leningrad
A. ventricosus Pers.	Phleeae	44.1	Dalnye Zelentsy
Poa alpina L.	Poeae	43.5	Khibin Mntns
P. bulbosa L.	Poeae	44.6	Dushanbe
P. pratensis L.	Poeae	44.8	Leningrad
P. trivialis L.	Poeae	43.8	Leningrad
P. tibetica Munro	Poeae	43.9	The Pamirs, Chechekty
P. silvicola Guss.	Poeae	43.4	Dushanbe
Arcthophila fulva (Tin.) Anderss.	Poeae	42.0	Taimyr
Festuca rubra L.	Poeae	44.1	Leningrad
Phippsia concinna (Th. Fries) Lindb.	Poeae	43.0	Taimyr
Catabrosa aquatica L.	Poeae	43.6	Leningrad
Puccinellia gigantea Grossh.	Poeae	42.5	Leningrad
P. capillaris (Liljebl.) Jansen	Poeae	42.5	Leningrad
P. hackeliana V. Kresz.	Poeae	43.3	The Pamirs, Chechekty
P. diffusa Kresz.	Poeae	42.4	Leningrad
Dactylis glomerata L.	Poeae	44.2	Leningrad
Arctagrostis latifolia (R. Br.) Griseb.	Poeae	43.1	Taimyr
A. arundinaceae (Trin.) Beal.	Poeae	41.5	Wrangel Island
Arundo donax L.	Arundineae	45.5	Ashkhabad
Phragmites communis Trin.	Arundineae	45.7	Leningrad
Molinia coerulea (L.) Moench	Molinieae	45.1	Leningrad
Aristida karelinii (Trin. et Rupr.) Roshev.	Aristideae	49.5	Ashkhabad
Aeluropus litoralis (Gonan.) Parl.	Aristideae	48.2	The Vaksh Valley

Table 12 (continued)

Species	Tribe	Temperature of 5-min heating which stops the protoplasmic streaming (° C)	Locality
Eragrostis curvula Ness.	Cynodonteae	48.3	Ashkhabad
Eleusina indica (L.) Gaerth.	Cynodonteae	49.3	Tashkent
Cynodon dactylon (L.) Pers.	Cynodonteae	46.9	Dushanbe
Chloris uliginosa Hack.	Cynodonteae	48.1	Ashkhabad
Panicum miliaceum L.	Paniceae	48.5	Leningrad
P. autidotale Retz.	Paniceae	48.4	Ashkhabad
Paspalum dilatatum Poir.	Paniceae	48.2	Ashkhabad
Saccharum spontaneum L.	Andropogoneae	47.6	Dushanbe
Imperata cylindrica (L.) Beauv.	Andropogoneae	48.3	Dushanbe
Sorghum halepense (L.) Pers.	Andropogoneae	47.9	Dushanbe
S. techicum (Koern.) Roshev.	Andropogoneae	48.8	Dushanbe
S. almum Parodi	Andropogoneae	49.7	Ashkhabad
S. sudanense (Piper) Stapf	Andropogoneae	49.0	Ashkhabad
Zea mays L.	Andropogoneae	48.9	Ashkhabad

and it was not always lower than the heat resistance of the Mediterranean and tropical plants. Biebl believed the excessive thermoresistance of Arctic plants to be related to their high cold resistance, and referred to the ideas advocated by Levitt (1958) that plants are able to exhibit general, non-specific resistance. Besides, Biebl (1964, 1967b) repeatedly indicated that in comparing the temperature ecology with the thermostability, it is important to take into consideration not the macroclimate, but local temperature conditions of respective habitats.

In the author's laboratory, particular attention has been directed to assessing the primary thermostability of cells. Table 12 summarizes the data on the effect of a 5 min heating that arrested the streaming of protoplasm in adaxial epidermal cells in the leaf sheath of 59 grass species. The data for non-hardened plants are included in this table. The curves of the heat resistance of 12 grass species have also been obtained (Fig. 32). The grasses shown in Table 12 were arranged according to the system proposed by Tsvelyov (1969). All of the species belonging to the tribes Aristideae, Cynodonteae, Paniceae and Andropogoneae exhibit higher cellular thermostability (47.6–49.5° C) than do those of the festucoid tribes—Triticeae, Bromeae, Aveneae, Phleae, Poeae (42.2–45.9° C); Arundineae and Molinieae (45.1–45.7° C) occupy an intermediate position. The species in the first four tribes live in tropics and subtropics and vegetate during the hot season. The species of the festucoid tribes are found, as a rule, in temperate and cold geographical belts. Those species of the festucoid tribes which are found in the tropics are the ephemeres or ephemeroids there and develop during the cool season. Some species in these tribes typical of the tropics and subtropics ascend the mountains. Among the grasses studied, the reed *Phragmites communis* is a very surprising one. It can be found in the Khibins and the Kara-Kum desert alike.

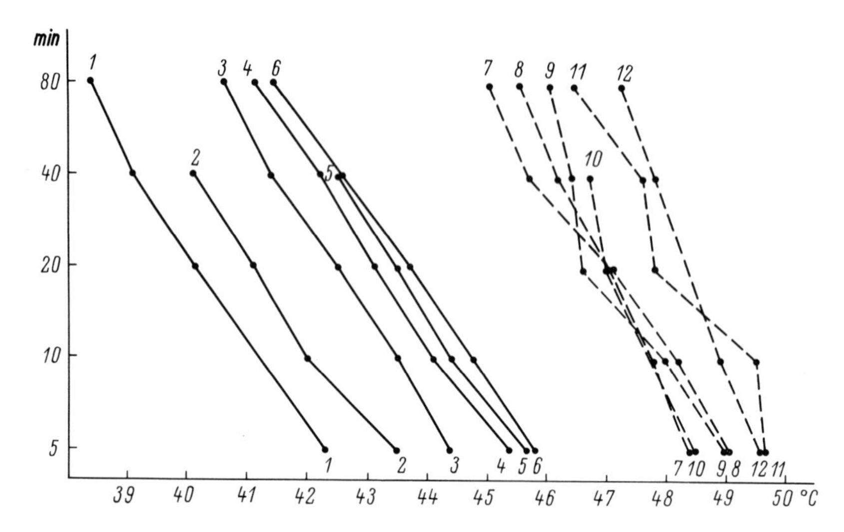

Fig. 32. Plots of heat stability of epidermal cells of leaf sheath in various grasses. (Original data of Alexandrov, Shukhtina and Yazkulyev). *Abscissa:* temperature of heating; *ordinate:* time length during which protoplasmic streaming persisted (log scale). *1: Arctagrotis arundinacea*, *2: Poa alpina*, *3: Dactylis glomerata*, *4: Elymus arenarius*, *5: Phragmites communis*, *6: Hordeum leporinum*, *7: Panicum miliaceum*, *8: Zea mays*, *9: Eleusina indica*, *10: Imperata cylindrica*, *11: Aristida karelinii*, *12: Sorghum almum*. *Solid lines:* grasses of temperate belt tribes *dashed lines:* grasses of subtropical and tropical tribes

Apparently, its low thermostability allows it to move to the north, though the price of this translocation is paid for by loss of the grasses' capacity for generative reproduction. In those regions the plant survives thanks to its capacity for vigorous vegetative reproduction. Life in a hot climate, on the other hand, becomes feasible due to intensive transpiration and the preference of the grass to grow near water bodies as well as in those localities where the ground water is high (Myalo, 1962).

Similar evidence was provided for various representatives of the Amaryllidaceae family (Feldman et al., 1970; Table 13; Fig. 33), as well as for the *Ornithogalum* (Feldman et al., 1975a) and *Allium* (Feldman and Kamentseva, 1974; Feldman et al., 1975; Kamentseva, 1974b) genera. The data depicted in Table 13 are self-explanatory. A correspondence between the cellular thermostability and thermophily is evident. As for the *Ornithogalum* genus, some 14 species living in the middle latitudes have been studied. The temperatures of a 5 min heat treatment that arrested the protoplasmic streaming, ranged for these plants from 44.7 to 46.7° C; for the single tropical species studied in this work, the temperature was found to be 48.4° C (Fig. 34). Considering the *Allium* species, it can be said that in the ephemeroids, which terminate vegetating prior to the onset of heat, the temperature which arrests the protoplasmic streaming is lower than that in the long-vegetating species. Likewise, this difference can be noted when either plasmolysis or respiration are used as indicators.

In addition, correspondence between the cellular thermostability and the temperature ecology of a species can be noticed frequently when other taxonomically

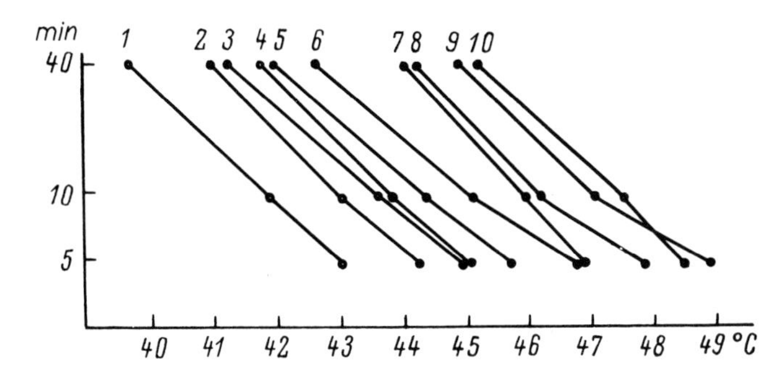

Fig. 33. Effect of heat on protoplasmic streaming in leaf epidermal cells in different Amaryllidaceae (Feldman et al., 1970). *Abscissa:* temperature of heating; *ordinate:* time length during which protoplasmic streaming persisted (log scale). *1: Galanthus platyphyllus, 2: Leucojum vernum, 3: Ixiolirion montanum, 4: Sternbergia fischeriana, 5: Ungernia sewerzovii, 6: Haemanthus albiflos, 7: Hippeastrum* sp., *8: Crinum commelini, 9: Pancratium maritinum, 10: Zephyranthes rosea*

Table 13. Cellular thermostability in various representatives of the Amaryllidaceae family. (From Feldman et al., 1970)

Species	Temperature of 5-min heating which stops protoplasmic streaming, °C	Distribution
Galanthus platyphyllus	43.0 ± 0.1	Middle latitudes
G. nivalis	43.2 ± 0.1	Middle latitudes
G. plicatus	43.4 ± 0.2	Middle latitudes
Narcissus triandrus	44.1 ± 0.2	Middle latitudes, northern subtropics
N. angustifolius	44.9 ± 0.2	Middle latitudes
N. pseudo-narcissus	45.3 ± 0.3	Middle latitudes
Sternbergia lutea	44.6 ± 0.2	Middle latitudes
S. fischeriana	45.0 ± 0.3	Northern subtropics
Leucojum vernum	44.2 ± 0.2	Middle latitudes
L. aestivum	45.7 ± 0.1	Middle latitudes
L. autumnale	46.3 ± 0.2	Northern subtropics
Ixiolirion montanum	45.0 ± 0.2	Northern subtropics
Ungernia sewerzovii	45.7 ± 0.2	Northern subtropics
Clivia miniata	45.9 ± 0.3	Northern subtropics
Nerine bowdenii	46.3 ± 0.3	Southern subtropics, tropics
Haemanthus katherinae	46.7 ± 0.2	Southern subtropics, tropics
H. albiflos	46.8 ± 0.2	Southern subtropics, tropics
Habranthus carinata	47.5 ± 0.2	Southern subtropics, tropics
Crinum commelini	47.8 ± 0.1	Southern subtropics, tropics
Hippeastrum hybridus hort.	48.1 ± 0.2	Southern subtropics, tropics
Vallota speciosa	48.5 ± 0.5	Southern subtropics, tropics
Zephyranthes candida	48.3 ± 0.1	Southern subtropics, tropics
Z. rosea	48.5 ± 0.5	Southern subtropics, tropics
Z. treatiae	48.8 ± 0.2	Southern subtropics, tropics
Pancratium maritimum	48.8 ± 0.2	Northern and southern subtropics

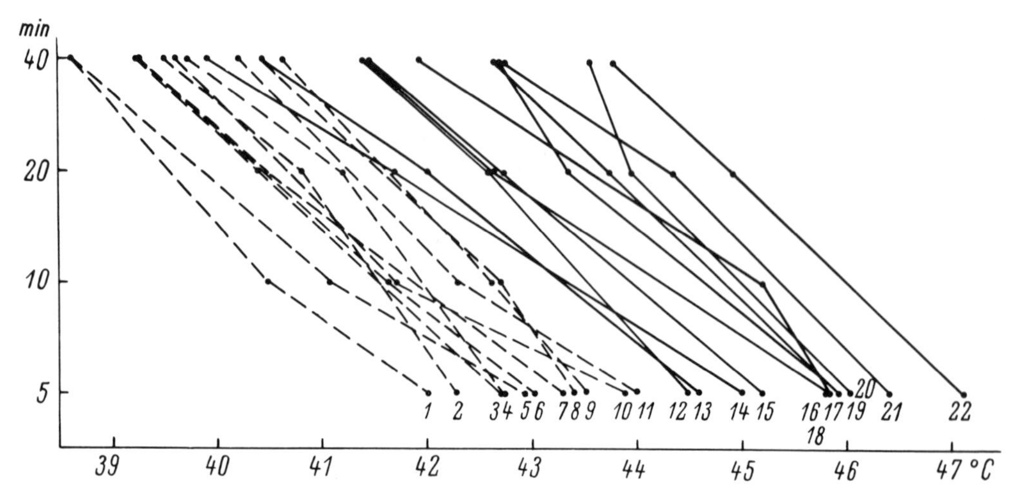

Fig. 34. Thermostability of protoplasmic streaming in leaf epidermal cells in various *Allium* species (Feldman and Kamentseva, 1974). *Abscissa:* temperature of heating; *ordinate:* time length during which protoplasmic streaming persisted (log scale). *1: A stipitatum, 2: A. alexejanum, 3: A karataviense, 4: A. alfatunense, 5: A. giganteum, 6: A. paradoxum, 7: A. gultschense, 8: A. christophii, 9: A. winklerianum, 10: A. suvorovii, 11: A. altissimum, 12: A. oschaninii, 13: A. hymenorrhizum, 14: A. longicuspis, 15: A. barszczewskii, 16: A. pscemense, 17: A. vavilovii, 18: A. filidens, 19: A. galanthum, 20: A. senescens, 21: A. schoenoprasum, 22: A. brevidens. Solid lines:* ephemeroids

related species are compared. Thus, for instance, *Euphorbia helioscopia*, which grows in the vicinities of Ashkhabad, is an ephemere, whereas *E. cheirolepis* vegetates during the whole summer. The leaf epidermal cells in the latter are 2° C more heat resistant than corresponding cells in the ephemere plant (Yazkulyev, 1969). The marine grasses *Zostera marina* and *Z. nana* occupy littoral areas, *Z. marina* growing at greater depths than *Z. nana*. Moreover, an area of distribution for the former reaches up to 70° North, whereas *Z. nana* never goes farther than 60° North. On the other hand, *Z. nana* can be found much farther to the south than *Z. marina*. The heat resistance of these grasses differs by 6° C (Fig. 35).

Thermostability of various functions has been studied in great detail in two *Leucojum* species (Feldman, 1964, 1969; Feldman and Kamentseva, 1967, 1971). The spring snowflake *(L. vernum)* vegetates in Leningrad from early April till May; the summer one *(L. aestivum)*, from mid-April till mid-June. Accordingly, the summer snowflake showed a higher heat resistance of all the indicators studied: protoplasmic streaming, respiration, photosynthesis, the ability to reduce tetrazolium salts to formazan.

The same regularity is evident in one plant when the heat resistance of the various plant parts, exposed to contrasting temperature conditions are being compared. For instance, the protoplasmic streaming in the cotton leaf epidermal cells is arrested by a 5-min heating to 46° C, whereas for the same effect to be achieved in the septum cells of a cotton boll, the temperature must be elevated to 50° C. This difference can be explained by studying the data reported by Aizenshtat (1952). This author showed that the temperature of a cotton leaf plate

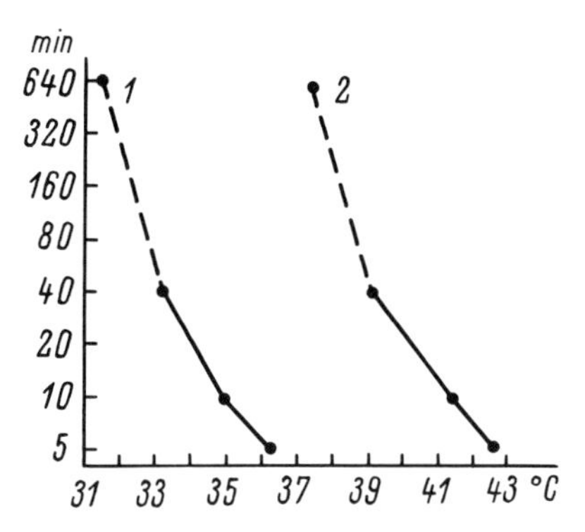

Fig. 35. Effect of temperature on protoplasmic streaming in leaf epidermis cells in two species of marine grasses: *Zostera marina (1)* and *Z. nana (2)* (Feldman and Lutova, 1962). *Abscissa:* temperature; *ordinate:* time period during which protoplasmic streaming persisted (log scale)

exposed to sunlight in daytime is 1–5° C lower than the air temperature. The means for cooling of the sphere-shaped boll through transpiration are severely limited, thus resulting in elevation of the temperature of its tissues on the illuminated side by 5–7° C, and of those in the shade, by 2–3° C over the air temperature.

In the majority of grasses (Yazkulyev, 1964b), thermostability of the adaxial epidermis in the leaf sheath is higher than that of the leaf plate epidermis. This can apparently be ascribed to the difficulty encountered in lowering the temperature of the sheath surface contiguous to the stem through cooling by transpiration. Comparing the heat stability of the spring and summer leaves of the lungwort *Pulmonaria obscura* (Feldman et al., 1966a; Kamentseva, 1969, 1974a) data of considerable interest could be obtained. The spring leaves appear early in spring and function at appreciably lower ambient temperatures than the summer leaves, which, eventually, take the place of the spring ones. Accordingly, the spring leaves displayed lower resistance of the protoplasmic streaming to a 5-min heat shock (42.5° C versus 44.1° C for the summer leaves), of the ability for plasmolysis (53.8 vs. 55.3° C), of photosynthesis (39.5 vs. 42.0° C), respiration (50.0 vs. 53.0° C) and of the ability to reduce tetrazolium salts to formazan (53.8 vs. 55.3° C).

Data are also available in the literature concerning intraspecific variations in the heat resistance of plant cells related to the differences in temperature habitats. Mooney and Billings (1961) conducted an extensive research on a number of geographically isolated populations of *Oxyria digyna* growing from Colorado and California to Alaska and Greenland. They found thermostability of the southern populations to be higher. Oleynikova and Uglov (1962) and Oleynikova (1964a) found that the cellular thermostability of the drought-resistant southern varieties of wheat, beans from Spain and lentils from Iran was higher than that of the varieties distributed farther to the north. The temperature of heating that arrested the protoplasmic streaming or enhanced leakage of electrolytes from the leaves was taken as the criterion for comparing the heat resistance.

Table 14. Thermostability of sublittoral red algae from various climatic zones. (From Biebl, 1967b)

Habitat	Thermostability (° C)	Sea water temperature (° C)
Puerto-Rico	32–35	28
Roskow	27–30	17
Reykjavik (Iceland)	26	10
Godhavn (W. Greenland)	22–24	6

In spite of the diversity of the experimental designs reported, and the difficulties encountered in assessing the temperature conditions under which plant tissues function, the observations accumulated confirm that in the evolution of plants, as in the animal kingdom, the level of primary thermostability of tissue cells has been brought into a correspondence with the temperature ecology of the species in an active period of a plant's life.

It remains to be seen if this is also the case in algae capable of temperature adjustment of their cells; whether it is feasible to detect the genetically established differences in their heat resistance which reflect variations in the temperature conditions of their life.

Monfort et al. (1955, 1957) studied heat stability of the gas exchange in 17 species of the green, brown and red algae from the Kiel fjord. In order to evaluate the genetically determined nature of the level of resistance, the algae have been kept before test experiments for long periods of time at a constant temperature of 14° C. The results obtained pointed to a direct relationship between the heat resistance of the algae and the habitat temperatures. The highest resistance was found in the algae growing in the supralittoral zone, and, particularly, in shallow and quiet bays; the algae in the upper and middle littorals were less resistant, still more labile were the plants in the lower ebb-tide zone. The sublittoral plants were found to exhibit the highest sensitivity to heat. A difference between thermostabilities of the extreme representatives amounted to 13° C.

Temperature ecology of the algae was the subject matter of a continuous series of investigations carried out by Biebl over a period of many years (Biebl, 1939, 1958, 1962a, 1962b, 1964, 1967d, 1969a, 1969b, 1969c, 1970). Scores of the green, brown and red algal species in various regions of the globe have been studied. It was the response of the algae to a 12 h heating that was mostly evaluated. Either macroscopic examinations or microscopic tests (plasmolysis, vital staining) were used as criteria. The main issues of the ample evidence accumulated by Biebl may be summarized as follows: the sublittoral algae always exhibit greater sensitivity to heat than do the littoral forms in the same region. The littoral algae in Brittany showed thermostability that was lower than that of the algae in Naples. No difference in thermostability has been found for the sublittoral algae in those regions. The sublittoral algae in Greenland are more sensitive to heat, whereas the tropical zone algae display higher resistance, than the sublittoral algae in the middle latitudes. Table 14 summarizes the data obtained by Biebl for the red algae. The freshwater algae *Ulothrix* and *Tribonema* in Alaska are more thermolabile than the same algae in Austria; no such difference has been detected for *Spirogyra*. A colleague of Biebl, Dr. Schölm (1968) estimated the heat resistance of various freshwater algae and reported that in summer the algae found close to the surface exhibited higher thermostability than the benthic algae. Again, thermostability was higher in the algae growing in stagnant waters than in those growing in flowing streams.

These observations may be supplemented by the findings of Feldman et al. (1963). These authors have found that thermostability of the *Fucus* algae diminished in the following order: *Fucus filiformis* > *F. vesiculosus* (the littoral form) = *F. disticus* > *F. vesiculosus* (the sublittoral form) > *F. serratus*. This list corresponds to the zonal distribution of the *Fucus* algae along a littoral. The algae growing in the upper horizons of a littoral are exposed to more intensive heat and possess higher thermal resistance of their cells.

To recapitulate, despite the abundance of data obtained on algae, all the doubts that the majority of the differences in heat resistance referred to above are being genetically determined cannot be discarded. The ability of algae to achieve a modificational temperature adjustment must be taken into account. The doubts mentioned may be disregarded completely only when considering the findings reported by Monfort and his co-workers, who kept all the algae before evaluating their thermostability at a constant temperature for prolonged periods of time.

2.3 Microorganisms

Various microorganisms may be found in practically all temperature zones where life is conceivable. The resistance of microorganisms to heat is brought into a correspondence with their temperature ecologies. High stability of the thermophiles is a prerequisite for their normal existence at temperatures in the order of 50–70° C and higher. In many a thermophilic species, a temperature optimum for life is found close to a temperature maximum. For instance, one thermophilic alga (*Oscillatoria* sp.), which lives at 64° C, is killed in 10 min on exposure to 68° C (Bünning and Herdtle, 1946). Lower resistance is typical of the mesophilic organisms, which, in the vegetative state, are killed at temperatures optimum for life of the thermophiles. The psychrophiles, in their turn, which are quite comfortable at temperatures of around 0–15° C, are much more heat labile than the mesophiles. Heating of sea water with a sediment containing psychrophilic bacteria for 100 min to 40° C results in death of 80% of the organisms. The psychrophilic bacterium *Vibrio marinus* is killed in 1 min if transferred to 42° C (Morita, 1966). A 20 min exposure to 35° C completely suppresses the activity of the enzymes in the psychrophilic yeast *Candida* sp. (Stokes, 1967), Hagen et al. (1964) isolated from flounder eggs a psychrophilic Gram-negative rod, which could not grow at temperatures of above 19° C, so that at 25° C approximately 90% of the cells were killed in 15 min. The necessity of an elevated thermostability in thermophiles is readily explainable, but evidently the rather low thermostability in psychrophiles is of no lesser importance.

Summary. The preceding two chapters dealt with alterations of the level of cellular thermostability. Such changes have occurred in the following instances. Modificational changes: (1) The temperature adjustment in *Protozoa* and algae caused by temperature fluctuations within the limits of the tolerant zone; (2) the cyclic variations of thermostability of animal tissue cells effected by endocrine factors; (3) the variations in heat resistance of plant tissue cells, related to activation and decline of growth processes; (4) an augmentation of thermostability of plant cells produced by heat and cold hardening, by water deficit, traumatic irritation; (5) an increase in thermal resistance of animal tissue cells produced by heat hardening and elevation of salinity; (6) variations of thermostability of leaves in some plants related to variations in the length of day. Genotypic variations: adaptation of the level of primary thermoresistance of animal and plant cells to the habitat temperature, which leads to interspecific and frequently also to intraspecific differences in thermoresistance. This list is obviously not exhaustive as regards the modificational and genotypic changes of cellular thermoresistance. In fact, one is often confronted with considerable individual variations in the level of cellular heat resistance, which, at this time, can hardly be related to this or that external or internal factor.

Chapter 3
Variations in Thermostability of Protoplasmic Proteins as a Basis for Changes in the Level of Primary Cellular Thermoresistance

In Section 1.1.2 arguments were offered in favor of the concept that the primary thermostability of a given cellular function is an indirect indicator of stability of certain proteins, which are related to the function concerned and are, at the same time, the most sensitive to the denaturing action of heat. For the purposes of further analysis it is important to realize whether modificational or genotypic changes in the resistance of cells to heat are accompanied by stabilization or labilization of the cells in regard to other injurious agents. It was shown earlier (Nasonov and Alexandrov, 1940) that not only elevated temperature, but also numerous other injurious agents, when applied to the cell, cause denaturation of the protoplasmic proteins. It can be expected, therefore, that stabilization of the proteins to heat will also make the latter more stable against other denaturing agents.

3.1 Are the Shifts in Cellular Thermoresistance Accompanied by Alteration of the Cellular Resistance to Other Injurious Agents?

Although relevant data are as yet scarce, a discussion of these is nonetheless worthwhile. First of all, the degree of unspecificity of modificational changes of the cellular thermostability must be evaluated.

Polyansky and Irlina (Polyansky, 1957; Irlina, 1963a, 1963b; Polyansky and Irlina, 1973) have shown that adjustment of *Paramecium caudatum* to elevated temperature is accompanied by an increase in the resistance of the *Protozoan* not only to heat, but also to ethanol and glucose. Concomittantly, however, the resistance to certain salts (LiCl, NaCl, KCl, $CaCl_2$, $MgCl_2$, KCN) decreases. Maintaining the amoebae at elevated temperatures as well as starvation also resulted in elevation of their resistance to the first two factors. On the other hand, the temperature adjustment of three marine algae (*Ascophyllum nodosum*, *Fucus serratus* and *F. vesiculosus*) increased the resistance of the cells to heat, did not affect their resistance to ethanol and KCNS, and reduced their stability against cold (Lutova et al., 1968). The *Escherichia coli* cells adapted by Zhestyanikov

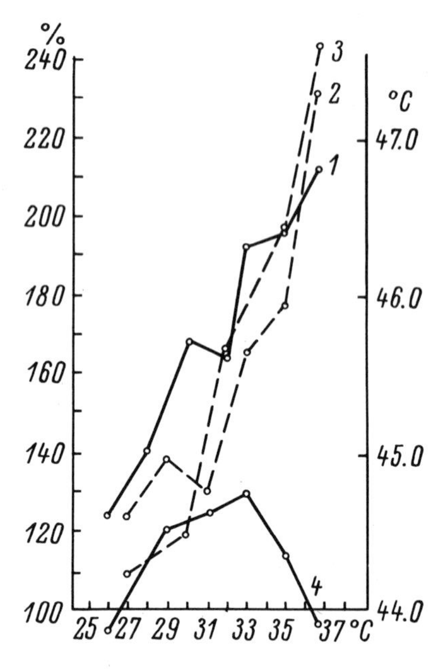

Fig. 36. Dependence of resistance of *Tradescantia* leaf epidermal cells to various agents on temperature of an 18 h heat hardening (Alexandrov and Feldman, 1958). *Abscissa:* temperature of hardening; *ordinate left:* time length during which protoplasmic streaming persisted in hardened cells (percent of that in non-hardened cells) during ethanol, acetic acid and ammonium hydroxide treatment; *ordinate right:* temperature of a 5-min heating that stops protoplasmic streaming. *1:* heating, *2:* 12% ethanol, *3:* 0.02 N acetic acid, *4:* 0.004 N ammonium hydroxide

(1963, 1964) to growing at 50° C did not acquire any resistance to X-rays, but enhanced their resistance to ethanol.

In the studies of heat hardening of plants, the rather non-specific stabilization of cells has been repeatedly observed. After 18-h heat hardening at supraoptimum temperatures, the protoplasmic streaming in *Dactylis glomerata* and *Tradescantia fluminensis* cells became less sensitive to heat, ethanol, acetic acid, and high hydrostatic pressure (HHP); at the same time, the resistance to ammonia showed no changes (Fig. 36). Lomagin et al. (1963) succeeded in obtaining, after an 18-h heat hardening, an elevation of the resistance of *Tradescantia* cells to ethyl ether, KCNS, $CaCl_2$ and sodium azide, applied in the doses which were effective in producing an overall injury of the cell. Susceptibility of the cells to the action of sodium azide and 2,4-dinitrophenol, administered in specifically inhibiting concentrations, remained the same. An increment of the resistance of protoplasmic streaming to ethanol and HHP could be also observed following a 10 sec(!) hardening of *Tradescantia* leaves at 50.5° C (Lomagin, 1961).

In the experiments on heat hardening, apart from an enhancement of heat stability, one could also observe an increase in the resistance of the protoplasmic streaming in the *Zostera marina* cells to ethanol (Feldman and Lutova, 1962), as well as an elevation of the resistance of photosynthesis in the *Podophyllum peltatum* leaves (Lutova, 1962). Lomagin and Antropova (1968) reported the resistance of the myxomycete *Physarum polycephalum* plasmodia to HHP to be raised by heat hardening, whereas the resistance of the slime mold to ethanol remained at the original level. By exposing air-dry or swelled wheat and barley seeds to heat hardening temperatures, Babayan (1972) detected an increment in the resistance of the seeds to heat, X-rays, ethanol, chloroethyleneimine, hydroxylamine, ethyleneamine and to acetone vapors. The plasmolytic method was used by Antropova (1971) to observe an increase in resistance to HHP in a heat-hardened moss. Kiknadze (1960) demonstrated the quenching of the chloroplast intrinsic fluorescence in *C. persicifolia* mesophyll cells that had been produced by the action of heat, ethanol and acetic acid. The quenching was accompanied by partial leakage of chlorophyll from the chloroplasts, followed by dissolution of the chlorophyll in large drops of fat located in the cytoplasm. As a result, the drops started to emit a red fluorescent light. In heat-hardened cells this phenomenon occurred at higher doses of heat and at higher concentrations of ethanol. The resistance of the association of chlorophyll with the chloroplast structures to the action of acetic acid was not elevated by heat hardening.

At first glance, the results obtained by Lomagin (1964) in his studies concerning the effects of heat hardening on the sensitivity of the *Tradescantia* leaf epidermal cells to ultraviolet irradiation (UV), seemed to be rather strange. The injurious action of the UV rays is notable for the explicitly manifested after-effects. The result of irradiation was estimated next day by measuring the percentage of the cells which had lost their antocyan. It was found that in the leaves heat-hardened by a 10-sec immersion in water heated to 54° C, followed by exposure of the leaves to UV at a certain dose, all of the cells retained their antocyan 24 h later, whereas in non-hardened leaves, the percentage of such cells after an irradiation treatment was less than 10%. A hasty inference might be drawn that heat hardening increases the resistance of cells to UV. The same result, however, was obtained when leaves had been irradiated prior to heat hardening. Subsequent experiments have shown that heat hardening does not elevate the primary thermoresistance of cells to UV. It is effective only in diminishing the rate of destructive after-action, apparently by inhibiting the entire cellular activity.

Studies of heat hardening of plants in the authors laboratory have never shown any elevation of the resistance of plant cells to cold (Alexandrov, 1963; Antropova, 1971). Lange (1961) has also reported that heat hardening of plants in summer was not accompanied by elevation of their cold resistance.

A 10-min heat hardening of *Paramecium caudatum*, together with elevation of the heat stability of cells, also increased their resistance to ethanol (Polyansky and Irlina, 1967). Shlyakhter (1959) increased the resistance of frog muscles to heat, ethanol and quinine by heat hardening. Sensitivity to KCl and $CaCl_2$ was not affected. Thörner (1922) observed that a heat-hardened frog nerve retained its excitability for longer periods of time in the atmosphere of nitrogen. Presumably in that case, similar to the experiments with UV irradiation communicated by Lomagin, the result obtained was a consequence of an overall suppression of metabolism.

When cold resistance of plants is accompanied by an increase in their heat stability, a rather non-specific stabilization of the cells evidently takes place. It has been shown in a number of plants that in winter one could obserce not only the

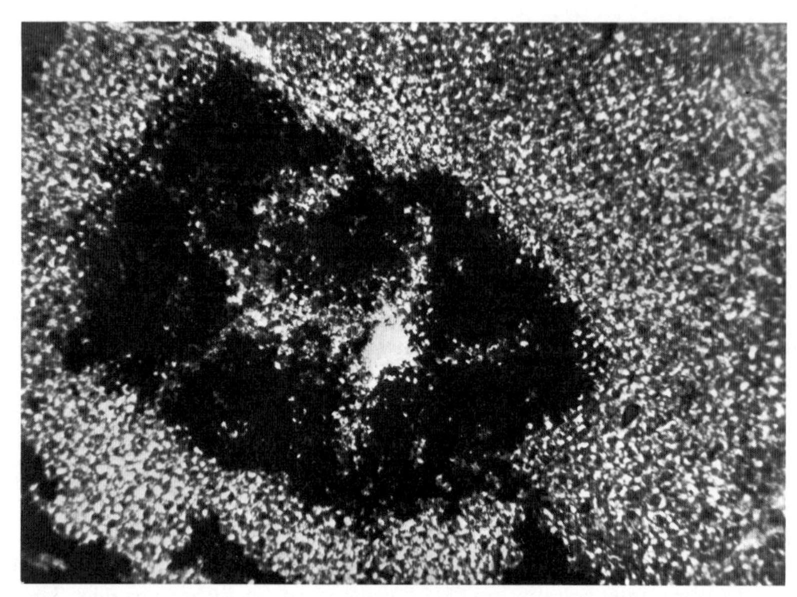

Fig. 37. Wound injury increases resistance of the *Campanula persicifolia* leaf epidermal cells to ethanol (Feldman, 1960). A cut-out *Campanula* leaf piece after a pinprick was treated with 8% ethanol for 48 h followed by staining in neutral red. Cells continguous to the wound are alive and accumulate the dye in their vacuoles; cells remote from pinprick area are dead, showing a diffuse pattern of staining of their protoplasm and nuclei

maximal resistance of the cells to heat and cold, but also to HHP (Fig. 25). The resistance to ethanol also dramatically increased (Alexandrov et al., 1959; Alexandrov et al., 1964b). In the experiments of Denko (1964), cultivation of *Cabomba* at low temperatures increased the resistance of the cells to cold, heat, HHP and ethanol. In winter, when *Dactylis glomerata* is taken indoors from the cold, a parallel decrease in the resistance of cells to both heat and HHP can be observed, along with a reduction of cold hardening (Fig. 27).

An enhancement of cellular thermoresistance, produced by prolonged drought in a number of plants was accompanied by an augmentation of their resistance to HHP and ethanol (Zavadskaya and Denko, 1966, 1968).

Response of the *Campanula persicifolia* and *Gagea lutea* leaves to a traumatic injury elevated the resistance of the cells to both heat and ethanol (Fig. 37).

Earlier data was discussed concerning an increase in thermostability occurring concordantly with a decline of the growth of plant cells. The latter process is accompanied also by an increase in the resistance of cells to HHP (Alexandrov et al., 1959; Alexandrov et al., 1964b; Gorban, 1968). Gorban (1972) has observed in the *Zebrina pendula* leaf cells and in wheat coleoptiles that the older the cells are, the higher is their resistance not only to heat, but also to ethanol administered in concentrations of below 9%, to KCNS, acetic acid, and HHP. The development of resistance to different agents, however, does not proceed concordantly in all cases. At different ages the cells manifest different responses to ammonia, so that the cells of young coleoptiles prove to be less sensitive.

Modificational variations of primary thermostability in animal cells associated with hormonal shifts occurring in the development cycle are also evidently to a considerable extent non-specific. Pashkova (1962a) has detected that during a period of a reproduction of *Rana temporaria*, there occurs, apart from a decrease in the heat stability of the muscles, a reduction in their resistance to ethanol. This was also the case if a suspension of hypophysis was administered to the frogs. An injection of thyrotropin or of an extraction of the thyroid gland enhanced the resistance of the frog muscles to both heat and ethanol (Pashkova, 1963a). In the experiments reported by Suzdalskaya and Kiro (1963), the turtle muscles from animals kept in the cold exhibited higher sensitivity to both heat and ethanol as compared to the muscles from the "warm" animals. Ilyinskaya (1966) showed that with transition of the coddling mot pronymphs to the diapause, their muscles acquire higher resistance not only to heat, but at the same time also to ethanol, dinitrophenol, $HgCl_2$, KCNS, malachite green and methylene blue. Consequently, an arrest of the pronymph development is accompanied largely by a non-specific stabilization of the cells.

The degree to which the genotypic differences in thermostability of cells are accompanied by corresponding differences in the response of the cells to other injurious agents should now be considered. Zhestyanikov (1964) found that thermoresistant *Escherichia coli* strains showed an elevated resistance to ethanol. Those authors who studied various amebal clones (Polyansky et al., 1967; Sopina, 1968a; Yudin and Sopina, 1970) have also noted that the clones conspicuous for their high thermoresistance displayed an elevated resistance to ethanol. *Opalina ranarum* and *Nyctotherus cordiformis*, parasites in the *Rana temporaria* intestine, possess higher thermo and ethanol resistance as opposed to those living in toads. Lozina-Lozinsky (1961) compared the response to various injurious agents in the *Paramecium multimicronucleatum* and *P. aurelia* populations taken from a hot radioactive Slavyanovsky spring with that observed in the populations of these same species found in an ordinary, non-radioactive pond. The populations from the Slavyanovsky spring were found to be more resistant to elevated temperatures, X-rays, dinitrophenol (administered in concentrations exceeding those which specifically inhibit the coupling of oxidative phosphorylation to respiration), hydroxylamine and methylene blue. At the same time, in *P. aurelia* from a hot spring, the resistance to HCN and H_2O_2 was found to be decreased, whereas *P. multimicronucleatum* from both sources displayed equal sensitivity to both UV irradiation and HHP. Muscles taken from related species of frogs, toads and mussels which differed in their thermostability, exhibited similar sensitivity to ethanol (Ushakov, 1955, 1956b). Braun and Fizhenko (1963) detected no difference in the resistance of *Rana ridibunda* and *R. temporaria* red blood cells to either HHP or acid, in spite of the difference in their heat stability. On the other hand, the *R. ridibunda* ciliary epithelium cells, as well as glycerol models of these, were more resistant than those of *R. temporaria*, not only to heat but to high hydrostatic pressure too (Arronet, 1966). Consonant with the latter data are the findings reported by Yazkulyev (1969), for three pairs of related plants which differed in their thermophily. The cells in the more warm-loving species combined an elevated resistance to heat with a greater resistance to HHP.

The evidence referred to above shows that modificational variations of thermostability of plant and animal cells, in the majority of cases, are accompanied by corresponding shifts in the cellular resistance to other injurious agents, in particular to ethanol and HHP—the agents studied in greater detail in this regard. As to the genotypic variations of thermostability, the problem of their specificity has been studied in much less detail and the results obtained are considerably less definitive.

What explanation can be provided as to the simultaneous and unidirectional changes in the sensitivity of cells to elevated temperature, ethanol, acid, HHP, etc.? An hypothesis has been advanced stating that protein molecules in the cell are the targets for these injurious agents. More specifically, the point of application of the injurious action is located in the bonds supporting the secondary, tertiary and quaternary structures of protein macromolecules. Disruption of these bonds results in denaturation, i.e. transition of a unique native conformation of a polypeptide chain into a disordered state. An enhancement or reduction of the protein thermostability is accounted for by strengthening or weakening of the bonds, and interactions responsible for both maintaining a native conformation of a polypeptide chain and for keeping the macromolecular subunits together. This may be achieved by various means: by altering the primary structure of a protein, by fixation or release of the stabilizing ligands (a ligand is a small molecule bound to a protein macromolecule through weak, non-covalent bonds), by changing the composition of the milieu surrounding the macromolecule, by changing the structure of water, etc. Various denaturing agents, depending on their physico-chemical properties, may produce destruction that may vary in degree, of certain elements supporting a native structure of a protein macromolecule (Frenkel, 1962). Strengthening of particular bonds, therefore, can simultaneously enhance the resistance of a protein to the action of a number of denaturing agents which specifically damage those bonds. At the same time, the resistance of this protein to the action of other agents, which mostly destroy other bonds, may remain either unaltered or affected to a lesser extent.

Most frequently, a parallelism can be observed in modificational changes of resistance to heat and ethanol. It might be inferred, therefore, that the loci minoris resistentia regarding these denaturing agents are very closely situated in the protein molecules. In some cases, however, simultaneous changes in the resistance of cells to two agents can be produced not by a single factor, but by two separate ones.

Thus, for instance, the factors responsible for the resistance of cells to HHP have been studied (Alexandrov and Denko, 1971). It was found that this resistance often shifts in the same direction as the stability of the cells against heat. Under experimental conditions, however, it proved feasible to elicit oppositely directed changes in the resistance of cells to these two agents. It was shown in plant and animal cells that an osmotically produced dehydration of cells enhanced their resistance to high hydrostatic pressure, whereas their resistance to heat simultaneously decreased. Polyploidization in *Campanula persicifolia* and *Philadelphus lemoinei* elevated the osmotic pressure of the cell sap and increased the resistance of the cells to HHP. This, however, did not affect the sensitivity of these cells to

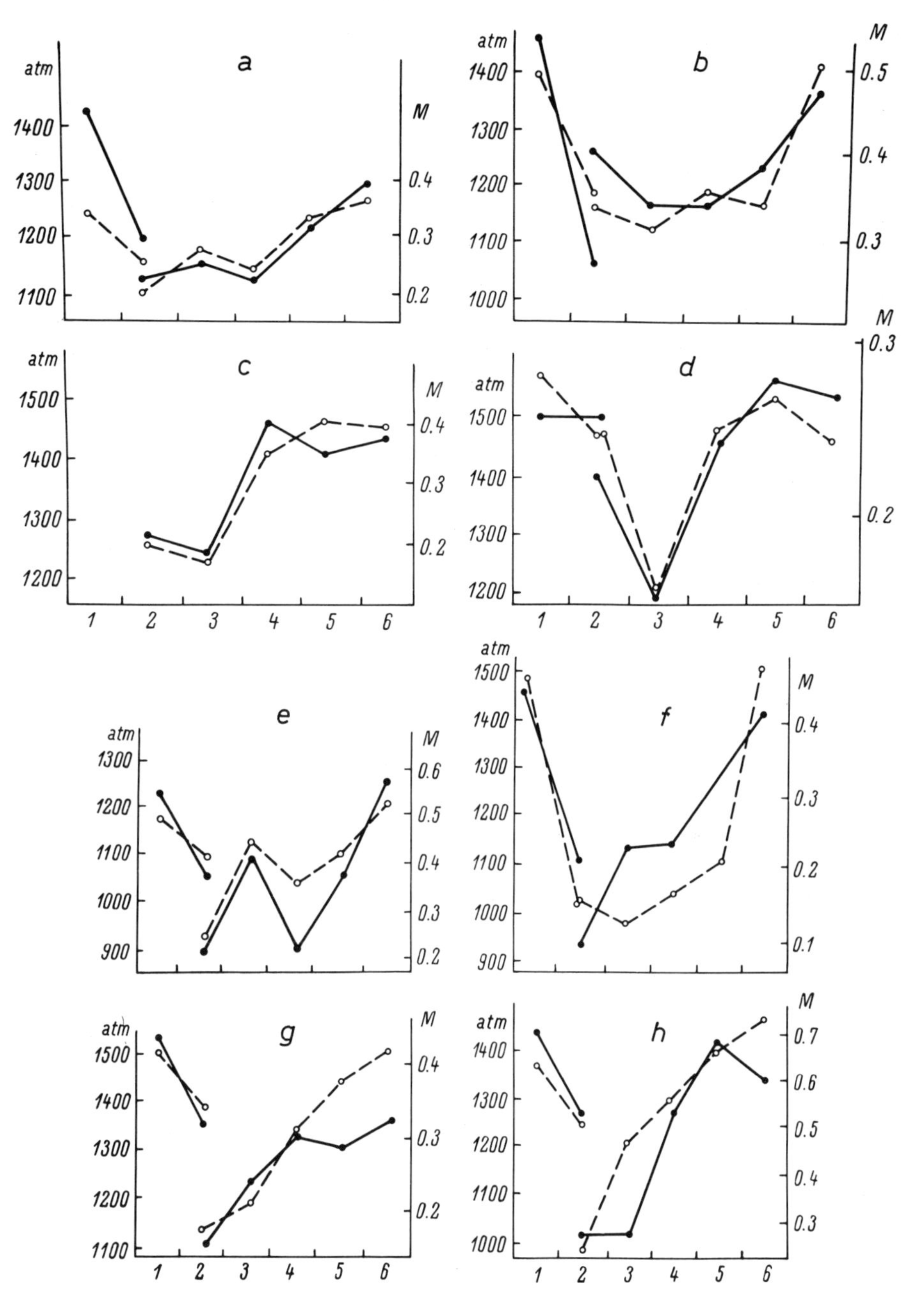

Fig. 38 a–h. Seasonal variations of stability of leaf epidermal cells against high hydrostatic pressure *(solid lines)* and variations of osmotic pressure of cell sap *(dashed lines)* (Alexandrov and Denko, 1971). *Abscissa:* seasons: *1:* winter, *2:* spring, *3:* summer, *4:* autumn before first autumn frosts, *5:* autumn after the onset of frosts, *6:* winter; *ordinate left:* hydrostatic pressure that stops protoplasmic streaming in 5 min; *ordinate right:* minimal concentration of KNO_3 producing plasmolysis in cells. (a) *Leocanthemum vulgare*, (b) *Dactylis glomerata*, (c) *Campanula persicifolia*, (d) *Chelidonium majus*, (e) *Geum rivale*, (f) *Sedum spurium*, (g) *Saxifraga cuneifolia*, (h) *Vinca minor*

heat (Ashraf, 1971). A close correlation has been observed between seasonal variations of the osmotic pressure in the cells of eight plant species and the variations in their resistance to HHP (Fig. 38). In only five species were these changes accompanied by corresponding changes in their cellular thermostability, whereas in other species *(Sedum spurium, Saxifraga cuneifolia, Chelidonium majus)* practically no seasonal variations of thermostability could be detected. On the other hand, we have seen that with heat hardening of plants, their resistance to both heat and HHP, as a rule, increased, though elevation of the osmotic pressure could never be observed in the cells. Moreover, as will be shown later, certain anti-denaturing agents, such as sugars, glycerol, and calcium ions, in the concentrations below those effective in osmotic dehydration of cells, produce a simultaneous increase in the resistance of cells to both agents (heat and HHP). An in-depth analysis revealed that the level of resistance of cells to high hydrostatic pressure may be altered by factors, which either do not affect cellular thermostability, or produce oppositely directed shifts in the latter. On the other hand, the influences which stabilize cellular proteins can, simultaneously and unidirectionally, modify the sensitivity of the cells to both factors. This phenomenon may explain the absence of complete uniformity in the evidence obtained.

The experiments reported by Zhestyanikov (1963, 1968) pointed to the absence of any relationship in the resistance of bacteria to heat and X-rays. Lozina-Lozinsky (1961) found that infusoria from a hot radioactive spring displayed an increased resistance to both factors. It is highly likely that the resistance to those two factors was determined by two different mechanisms.

The cause of an unspecific modification of cellular resistance to agents of different physico-chemical nature may lie, aside from stabilization of the proteins, in either an increase or decrease in the cellular permeability. Unfortunately, the effects of modificational and genotypic variations of thermostability on the cellular permeability have not been approached at all. By using heat and HHP to produce a cellular injury, an investigator is free of difficulties of this kind. If, however, he is inclined to extend the range of the injurious agents so as to include dissolved substances, he is obliged to provide answers to the question of whether the resistance of intracellular components has undergone modification, or whether the permeability of the cells toward that particular agent has been modified.

The complex nature of the problem in question and the insufficiency of pertinent observations accumulated to date are obvious. Nevertheless, it is appropriate to infer that the repeatedly noted simultaneous and unidirectional modifications of the resistance of cells to agents of essentially different nature are, rather frequently, accounted for by the changes in the level of stability of the protein macromolecules in the cell. Consequently, if modificational or genotypic adaptations of organisms to any natural factor are accompanied by changes in the resistance of their cellular proteins to that factor, this may result in an analogous modification of the organisms' response and in the response of their cells to other, non-ecological factors which have never had a chance to come into contact with the organisms during the whole evolution period of the organisms concerned. Relevant findings can frequently be met in literature, and they are considered as non-specific shifts in resistance.

3.2 Changes in Protein Thermostability Caused by Isolation of the Proteins From Cells

The preceding sections discussed the changes in the primary thermostability of cells, regarding them as an indicator, though an indirect one, of corresponding shifts in the resistance of the protoplasmic proteins to the denaturing action of heat. An attempt was made to substantiate this contention by resorting to a number of arguments referred to in Section 1.1.2. The reader may find the arguments offered unsatisfying. A direct proof for the validity of utilizing the primary thermostability of cells as a criterion of the heat stability of cellular proteins may be seen in those experiments, which would have shown that a difference in the cellular primary thermoresistance (of the cells taken either in different states or from different study objects) corresponds to unequivocal differences in the heat resistance of the proteins isolated from those cells. Such proofs will be offered in abundance; a presentation of these should be, however, preceded by an analysis of the principle difficulties and traps which continually interfere with relevant experiments.

The difficulty is that proteins, in particular the enzymes[3] inside the cell, and those isolated from it, may prove to be absolutely different objects, both in regard to their biochemical characteristics (Konev et al., 1970) and in their response to various agents.

Most often a protein displays much greater stability inside the cells than in a tissue homogenate, or when isolated in a preparation. As early as 1933, Pauli and Valkó wrote in their monograph on proteins, that during isolation and purification procedures the proteins are separated from the physico-chemical milieu which maintains them in a native state; thereby the proteins, as a rule, become more labile and denature more readily. For instance, bovine liver glutamate dehydrogenase, when retained in the liver tissue, shows 12–14° C higher heat stability than the enzyme which had been purified and assayed for its resistance in vitro (Yakovleva and Gubnitsky, 1970, 1971, 1973). Aldolase (Makhlin, 1965) and actomyosin (Arronet, 1966) are more thermostable when they are in the frog muscles than when isolated from tissues. Hemoglobin is more stable against heat in the red blood cells than in solution (Kusakina and Skholl, 1963). The ATPase activity of the frog muscle actomyosin is inhibited in vitro by 1 M urea, whereas in the muscles 10 M urea does not completely inhibit the ATPase activity (Konev et al., 1968; Stabrovskaya and Braun, 1969). The same regularity also holds true for plant cells. Acid phosphatase is more stable against heat in the snowflake (Feldman 1969) and in wheat cells (Shcherbakova, 1971 b) than in extracts of these cells. Higher thermostability of enzymes in situ than in vitro has been reported also for thermophilic, mesophilic and psychrophilic bacteria (Maas and Davis, 1952; Militzer and Burns, 1954; Burton and Morita, 1963; Morita and Burton, 1963). Figure 39 shows the plots of inactivation of the *Bacillus stearothermophilus* pyruvate oxidase in intact cells and in particles derived from the cells (Militzer and Burns, 1954).

---

[3] Designations of enzymes in this book will be given as indicated in the original publications.

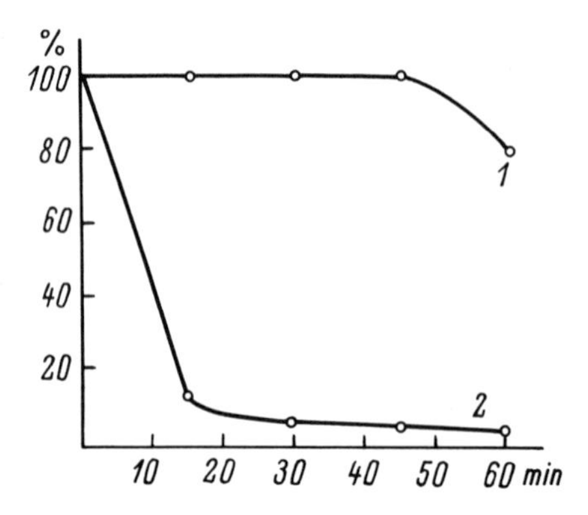

Fig. 39. Activity of *Bacillus stearothermophilus* pyruvate oxidase as function of length of heating of intact cells *(1)* and isolated particles *(2)* to 60° C (Militzer and Burns, 1954). *Abscissa:* length of heating; *ordinate:* enzyme activity (% of control)

Sometimes thermostability of the enzymes in situ and in vitro obeys an opposite regularity. Gorban (1972) has found that the heat stability of urease was appreciably lower in cucumber cotyledons than in their homogenates. Certain enzymes enhanced their thermostability when isolated from bacterial cells (Grisolia, 1964). Peroxidase in intact vegetables was more readily inactivated by an intensive head treatment of the plants (12 min at 98° C) than in a purified state. If, however, a prolonged exposure to a somewhat lower temperature was used, the regularity proved to be the opposite (Herrlinger and Kiermeier, 1948).

Finally, examples are also known when isolation of a protein from the cell does not modify its thermostability: homoserine transsuccinylase within the cell, in extracts, and in a purified state displayed the same temperature dependence in its activity and the same thermostability (Ron and Shani, 1971b). Thus, no general regularity exists in this respect.

The modification of the response of proteins to injurious agents, produced by isolating the proteins from cells, may be not only of quantitative, but of qualitative nature as well. According to Konev et al. (1968), the denaturing of myosin with sodium dodecylsulfate in the muscles and in solution produced oppositely directed shifts in the fluorescence spectrum. Liver glutamate dehydrogenase in vitro exhibits higher sensitivity to heat with respect to its ability to catalyze the reaction of the reductive amination, as compared to that of the reaction of the oxidative deamination. When the enzyme preparation is heated in situ the reverse response can be observed (Yakovleva and Gubnitsky, 1970, 1971, 1972, 1973).

There are many causes to account for the contrasting behavior of proteins in situ and in vitro in regard to injurious agents. First, protein is found in essentially different concentrations in the cell and in a solution. Second, a protein molecule in the cell and in a solution is surrounded by appreciably different media having extremely dissimilar compositions. Furthermore, by isolating the protein from the cell it is often dissociated from the complex in which it is incorporated (other proteins, lipids, polysaccharides). In addition, the ligands are split from it. All of this taken together can lead to either stabilization or labilization of the protein structure. In studies of the heat stability of enzymes in tissue homogenates the destruction of the compartmentalization of the cells creates an artificial medium that may considerably affect the state of the protein macromolecule.

On the other hand, when testing the heat stability of enzymes in situ, frequently tissues must be heated to a temperature that is disastrous to the cell system. In this case, the denaturing action of heat upon the enzyme concerned may be complemented by inactivation of the enzyme by proteases or other toxic products, which in vivo had been spatially separated from the enzyme. An injurious agent, when applied to the cell, may affect not the activity of a protein, but its extractability by either facilitating it or making it more difficult. This, in its turn, may affect the result of an experiment and complicate explanation of the latter. Therefore, it is not the discrepancies in the observations made on the stability of proteins in situ and in vitro that are surprising, but a coincidence in the results of such comparisons that can be observed.

Apart from comparing the thermostability of cells with that of the proteins isolated from them, it may be helpful to compare the response to heat of the contractile proteins in muscles and ciliary cells, on the one hand, and in glycerol cell models, on the other (Alexandrov and Arronet, 1956; Arronet, 1971). Such models may be regarded as an intermediate link between the contractile proteins in a living cell and in solution. To prepare the models, muscle fibers or pieces of tissue with the ciliary cells are killed in the cold by a 45% glycerol solution. After several days, glycerol is substituted for a salt solution. The muscle fibers treated thus react by contraction to the addition of ATP, whereas the ciliary cells, by vigorous ciliary beating. In the course of preparation of the models, all soluble substances are extracted from the cells, including soluble proteins, carbohydrates, salts and various metabolites. Thereby, the cellular systems responsible for the protein synthesis and energy production are being destroyed. The nucleus and protoplasm content is coagulated. And, in spite of this treatment, the contractile proteins retain their ATPase activity as well as their ability to contract upon addition of ATP. That is possible only in the case of the native arrangement of actomyosin being preserved. The state of actomyosin in glycerol models is very much the same as in vivo rather than in solution.

3.2.1 Modificational Variations of Protein Thermostability

3.2.1.1 Plants

The rather ample evidence concerning modificational variations in primary thermostability of plant cells (see Chap. 1) can be supplemented with only scanty data on the state of isolated proteins. There appear to be no reports dealing with the resistance of proteins as a function of temperature fluctuations within the tolerant zone, of salinity, and of growth processes. Some findings are available, however, pertinent to heat hardening.

Novoselova et al. (1966) subjected wheat seedlings to repeated heat treatments, alternating these with exposure of the plants to 22° C for several days. The activities of a number of enzymes were then estimated in the treated and control plants. It was found that a preliminary heating of the plants increased the heat stability of their enzymes. Unfortunately, this publication lacks any coherent description of the experiments reported and the methods used, a statistical evaluation of the results is missing, and, more important, it is unclear whether the authors were dealing with the effects of heat hardening on the heat stability of the enzymes, or with the effects on the amount of the enzymes in the treated cells. Similar results can be found in a later publication by Novoselova et al. (1971).

This problem has been taken up in a series of studies performed in the author's laboratory. Initial experiments along these lines have been accomplished by Feldman (1966) on dialyzed and non-dialyzed homogenates of hardened (18 h at 36 or 40° C) and non-hardened cucumber leaves. After heating the homogenates for 15 min at different temperatures (40–80° C) a decrease in the urease activity was assessed. The activity was found to be reduced to a lesser extent in the homogenate of hardened leaves. The effect was not infallibly reproducible and, in subsequent years, the same approach failed to reveal any differences in the heat stability of urease from heat hardened and non-hardened leaves. An enhancement of thermostability following heat hardening has been observed in acid phosphatase from the cucumber leaves and wheat as well as in ATPase in *Caragana arborescens* leaves (Feldman et al., 1966b; Feldman, 1968; Shcherbakova et al., 1973). It should be indicated, however, that activity of these enzymes markedly decreased after heat hardening. This reduction of activity could be a consequence of either inactivation of a fraction of the most sensitive molecules, or of a partial suppression of the activity of all the molecules of an enzyme during heat hardening. It can be seen, however, from the plots given in Figure 40, that although the curves of the temperature-dependent inactivation of enzymes isolated from the hardened leaves tend to lie below those of the non-hardened leaves over a wide range of temperatures, the curves intersect at higher heating temperatures. This indubitably points to the presence of enzyme molecules with elevated resistance in extracts of the hardened leaves.

In crude extracts of non-hardened wheat leaves, glucose-6-phosphate dehydrogenase proved to be more heat-stable than in extracts of the control leaves (Fig. 41). Spontaneous inactivation of the enzyme at 24° C in the extract of hardened leaves also proceeded at a lower rate (Shcherbakova, 1972). Heat hardening of pea leaves (18 h at 35–37° C) was found to elevate the heat resistance of glucose-6-phosphate dehydrogenase, acid phosphatase and ferredoxin (Feldman et al., 1975b). These data on heat hardening are fairly consistent with the findings reported by American authors. Kinbacher and his associates (Kinbacher et al., 1967; Sullivan and Kinbacher, 1967) heat hardened entire *Phaseolus* plants by exposing them for several days to 44–47° C for only 2 h per day. They isolated semi-purified preparations of malate dehydrogenase and of the protein fraction 1 from the leaves of the heat-hardened and untreated plants. The preparations were then exposed to a test heating. In all cases the proteins from the heat-hardened plants proved to be more resistant to heat. As a result of heat hardening, malate dehydrogenase increased its thermostability by almost 3° C; in addition, its resistance to pH shifts also enhanced, especially so in regard to shifts into acidic pH range. The heat stability of the kidney bean leaves themselves was elevated by heat hardening by 3.6° C on the average. The authors concluded that heat hardening augmented the stability of the proteins toward the denaturing action of heat by modifying the protein configuration.

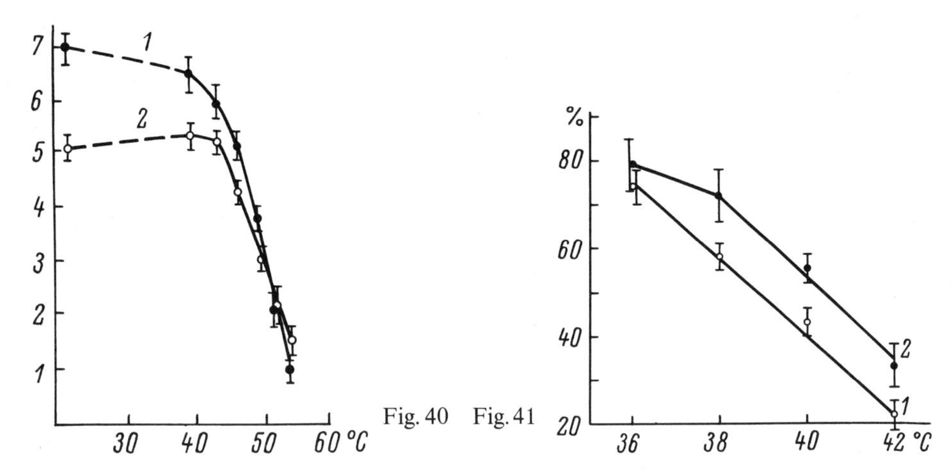

Fig. 40. Change in the ATP-ase activity in extracts of non-hardened *(1)* and hardened *(2)* acacia *Caragana arborescens* leaves after 30-min exposure to different temperatures (Feldman, 1968). *Abscissa:* temperature of heating; *ordinate:* ATP-ase activity (μg$P_{inorg.}$ per mg protein nitrogen per 5 min at 30° C, × 10^3). Hardening of leaves for 18 h at 42° C

Fig. 41. Thermostability of glucose-6-phosphate dehydrogenase in extracts of non-hardened *(1)* and hardened *(2)* wheat leaves. (Shcherbakova, 1972). *Abscissa:* temperature of 10-min heating of extracts; *ordinate:* enzyme activity (% of initial activity). Heat hardening for 3 h at 40° C

The resistance of crude protein preparations or of partially purified fractions was assessed in the studies cited. An attempt to detect a difference in purified proteins was reported by Feldman and her colleagues (1973). These workers heat hardened entire pea plants for 18 h at 35–37° C. Ferredoxin was then isolated from the leaves followed by an exhaustive purification that yielded a preparation giving a single band in polyacrylamide gel electrophoresis. An evaluation of the resistance of a purified ferredoxin preparation to heating for 15 min consisted of two assays: one measuring the relative absorption at 422 and 277 nm, and another one of estimating the capacity of the preparation for photoreduction of NADP. The heat stability of ferredoxin from the heat-hardened pea plants, estimated by both methods, proved to be higher (Fig. 42). The difference was as large as 1.3° C and corresponded to the heat stability increase of photosynthesis and photosynthetic reaction provoked by heat hardening of the plant (Ageeva and Lutova, 1971). Semi-purified ferredoxin is considerably more resistant to heat than is the purified protein; nevertheless, the difference between the heat stability of semi-purified ferredoxin preparations from the heat-hardened and control leaves was found to be the same as in the case of purified preparations. Modification of ferredoxin resistance brought about by heat hardening was obtained with two pea varieties—"Henry" and "Pozdny spely". For unknown reasons no difference could ever be detected in the third variety—"Prevoskhodny".

The evidence obtained in the studies on test heating of proteins isolated from cells can be extended by the observations concerning thermal effects on the proteins in heat-hardened and non-hardened cells in situ. Shcherbakova (1971a) estimated a decrease in the amount of water-soluble proteins that can be ex-

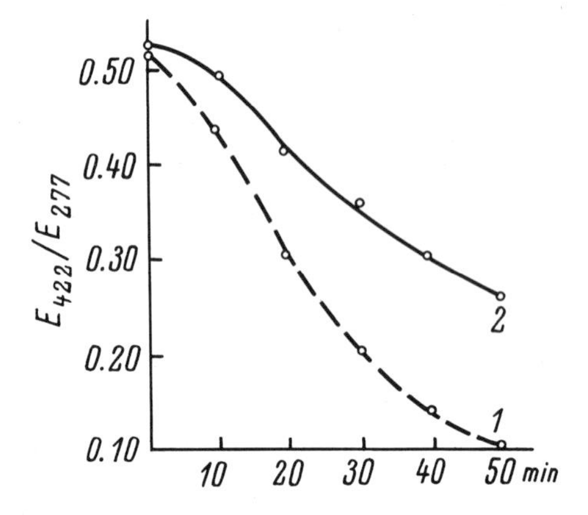

Fig.42. Effect of heat hardening on thermostability of ferredoxin isolated from pea leaves (Feldman et al., 1973). *Abscissa:* length of heating at 48° C; *ordinate:* extent of denaturation expressed as ratio of light absorption at 422 and 277 nm. *1:* ferredoxin from non-hardened leaves, *2:* ferredoxin from hardened leaves

tracted from the heat hardened and control wheat leaves after a 30-min exposure of the leaves to 50° C. After heating the leaves, which had been heat hardened for 3 h at 36–43° C, more leaf protein could be extracted than after heating of the control, non-hardened leaves. That testified to a lesser extent of thermal denaturation of the proteins in the heat hardened cells.

It is known that petroleum ether cannot extract chlorophyll from a native complex of chlorophyll with proteins and lipids. If, however, the protein has been denatured by heat, chlorophyll can be extracted readily with petroleum ether. After heat hardening, the extraction of chlorophyll from *Tradescantia fluminensis* and *Taraxacum officinalis* leaves requires higher temperatures of heating than in the case of non-hardened plants (Lutova, 1963a). These findings have been corroborated by Novoselova et al. (1971). Kiknadze (1960) has found also, using fluorescent microscopy, that in the case of heat-hardened leaves, for the complex of chlorophyll with the proteins and lipids (in the cell) to be destroyed, more intensive heat treatment must be applied. And, finally, Shcherbakova (1969) demonstrated that the heat stability of dehydrogenases was elevated by 2° C in the leaf cells of heat-hardened wheat seedlings. Reduction of tetrazolium salts to formazan served as an indicator of thermal resistance in her experiments.

In Section 1.2.2.1 b, data were referred to which unequivocally testify that heat hardening elevates the primary thermoresistance of various cellular functions. So far any one function that we have studied has proved to acquire an elevated resistance as a result of heat hardening. Considering primary thermostability as an indirect indicator of thermal stability of cellular proteins, it may be inferred that stabilization involving all of the proteins in the cell underlies heat hardening. The latter can be obtained with treatments as brief as several scores of seconds (Lomagin, 1961; Zavadskaya, 1963a), and, consequently, cannot be a result of the substitution of cellular proteins for new, more thermostable proteins. It is only natural to suppose that this stabilization affects the already existing proteins in the protoplasm. Later in this book I shall discuss the possible mechanisms of such stabilization. At this point, the primary concern is to prove firmly that it does occur in heat hardening. Apart from an indirect criterion—the primary thermostability of various cellular functions—this conclusion is corroborated by the above-mentioned evidence accumulated in the studies on the heat resistance of proteins isolated from hardened and untreated leaves. Unstable reproducibility of the results reported in evaluating heat resistance of isolated proteins may be ascribed to the possibility that a stabilizing factor can form bonds, varying in their strength, with different proteins in the cell. Isolation and purification procedures may destroy these bonds with the protein under study, which fact will be manifested in a loss of the elevated thermostability of the protein.

Taking into consideration the above probability, the data demonstrating the enhanced thermostability of the proteins isolated from hardened cells must be regarded as of higher supporting value as opposed to the findings reporting no differences in the heat stability of the proteins from hardened and control cells.

As referred to earlier (see Section 1.2.2.3), numerous workers have detected an increase in cellular thermostability elicited by cold hardening in various plants. The problem of how it affects the response of isolated proteins to heat has not been practically dealt with. In 1934, Tysdal, when heating an alfalfa root extract, found thermostability of diastase increases in winter. The next study of this phenomenon was performed by Shcherbakova in 1969. She found that in experimental cold hardening of wheat seedlings, as in heat hardening, the heat stability of dehydrogenases, assayed by reduction of tetrazolium salts, grew higher. Cold hardening of wheat seedlings both in nature and under experimental conditions, much like heat hardening, elevates thermostability of glucose-6-phosphate dehydrogenase as revealed by measuring the enzyme activity in a crude extract (Shcherbakova, 1972). However, neither artificial nor natural cold hardening could ever produce any modification of the heat stability of acid phosphatase in this plant (Shcherbakova, 1971 b).

3.2.1.2 Animals

In Section 1.2.3.3, findings were discussed relevant to regular seasonal variations of the primary thermostability of frog cells associated with hormonal shifts. Based on the fact that these variations can be detected immediately after the end of an intensive pulse-heat treatment, it is believed that they reflect a change in the level of resistance of cellular proteins to the denaturing action of heat. No seasonal variations could be observed with isolated preparations of actomyosin (Vinogradova, 1965a) and succinate dehydrogenase (Makhlin, 1965). In contrast to the response of intact cells, succinate dehydrogenase isolated from muscles does not lower its thermostability after administration of a hypophysis preparation. The controversy is resolved in the experiments with the glycerol models of muscle fibers and of the ciliary cells. In this case the models behave as living cells: they vary their thermostability in various seasons synchronously with the cells from which they have been made (Chernokozheva, 1967b; Dregolskaya and Chernokozheva, 1970). Consequently, a hormonal shift alters the level of the protein heat stability by means of certain influences which are also retained in the cell model, in which the native conformation and arrangement of contractile proteins, and, presumably, their association with other cellular components surviving the glycerolization procedure, have not been distorted. These interactions, however, break up completely upon isolation of the proteins from cells followed with solubilization of these proteins. Likewise, in the warm-blooded animals, Skholl (1965) has detected seasonal variations in the heat stability of muscle fibers of the siberian marmot; such variations could not be observed, however, in thermal resistance of actomyosin preparations. On the other hand, seasonal fluctuations of thermostability could be found in both muscles and muscle fiber models derived from the field mouse (Skholl, 1970).

These findings, at any rate, warrant the conclusion that the seasonal variations of thermostability of protoplasmic proteins, and thereby those of the cells themselves, are not accounted for by the changes in the primary structure of protein molecules. During growth and development of tadpoles at periods of pre- and prometamorphosis, the heat stability of muscle tissues increases, whereas during metamorphosis it dramatically decreases. Thermostability of actomyosin shows no fluctuations; unfortunately, cell models were not studied in those experiments (Chernokozheva, 1970). Chernokozheva observed somewhat different relationships during growth of immature frogs: in this period of the frog ontogenesis the heat stability of the frogs' muscles appreciably increases, whereas that of the muscle models remains unchanged. Ushakov and Glushankova (1970) noticed a decrease in the heat resistance of the muscles after transfer of the bellied toads to 5–8° C in summer. The heat resistance of the glycerol models, aldolase and adenylate kinase remained unaffected.

Of direct relevance to the problem in question are the data obtained by Andreyeva (1970) on age-dependent variations in the heat stability of the collagen preparations from the frog *Rana ridibunda*. It is known that collagen fibers contract when heated to certain temperatures. This contraction is accounted for by a transition of collagen into a denatured state. The temperature range for this transition depends on both the primary structure of collagen, which specifies the pattern of intra and intermolecular bondings supporting the conformation of the macromolecule, and on the stabilizing interactions of collagen with the natural ligands, in particular with polysaccharides. The temperature at which contraction of collagen fibers reaches half the maximally attainable magnitude, is designated T_S. It has been found that T_S of the tendons from two-year-old frogs is 2.1° C, and that of the skin is 1.3° C, lower than that value observed for 3–4 year old frogs. The findings communicated by Andreyeva are in fair agreement with those reported by other workers, who have also found T_S of collagen tissue from man (Briscoe et al., 1959) and rat (Joseph and Bose, 1962) to increase with age. A native protein—procollagen or tropocollagen—can be extracted from collagen fibers and then solubilized; this protein also denaturates on heating. The temperature at which 50% denaturation of tropocollagen occurs is designated T_D. This transition, however, takes place at a temperature approximately 25° C lower than that required for a similar effect to occur in the collagen incorporated in the system of a fiber. The fact that denaturation of solubilized tropocollagen takes place at lower temperatures is readily explained by elimination, during extraction and purification procedures, of the components responsible for stabilization of the protein in situ. Andreyeva has shown that, in the course of these procedures, not only the resistance of collagen decreases, but also the age-dependent differences of thermostability are lost. Consequently, the absence of age-dependent differences in the stability of collagen, observed in in vitro assays, is an artifact. It will be later shown that when differences in the heat stability of collagenous tissue are determined by the differences in the primary structure of the protein, these differences can also be seen when studying solubilized tropocollagen.

In the foregoing, reference was made to the data obtained by Friedrich (1967) on increase in thermostability of the ciliary epithelium of the mussel gills following a 20-min heating to supraoptimal temperatures. A decrease in aldolase activity could be detected in homogenates of the gills, which was accompanied by an elevation of the stability of the enzyme against heat. Mews (1957) kept *Helix pomacea* at 25 and 8° C for a fortnight. As a result, the heat resistance of the gastric protease was elevated in the "warm" snails, whereas that of lipase showed no dependence on the cultivation temperature. Tsukuda and Ohsawa (1971) acclimated the goldfish for 40 days to 18 and 28° C. Then the serum proteins, the soluble proteins from muscles and lens were analyzed by cellulose membrane electrophoresis to reveal that the pattern of serum and muscle proteins was slightly different in the "warm" and "cold" animals. If, prior to an electrophoresis run, these protein preparations were heated for 30 or 60 min at 46° C, a portion of the proteins did not move from the start line. In each of the three protein preparations obtained from the goldfish acclimated to 18° C, more protein sedimented at the start line during an electrophoresis run than in the preparations from the fish kept at 28° C. The authors believe that acclimation to elevated temperature is either accompanied by accumulation of thermostable types of proteins or can be accounted for by the appearance of a factor, which stabilizes certain proteins in the organism. Johnston et al. (1975a) found that the heat stability of the Mg^{2+}-activated myofibrillar ATP-ase from the goldfish acclimated to 1° C was appreciably lower than that of the enzyme from the fish acclimated to 26° C. The authors believe that this is a result of critical changes in the higher orders of structure of the contractile complex in the region of the active site of the enzyme concerned.

3.2.1.3 Microorganisms

Christophersen and Precht (1950a, 1950b, 1951, 1952) cultivated *Torulopsis kefyr* at 20 and 41° C. The "warm" yeasts were 2–3° C more stable toward heat treatment. It is difficult to decide whether that shift in their resistance was of modificational nature or due to selection. What is important, however, is that the modification of the cellular thermostability was accompanied by an elevation of the stability of some enzymes (catalase and dehydrogenases) against heat. Similar findings have been reported for the thermophilic bacteria. Campbell (1954, 1955) cultivated thermophilic bacteria at 35 and 55° C; α-amylase isolated from the "cold" bacteria was 90% inactivated by exposure to 90° C for 1 h, whereas the enzyme from the "warm" organisms was only 6–10% inactivated. Essentially similar results have been obtained by Isono (1970). Brown et al. (1957) assessed the heat stability of the *Bacillus stearothermophilus* pyrophosphatase in the organisms grown at different temperatures: the enzyme from those kept at 50 and 60° C displayed similar thermoresistance, whereas pyrophosphatase from the cells grown at 70° C was more thermostable than that from the bacteria kept at 40° C. On the other hand, Maas and Davis (1952) were unable to detect any dependence of the resistance of pantothenate synthetase on the cultivation temperature of *Escherichia coli.* Thompson and Thompson (1962) likewise could not observe any effects of cultivating *B. stearothermophilus* at 45 and 60° C on the heat stability of its aldolase. Crabb et al. (1975) studied the proteins from the facultative thermophile *Bacillus coagulans* grown at 37 and 55° C. The authors concluded that, in contrast to the obligate thermophiles in which the ability for life at high temperatures is associated with the intrinsic thermostability of their proteins, in the facultative thermophiles the intrinsic thermostability of their proteins is similar to that in the mesophiles. Cultivation of the facultative thermophiles at high temperatures, however, results in modification of the intracellular milieu that induces an elevation of thermostability of their proteins. Among the changes accounting for this modification is, in particular, increase in the ionic strength. Proteins isolated from the facultative thermophiles exhibit the same sensitivity to heat as do the mesophilic proteins.

Paramecia living at 28° C are able to survive lethal heating for appreciably longer periods of time than the organisms cultivated at 15° C. Seravin et al. (1965), using cytochemical approaches, detected higher thermoresistance of succinate dehydrogenase, DNP-diaphorase and glucose-6-dehydrogenase in the "warm" infusoria. Acid phosphatase and non-specific esterase showed no differences.

Summary. The evidence presented in this section demonstrates that various mechanisms of stabilization and labilization of protein molecules are responsible for modificational changes in the level of the primary thermostability of various functions in plant and animal cells. Variations can be observed in the degree of universality of the stabilization: a change in the resistance to the temperature factor may be accompanied by a uniform modification of the resistance of cells to a number of factors, essentially dissimilar in their physico-chemical nature. Stabilization of the protoplasmic proteins may be of varying strength. It can either withstand isolation of the protein from the cell and persist even after a rigorous purification, or the stabilizing factor may be lost in the course of such treatments, and a shift in thermal stability of the protein (revealed indirectly by a change in the primary thermostability of cellular functions) will not be observed in the protein solution. Shifts have been described which could not be detected in solubilized proteins, whereas in the glycerol cell models they were recognizable. And, finally, shifts in the level of protein stability may be produced by those factors which are being eliminated during preparation of the cell models. Thus, the unavoidable interaction of the "tool" with the study object explicitly emerges in such experiments: a cytophysiological assessment of the heat stability of a function reveals that the level of the protein resistance has been changed in the cell, while the protein involved remains unknown. The only thing that is known is that this is a protein, whose stability of against heat is a limiting factor preventing during a thermal treatment the performance of the function chosen as an indicator (be it protoplasmic streaming, respiration, selective permeability, etc.). On the other hand, a particular protein can be isolated with the aid of the biochemical technique. The level of thermostability of that protein may, however, undergo such dramatic changes that the data eventually obtained can hardly be used for the purpose of comparative assessment. These two methodological approaches prove to be mutually complementary sensu Bohr (1963). The discovery of that stage of a physico-chemical treatment, at which the level of the protein stability is altered and the differences that existed in vivo disappear, may prove of considerable help in the search for the mechanisms responsible for regulating stability of protein macromolecules during an individual life of an organism. The above, to a considerable extent, holds true also for the data referred to in the section that follows, which deals with genotypic variations of protein resistance.

3.2.2 Genotypic Variations in Thermostability of Proteins

3.2.2.1 Plants

Khlebnikova has found (1932) that the watermelon leaves were able to achieve cooling through transpiration more effectively than were the vegetable marrow, pumpkin and melon leaves. This was accompanied by lower heat resistance of the watermelon leaf plates and by coagulation of water-soluble proteins, occurring at a lower temperature in that plant as compared with the other three plants.

Earlier (p. 86) mention was made of the higher thermostability of some functions of the summer snowflake compared to that of the less warm-loving spring one. The studies on the proteins in these two closely related species revealed the following facts (Feldman, 1969, 1973; Feldman and Kamentseva, 1971). If the leaves were heated, the heat stability of dehydrogenases, assayed by their ability to reduce tetrazolium salts to formazan as well as that of the total water-soluble protein fraction, ferredoxin, pyrophosphatase (the optimum activity at pH 6.0 and 4.7) and β-glycerophosphatase (Table 15) were reported to be higher in the summer snowflake. If, however, a homogenate of the leaves was heated, the following enzymes were found to display elevated thermostability in the summer snowflake: urease, pyrophosphatase (the optimum activity at 6.0) and β-glycerophosphatase (a difference in the latter two enzymes being somewhat lower in this case than when intact leaves were heated). Interspecific differences in the resistance of pyrophosphatase with the optimum activity at pH 4.7, detectable on heating intact leaves, were not observed if the enzyme thermostability was assayed in homogenates. In the case of acid phosphatase the difference disappears after a four-fold purification by ammonium sulfate precipitation.

McNaughton (1966) reported thermostability of malate dehydrogenase to be higher in the reed-mace platyphyllous populations living in hot climates than in those growing in cooler habitats; glutamate-oxalate transaminase and aldolase showed no difference in this respect. Heat stability was assayed by heating supernatant fractions of leaf homogenates. Prior to assays the plants of both populations had been kept under identical conditions for three months.

Unfortunately, these scanty data are, in fact, all that is known to date concerning thermostability of the proteins isolated from closely related plants which differ in their thermophily. Much more abundant evidence is available regarding animal proteins.

Table 15. Thermostability of cells and proteins in two *Leucojum* species. (From Feldman and Kamentseva, 1971)

Material and indicator of thermostability	The spring snowflake (*L. vernum*)	The summer snowflake (*L. aestivum*)
Leaf pieces, temperature of a 5 min heating (° C):		
effective in stopping protoplasmic streaming	44.6±0.1	46.1± 0.1
suppressing capacity for plasmolysis	52.6±0.3	55.3± 0.5
suppressing reduction of tetrazolium	55.7±0.3	58.0± 0.0
Leaves, intensity of photosynthesis following a 10-min heating at 44°	25 ±2	37 ± 3
Leaves, respiration intensity following a 5-min heating at 55°	17 ±5	46 ± 3
Leaves, amount of extractable protein following a 30-min heating at 55°	27 ±3	57 ±11
Leaves, ferredoxin activity following a 15-min heating at 48°	47 ±5	83 ±10
Leaves, pyrophosphatase activity following a 30-min heating at 55°	51 ±2	67 ± 3
Homogenates, pyrophosphatase activity following a 30-min heating at 50°	62 ±5	75 ± 3
Leaves, β-glycerophosphatase activity following a 30-min heating at 55°	46 ±4	71 ± 6
Homogenates, urease activity following a 40-min heating at 80°	41 ±4	55 ± 4

Note: Intensity of photosynthesis and respiration, enzyme activity and the amount of extractable protein are given as percent of the control values.

3.2.2.2 Animals

The comparative evaluation of thermostability of animal proteins evidently has been initiated by the work of Rossi (1905) who estimated the temperature at which coagulation of the serum, taken from mammals, birds and marine fishes, occurred. He could not find any correlation between this temperature and the animals' body temperatures, in spite of the fact that in various species the coagulation temperature varied from 54 to 60° C.

Further development of research along these lines is seen in the investigations reported by Korzhuev (1936) concerning thermostability of trypsin in various animals. The animals studied by him could be arranged in the following order in terms of the heat resistance of their trypsin: pigeon > dog > turtle > frog > perch = pike > Black Sea cod > Barents Sea cod. In this case, a correlation between the level of thermostability and the temperature at which the cells operate is obvious.

The third stage in these investigations was begun many years later, coincident with activation of cytoecological studies in the Soviet Union. Instead of discussing the comparative findings concerning very distant species (Connell, 1961; Makarewicz, 1965), data relevant to phylogenetically related species living in different climatic conditions will be emphasized. In studies of this kind, efforts have mostly been spent on comparing the proteins of *Rana ridibunda* and *R. tem-*

poraria. It has been seen that various tissue cells in the lake frog *R. ridibunda* are more thermostable than those in the grass frog *R. temporaria*. For a summary of pertinent data see Table 11, where, in addition, the results of comparative observations on the heat stability of the muscles, ciliary cell models and solubilized proteins are included. Almost all of the proteins studied so far in this regard proved to be more heat-stable in the lake frog, as can be seen from this table. This holds true for both intracellular and extracellular proteins such as collagen and serum albumin. The magnitude of the difference in the resistance to heat for the proteins of both frog species in the majority of cases is comparable to that reported in comparative studies on primary cellular thermostability of these frogs. This correlation can hardly be taken to suggest anything other than that the primary thermoresistance of cells does, in fact, reflect thermoresistance of the cellular proteins, and it may be used as an indirect measure of the latter.

According to B. Ushakov (1965a), purification of protein preparations does not affect the magnitude of interspecific differences in thermostability. The above data on collagen may serve as a good illustration of this view (Alexandrov and Andreyeva, 1967). A difference between the T_S values for the tendons from *R. ridibunda* and *R. temporaria* amounts to 6.1° C (T_S—the temperature effective in producing a 50% contraction of the maximally attainable contraction of the collagen tissue at a given rate of heating). After isolating collagen from tissues followed by appropriate purification, the protein decreased its thermostability by approximately 25° C. The difference in the heat resistance of purified collagen preparations derived from the two species of frogs remained the same, i.e. 6° C. Consequently, the interspecific differences in this case were not accounted for by those bonds which endow the protein with an elevated stability in situ, being destroyed upon purification. This, however, is not always the case. Rigby and Mason (1967) have estimated the T_S of collagen in a number of marine and terrestrial gastropod molluscs. They noted a good correlation between the T_S values and the temperature conditions of a species' life. T_S ranged from 50 to 60° C. After treating the tissues with polysaccharase, which breaks down the association of collagen with polysaccharides, the T_S values, found in some species to be above 50° C, dropped to 50° C and the interspecific differences disappeared. Tartakovsky et al. (1969) compared thermostability of myosin preparations from *Rana ridibunda* and *R. temporaria* and reported that the difference between the heat stabilities of the purified protein preparations from the two frog species grew less as compared with the differences observed in corresponding crude protein preparations.

Accordingly, genotypic interspecific differences in the heat resistance of proteins as well as modificational differences, may be retained after isolation of the proteins from the cells, may be reduced, or may disappear altogether. Considering interspecific comparisons too, we should rely on the primary thermoresistance of cells. This being an indirect indicator of protein thermoresistance in vivo, it should be remembered that by isolating proteins from cells, one may distort the relationships existing in vivo. Hence, one should consider those experiments on isolated proteins which do reveal interspecific differences in thermostability, to be more convincing than the observations yielding negative results.

Among the positive evidence, besides that discussed above, a mention should be made of a series of investigations carried out by B. Ushakov and his collabora-

Table 16. Findings from the laboratory of B. Ushakov on proteins, for which the specific differences in their thermostability corresponded to the differences in the thermophily of species

Enzyme	Animals	Number of species	Reference
Alkaline phosphatase	Crayfishes	2	Glushankova (1973a)
Adenylate kinase	*Polychaeta*	3	Glushankova (1973a)
	Bivalves	2+2	Glushankova (1973a)
	Fishes	2	Glushankova (1973a)
	Lizards	2+2	Glushankova (1973a)
Actomyosin	*Polychaeta*	2	Glushankova (1973a)
	Bivalves	2[a]+2+2	Vinogradova (1963b); Glushankova (1967a)
	Crabs	2	Vinogradova (1963a)
	Fishes	2+2	Vinogradova (1963a); Glushankova (1963b)
	Turtles	2[a]	Vinogradova (1963b)
Aldolase	*Polychaeta*	3	Kusakina (1967)
	Bivalves	2+2	Kusakina (1967)
	Fishes	2	Kusakina (1967)
	Lizards	2+2	Kusakina (1967)
Muscle acetylcholinesterase	*Polychaeta*	3	Kusakina (1967)
	Bivalves	2+2	Kusakina (1967)
	Cod fishes	2	Kusakina (1967)
	Batrachocottus baicalensis	2+5[a]	Kusakina (1962b)
	Lizards	2+2	Kusakina (1967)
	Toads	2[b]	Kusakina (1965b)
Brain acetylcholinesterase	Cod fishes	2	Kusakina (1967)
Hemoglobin	Toads	2[b]	Kusakina (1965b)
	Lizards	5[a]	Kusakina (1965a)

Numerals without (a) or (b) show the number of species of one genus. [a] Species belonging to related but different genera. [b] Subspecies of one species.

tors. In the majority of their studies, comparisons have been made on the species belonging to one genus, but which differed in their temperature ecologies. For numerous proteins and study objects a correspondence has been reported between the thermophily and the level of protein thermoresistance (Table 16).

Vasilyeva et al. (1969) found a difference in the stability of muscle aldolase from two sea-urchin species and two species of the scallops (*Pecten*) against heat that corresponded to a difference between their respective thermophilies. Rather thermolabile enzymes have been found in the Antarctic fishes which permanently live at temperatures of $-1.9°$ C, namely, aldolase, glyceraldehyde-3-phosphate dehydrogenase and fructose-2-phosphate aldolase. The latter was 6° C more sensitive to heat than analogous enzyme in the trout, and 17° C more heat-labile than that from the rabbit. Aldolase from Antarctic fishes was rapidly inactivated at 15° C (Feeney et al., 1967; Komatsu and Feeney, 1970). Licht (1964a, 1967) has found in eight lizard species, a close correlation between the temperature optimum for the muscle ATP-ase activity and its thermostability, on the one hand, and the preferred temperature for a species' existence on the other. However, in four lizard species, which considerably differed in their thermophily, the temperature optimum and the heat stability of alkaline phosphatase coincided. The author explained this finding by the fact that alkaline phosphatase is less sensitive to heat than is ATP-ase, and, therefore, the former is of lesser importance for the temperature adaptation of a species. As will be shown later, this explanation is not consistent. Moreover, the data reported by Licht are at variance with those obtained by Glushankova (1967a), who demonstrated differences in the heat stability of alkaline phosphatase from four lizard species. These differences were correlated with the differences in the thermophily of the animals. Abrahamson and Maher (1967) evaluated thermostability of pancreatic amylase from one frog and two lizard species. Leaving aside comparisons of phylogenetically distant animals—frog and lizard—and considering only the two lizards, it can be seen that in the more warm-loving lizard, the temperature optimum for the amylase activity as well as the resistance of the enzyme to heat is higher. Smith (1973a) assessed the resistance of succinate oxidase and monoamino oxidase activities to heating for 30 min at 37° C. These enzymes were assayed in the mitochondria isolated from the livers of the vertebrates belonging to 27 species. The activity of the enzymes was not inhibited by such treatment in the case of the reptiles, birds and mammals. In fishes and amphibia, the heat stability of the two enzymes varied considerably from species to species, and the differences were found to be in correspondence with the temperature conditions of a species' life, but not with the phylogenetic history of the species. In another investigation, Smith (1973b) used the same approach in determining the heat stability of succinate oxidase, malate dehydrogenase, NADH-dehydrogenase and cytochrome oxidase activities in the mitochondria isolated from 14 teleost species. These enzymic systems differed considerably in their thermostabilities; the highest stability was reported for cytochrome oxidase. The species-related differences in the heat resistance of all four enzymic systems corresponded to the differences between the respective temperature ecologies of the species. Johnston et al. (1973) studied thermostability of the myofibrillar ATP-ase isolated from the muscles of 19 species of the teleosts from the Northern and Mediterranean Seas, the Indian Ocean and from the equatorial lakes, where the temperature of water is around 37–40° C. A 50-fold difference was found in this group of organisms for the extreme values of the exposure, resulting in a 50% inactivation of an enzyme at 37° C (9.7 and 502.6 min). The specific differences between the heat stability values were found to correspond exactly to the differences in the respective temperature ecologies of the species.

Surprising results, unexpected even to the author himself, have been reported by Cowey (1967). He found that a 50% inactivation of α-glyceraldehyde-3-phosphate dehydrogenase from the rabbit muscles occurred in 3 min during exposure to 60 °C; similar inactivation of the enzyme from the lobster—in 5 min, and that from the cod—in 30 min. Comparison of such phylogenetically distant species as these, is, presumably, not valid for this enzyme.

Earlier evidence was presented in favor of a good correlation between thermophily of two species of frogs and some gastropods and the heat stability of their

Table 17. Thermostability of skin collagen fibers and of tropocollagen from some cod species. (From Andreyeva, 1971)

Species	Temperature during spawning (° C)	T_S (° C)	T_D (° C)
Cod	−0.4–+6.0	44.2±0.22	17.7±0.19
Green cod	3.0– 10.0	44.6±0.25	18.3±0.16
Haddock	5.5– 10.0	45.0±0.23	18.5±0.13
Merlang	9.5– 11.5	45.3±0.28	20.8±0.13

collagen. Similar findings have also been made in other species. Among these, a reference should primarily be made to the data reported by Gustavson (1953, 1955, 1956) and Takahashi and Tanaka (1953). These workers have shown that in the cold-water fishes, the T_S of their collagen were 37–45° C, whereas the T_S of collagen from the warm-water pelagic fishes were found to be within the range of 50–57° C; for the mammals these values were within 60–70° C. These observations have been corroborated by Eastoe (1957).

Privalov (1968) reported values for the temperature of a maximal absorption of the heat of denaturation of tropocollagen from the pike, merlang and cod as being 30.6, 21.3, and 20.5° C, respectively, that fully agrees with the degree of thermophily of these species. For the rat tropocollagen this value was about 40.8° C. Furthermore, Andreyeva (1971) obtained evidence on similar correspondence for four cod species as judged by the T_S and T_D of collagen (Table 17). Rigby (1968a) found in five species of Antarctic fish a surprisingly low temperature for the transition of skin collagen: T_S=23–27° C and T_D=5.5–6° C. According to Rigby, the corresponding values for the cod were 37 and 15° C, respectively. A direct dependence of the T_S of collagen upon the upper temperature limit for life of a species was shown by Rigby (1968b) in five worm species. T_D of tropocollagen of *Ascarida* living at the temperature of a warm-blooded host, is 52° C, whereas this value for the brandling is 21° C (Josse and Harrington, 1964). These findings can be supplemented by another series of investigations in which the authors were able to demonstrate a positive correlation between the T_S or T_D of collagen and the temperature ecology of a species, when taxonomically distant forms have been compared (Ewald, 1919; Esipova, 1957; Eastoe, 1957; Kazakova et al., 1958; Borasky, 1965; Rigby and Hafey, 1972; Rigby and Prosser, 1975; and others).

In order to show the extent to which the thermophily of an animal species affects thermostability of its collagen, frogs will again be discussed. Alexandrov and Andreyeva (1974) estimated the T_S of the *Rana terrestris* tendons to extend the observations made in *R. temporaria* and *R. ridibunda*. All three frogs belong to the same genus. According to Bannikov (1943) and Terentyev (1950), *R. terrestris*, in terms of its thermophily, occupies an intermediate position between *R. temporaria* and *R. ridibunda*. The values of T_S of the tendons, as measured by elevating the temperature at a rate of 1° C per min, were: 55.5, 56.9, and 60° C for *R. temporaria*, *R. terrestris* and *R. ridibunda*, respectively. The temperatures of 5-min heat exposures that arrested the ciliary beat of the frogs' ciliary epithelium were 41.9, 42.5, and 44.8° C, respectively; the temperatures of 10-min heatings that suppressed the excitability of the muscles isolated from these frogs were: 38, 39, and 42° C, respectively (these values have been derived from the plots obtained by B. Ushakov, 1961).

The evidence presented in this section unequivocally testifies to the existence in nature of a correlation between the thermophily of a species and thermostability of its proteins. It would, however, be a distortion of the style of biological literature reviews, if no evidence contradictory to the thesis being advanced could be found. In fact, it is apparent from the data in Table 11, that not all of the proteins in *R. ridibunda* exhibit an enhanced thermostability as compared with the corresponding proteins in *R. temporaria*. No difference could be found for acetylcholinesterase from brain, for the cathode fraction of aldolase, and for glutamate dehydrogenase. Read (1963, 1964a) studied aspartate-glutamate and alanine-glutamate transaminase from a number of mollusc species. He concluded that thermostability of these enzymes was found to be in better correspondence with the resistance of these molluscs to anaerobiosis than with their thermophily. Read believes that the properties of the enzyme which enable it to perform effectively under conditions of facultative anaerobiosis, make it, at the same time, more stable against heat.

Wilson et al. (1964) have carried out extensive research to assess the heat stability of the H_4-isozyme of lactate dehydrogenase in 55 species of vertebrates: birds, higher and lower reptiles, mammals, amphibia, teleosts and ganoid fishes, and cyclostomata. The authors aimed to find out, with the aid of this indicator, the evolution of H_4-lactate dehydrogenase. As a result, they were able to construct a diagram of the enhancement in the course of phylogenesis of the heat stability of this enzyme. The sequence commences with the fish followed by amphibia and lower reptiles (turtles). Thermostability of all these species is low. Beginning from the lower reptiles, the phylogenetic tree has two branches. One branch is terminated with the mammals, which retain thermolability of their H_4-lactate dehydrogenase. The other branch is represented by the higher reptiles (snakes, lizards, caymans) and birds, in which the enzyme is 10–20° C more heat-resistant than in the animals of the first group. In their work, the authors have ignored ecology of the animals, and in the choice of the species they have not aimed at comparing closely related species characteristic of different thermophilies. That is why the findings reported in this work are not suitable for the purposes of our discussion. According to Vinogradova (1970), however, lactate dehydrogenase may not conform to this regularity: no positive correlation could be detected between thermostability of this enzyme and thermophily of a species, when closely related species of lizards, frogs, fish and nereides were compared. A direct relationship between these indicators has been observed only for the crab.

Summarizing the evidence discussed in this section and that referred to in Sections 1.1.2 and 2.1, we have, nonetheless, to conclude that interspecific differences in thermostability of animal proteins do exist, and that they are determined to a considerable degree by the differences in the temperature ecologies of the species concerned.

As regards the intraspecific relationships, it must be admitted, provided the primary thermostability is used as the criterion, that the level of thermostability of proteins in many species is very conservative, i.e. the cellular thermostability of the organisms taken from the populations found under drastically different temperature conditions may be the same (Sect. 1.2.1.3). In other species, an intraspecific differentiation with respect to this character occurs (Sect. 2.1).

It is of interest to know whether a difference is retained in the heat stability of the cell models and of the proteins isolated from the cells in those species, for which intraspecific differences of primary thermostability of their tissue cells had been detected. In the European populations of the lake frog *R. ridibunda*, as compared to those living in Middle Asia, the muscle fibers (B. Ushakov, 1963b), ciliary cells (Glushankova et al., 1967), spermatozoa (Svinkin, 1959), glycerinated muscle fibers, adenylate kinase, hemoglobin (Glushankova et al., 1967), cholinesterase from liver and muscles, actomyosin (Vinogradova and Kusakina, 1963) have been found to show higher stability against heat. No differences have been observed for alkaline phosphatase (Glushankova et al., 1967), the anode fraction as well as the cathode fraction of aldolase (Pravdina, 1967), brain acetylcholinesterase (Konstantinova and Grigoryeva, 1969) and for collagen in both tissues and solution (Andreyeva, 1970). Opposite results have been obtained by Braun and Fizhenko (1963) for hemoglobin, i.e. the Middle Asian population of frogs was found to exhibit higher thermostability. In the lizard *Phrynocephalus helioscopus* from Armenia, its aldolase, adenylate kinase, alkaline phosphatase and liver cholinesterase were found to be more heat stable than the analogous enzymes in this subspecies from Kyzyl-Arvat; hemoglobin, on the other hand, was more thermolabile in the Transcaucasian lizards (Glushankova and Kusakina, 1967).

Comparison of two grayling subspecies from the Baikal Lake revealed that the myocardium, cholinesterase and muscle ATP-ase, aldolase, alkaline phosphatase and hemoglobin were more thermostable in the black grayling than they were in the white one. The curves of the heat stability of adenylate kinase and of actin from these subspecies eventually intersect and analysis of them is therefore infeasible (Ushakov et al., 1971). Higher thermostability was reported for hemoglobin, and, to a lesser extent, adenylate kinase and cholinesterase from the muscles of the northern Baikalian race of the Baikal omul, than from the Posol race. The other four the proteins studied—aldolase, actin, alkaline phosphatase and ATP-ase—showed no differences in their thermostability characteristics. In all these cases, it was difficult to relate the intraspecific divergence of cellular thermostability and that of the proteins to the temperature ecologies of the subspecies and races, because the putative differences in the ecologies of these were either insignificant or as yet poorly studied.

The findings of Andreyeva (1971) relevant to intraspecific differences in thermostability of collagens from cod are of considerable interest. We have seen that in cod the interspecific differences between the heat stabilities of collagen can be detected both in situ (T_S) and in vitro (T_D). These differences were found to be in correspondence with the differences in respective thermophilies of the fishes (Table 17). In addition, Andreyeva compared collagens from Atlantic cod and from those living in the Mogilny Lake on the Kildin Island. The temperature of water in the latter in summer is about 5–6° C higher than in the ocean. Accordingly, T_S of the collagen from the Kildin cod was 2° higher than that of the Atlantic cod. Tropocollagen preparations isolated from the skin, however, showed no difference in T_D. The same results were obtained by Andreyeva, when she compared collagens from the *Gadus merlangus* in the Barents Sea and from the variety of this fish living in the Black Sea. Collagen from the southern subspecies was 1.6° C more thermostable in tissues, whereas tropocollagen solutions did not exhibit any difference in T_D. In addition there were no differences found in thermostability of the muscle fibers (Ushakov, 1955), of hemoglobin and cholinesterase from the muscles and liver (Kusakina, 1965b) in the case of three subspecies of the toad *Bufo bufo*—*B.b.bufo* (Leningrad region), *B.b.verrucosissima* (Sukhumi vicinities) and *B.b.asiaticus* (islands in the Amur Bay).

The evidence accumulated concerning intraspecific differences reveals that when several proteins are being studied in this context some proteins show differences, whereas others do not. It may be inferred that, with the divergence of species, a shift in thermostability does not extend over all the proteins simultaneously, but that every particular protein is independently approaching a new level of heat stability.

3.2.2.3 Microorganisms

The relationship between the habitat temperature and the level of thermostability of the proteins revealed is most explicitly in the world of microorganisms. These latter, with extraordinary extensiveness, take advantage of those temperatures which are compatible with active life; the range of such temperatures in the case of microorganisms spans from −23 to +104° C. In microorganisms also, a correspondence exists between the primary thermostability of the cells and the temperature environment. This is manifested not only in the self-evident high level of heat stability of the thermophiles, but also in the less obvious, rather low level of the resistance to heat of the psychrophiles (see p. 89). A question arises whether the differences mentioned are actually determined by the different response of the proteins to high temperatures in the thermo, meso and psychrophilic microorganisms.

a) The Thermophiles. The astonishing capability of the thermophilic microorganisms for life and reproduction at temperatures not only effective in killing other organisms, but in denaturing most proteins in solution, requires that a consistent explanation be offered for this ability (Gaughran, 1947a; Allen, 1953; Koffler, 1957; Loginova et al., 1966; Brock, 1969, 1970). One of the first hypotheses was advanced by Gaughran (1947b) and Allen (1950, 1953). In the experiments of Allen, thermophilic bacteria were killed in a non-nutrient medium at 55° C at the same rate as were the mesophilic ones. These and other complementary observations made Allen believe that thermostability of thermophiles is determined not by an elevated thermostability of the cell's interior, but by a high rate of synthetic processes in the thermophiles at high temperatures, which exceeds that of degradation of the cellular components. Some evidence in favor of this concept is available. Zhestyanikov (1964) employed gradual adaptation, for instance, to obtain *Escherichia coli* strains capable of reproducing at 50° C. In the absence of nutrients, these strains did not differ in their thermoresistance from the non-adapted clones.

Nonetheless, the dynamic concept of Allen could not outweigh the data rapidly accumulating in the literature concerning elevated thermostability of thermophilic proteins. At present, there is no doubt left that the proteins of the thermophilic bacteria, as opposed to analogous proteins in the mesophilic forms, display higher stability against heat in the majority of the cases studied. The obligate thermophile *Bacillus stearothermophilus*, usually grown at 60–65° C, has been a favorite study object in relevant investigations. Here is a list of some proteins and protein systems of *B. stearothermophilus* conspicuous for their greater thermostability compared with that of analogous proteins from the mesophiles: the starch-liquifying enzyme (Stark and Tetrault, 1951), α-amylase (Campbell, 1954; Manning and Campbell, 1961, Manning et al., 1961; Campbell and Manning, 1961; Pfueller and Elliott, 1969; Isono, 1970), aldolase (Thompson and Thompson, 1962; Sugimoto and Nosoh, 1971), pyrophosphatase (Brown et al., 1957), glyceraldehyde-3-phosphate dehydrogenase (Amelunxen, 1967; Singleton et al., 1969; Amelunxen and Clark, 1970), ATP-ase (Hachimori et al., 1970), hexokinase, enolase, alcohol dehydrogenase, glutaminase, alkaline phosphatase, NADH-oxidase. Less high but still notably elevated heat resistance has been reported for pyruvate kinase and glutamate-oxalate transaminase (Amelunxen and Lins, 1968), leucyl-tRNA-synthetase (Vanhumbeeck and Lurquin, 1969), pyrimidine nucleoside phosphorylase (Saunders et al., 1969), DNA-dependent RNA polymerase (Remold-O'Donnel and Zillig, 1969), ribosomes (Saunders and Campbell, 1966; Friedman et al., 1967), the 50 S and 30 S ribosomal subunits (Altenburg and Saunders, 1971; Friedman, 1971), a cell-free system for protein synthesis (Friedman and Weinstein, 1964, 1966).

Taking *B. stearothermophilus*, for instance, it can be seen that a higher level of thermostability, compared with that of the mesophiles, is characteristic, evidently of the entire protein moiety of the thermophilic organism. A wealth of data are available concerning an enhanced thermostability of enzymes isolated from thermophilic strains of *B. subtilis*, *B. circulans* (Loginova and Karpukhina, 1968) and *B. coagulans* (Campbell, 1954, 1955)—α-amylase; from *B. thermoproteolyticus* (Endo, 1962; Matsubara, 1967)—thermolysin; the flagellar protein—flagellin—in some thermophilic bacteria (Koffler et al., 1957; Adye et al., 1957); maleic acid dehydrogenase in bacteria No. 2184 (Militzer et al., 1949) and pyrophosphatase in that bacterium (Marsh and Militzer, 1956), ferredoxin in *Clostridium tartarivorum* and *C. thermosaccharolyticum* (Fig. 43; Devanathan et al., 1969), ten glycolytic enzymes in these species of *Clostridia* (Nowell et al., 1969), fructose-1,6,-diphosphatase and phosphofructokinase in the extreme thermophile *Flavobacterium thermophilum*, which lives at 75° C (Yoshida et al., 1971; Yoshida and Oshima, 1971), fructose-1,6,-diphosphate aldolase in *Thermus aquaticus* (the enzyme is stable at 95° C—Freeze and Brock, 1970), glyceraldehyde-3-phosphate dehydrogenase in that organism (stable at 90°—Hocking and Harris, 1972), protease in *B. thermoproteolyticus* (Ohta et al., 1966; Ohta, 1967), phenylalanyl and isoleucyl-tRNA-synthetase in *T. aquaticus* (Zeikus and Brock, 1971), DNA-dependent RNA polymerase in *B. megaterium* (Nikiforov, 1970). Besides the above listed data, there are findings which report high thermal stability of mixed protein solutions obtained from thermophilic bacteria (Koffler and Gale, 1957; Samejima and Takamiya, 1958). It is practically impossible to give an exhaustive list of thermophilic

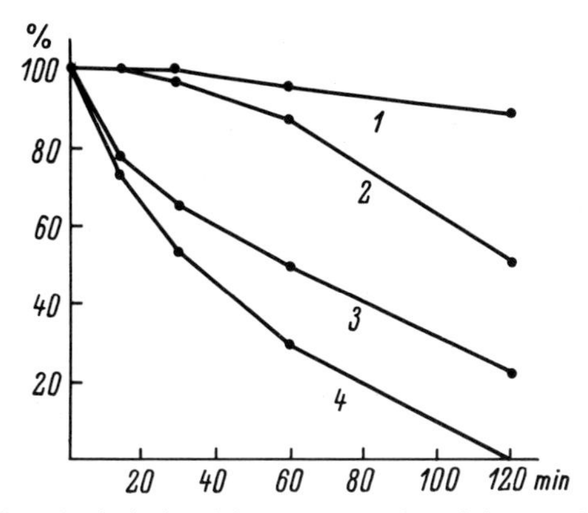

Fig. 43. Inactivation of ferredoxin isolated from two species of thermophilic bacteria, *Clostridium thermosaccharolyticus (1)* and *C. tartarivorum (2)* and two mesophilic species—*C. acidiurici (3)* and *C. pasteurianum (4)* by exposure to 70° C. (Devanathan et al., 1969). *Abscissa:* length of heating; *ordinate:* activity (% of initial activity)

proteins which are stable against intensive heat treatments, especially since relevant findings are being reported continually in increasing numbers.

Many algae, predominantly blue-green ones, are known to inhabit hot-water bodies, but their proteins have still been poorly studied. The observations on the thermostable cytochrome-*c*-reductase from *Aphenocapsa thermalis* are of interest in this regard (Marre and Servettaz, 1957; Marre et al., 1958).

In all these investigations, the proteins from the thermophiles have been compared with those from more or less related mesophilic species of bacteria. *Escherichia coli* and *Bacillus subtilis* have been most frequently used as mesophilic counterparts. Sometimes, more distant comparisons have been made with analogous proteins from animals. In the overwhelming majority of studies, the resistance of the proteins with enzymatic properties has been assessed by their catalytic activity. It has been found (Yoshida and Oshima, 1971; Yoshida, 1972; Yoshida et al., 1973) that the *B. thermophilum* fructose-1,6,-diphosphatase and phosphofructokinase as well as the *Flavobacterium thermophilum* fructose-1,6,-diphosphatase exhibit high thermoresistance of both their catalytic and allosteric sites.

The dependence of a bacterial enzyme's thermal stability upon the cultivation temperature has been reported in several papers, which were discussed in the above (p. 107). It is no easy task, however, to decide conclusively whether these findings reflect modificational or genotypic shifts.

In the works cited, protein preparations have been analyzed which were rather diverse in terms of their degree of purification—from crude extracts or homogenates to crystallized or homogeneous preparations as judged by ultra-centrifugation or polyacrylamide gel electrophoresis. It has been shown in some papers that high thermostability, typical of thermophilic proteins, can be detected in both crude preparations and highly purified ones. Many workers indicated that purification procedures did not affect the level of the heat stability of the proteins (Thompson and Thompson, 1962; Vanhumbeeck and Lurquin, 1969; Hachimori et al., 1970). Others, though, noted that thermoresistance decreased as proteins were progressively purified (Loginova and Karpukhina, 1968). The attention given to the problem of how the effects of purification modify the heat stability of proteins is explainable, because attempts are being made to elucidate the cause of the high thermostability of thermophilic proteins: whether it is associated with peculiarities in their primary structure, or whether stabilizers are involved. It will be shown later what a difficult problem this is. Whatever the answer is, there is no doubt that the thermophily of microorganisms is associated with a high (intrinsic or induced) thermostability of their proteins.

b) The Psychrophiles. This group of microorganisms is no less interesting. It was shown (p. 89) that the ability for active life at low temperatures is accompanied in these organisms by marked thermolability of their cells. The response of the proteins to the denaturing action of heat has been considerably less studied in the psychrophiles than in the thermophiles.

A series of papers (Morita, 1966; Stokes, 1967; Farrell and Rose, 1967) reported high thermosensitivity of various biochemical functions, revealed when psychrophilic microorganisms were taken to supraoptimal temperatures. Enzyme-dependent processes and syntheses of enzymes were arrested at temperatures optimum for the mesophilies, the difference in sensitivity frequently exceeding scores of degrees centigrade (see, for example: Upadhyay and Stokes, 1963a, 1963b; Quist and Stokes, 1972). Ingraham and Bailey (1959), however, were inclined to ascribe the peculiarities of the metabolism in psychrophilic bacteria to the cell organization, rather than to the differences in their enzymatic properties. Conversely, Hagen and Rose (1962), reporting their metabolic studies in the psychrophilic yeasts at supraoptimal temperatures, concluded that certain enzymes were strikingly thermolabile in these organisms.

The problem could be resolved in studies on the temperature response of proteins isolated from the cell. Unfortunately, research in this area has barely begun. Burton and Morita (1963; Morita and Burton, 1963) have studied malate dehydrogenase from the psychrophilic vibrion Ps 207. Exposure of lyzed cells to 35° C rapidly inactivated the enzyme. The authors believed that the inability of this vibrion to grow at temperatures above 30° C was associated with alteration of the cellular permeability and with the heat lability of the organism's enzymes. The psychrophilic bacterium Strain No. 82 grows at 0° C; at 35° C and above the growth ceases. This strain was compared with *E. coli*, the temperature maximum for which is 45° C. On heating the cell-free extracts to 46° C, $NADH_2$-oxidase and cytochrome-*c*-reductase of the psychrophile were inactivated at an apparently higher rate than analogous *E. coli* enzymes (Purohit and Stokes, 1967). Malcolm (1968a, 1968b, 1969b) has found the growth of *Micrococcus cryophilus* at temperatures exceeding 25° C to be limited by high thermolability of three aminoacyl-tRNA-synthetases: glutamyl-, histidyl- and prolyl-tRNA-synthetases. Notable suppression of their activity occurred already 10 min after transfer of a cell-free system to 30° C. Triose phosphate isomerase from the psychrophile *Clostridium* sp. strain 69 is much more thermolabile than this enzyme in the mesophilic *Clostridium pasterianum* species both in vitro and in vivo (Yuen Wan Shing et al., 1972).

Nash and Grant (1969) compared the response to temperature of the ribosomes isolated from the obligate psychrophiles *Candida gelida* and of those from the mesophilic yeasts of the same genus—*C. utilis*. The synthesis of ^{14}C-phenylalanine peptide on poly-U templates was taken as the criterion. A 5-min heating at 40° C fully inactivated the psychrophilic ribosomes. Similar treatment of the *C. utilis* ribosomes did not affect their activity. The authors explain their findings as due to higher termolability of ribosomal proteins in the psychrophiles. Besides, seven of the 13 aminoacyl-tRNA-synthetases proved to be highly thermosensitive. Leucyl-tRNA-synthetase was completely inactivated in 30 min at 35° C. In *C. utilis* this enzyme under identical conditions fully retained its activity. In addition, *C. gelida* has the soluble enzymes active in polypeptide synthesis on ribosomes, which exhibit high sensitivity to heat (Nash et al., 1969). Striking thermolability was noted in the extracellular proteinase from the psychrophile *Escherichia freundii* (Nakajima et al., 1974).

Summary. Here are some conclusions suggested by the data discussed in this and the first two chapters. Based on cytophysiological criteria and biochemical assessment of the response of isolated proteins to heat, the following statements may be formulated. The level of thermostability of proteins in vivo undergoes modificational variations which may be elicited by a change in the temperature of environment (temperature adjustments, heat and cold hardenings) or by other factors, some of which are already known (endocrine shifts, changes in water and salt content in the cell, length of day, wound injury, growth processes, age-dependent changes) and others which still remain undetected.

Comparison of the primary thermostability of cells, cellular models and of proteins isolated from the cells of closely related species suggests that, in the evolution process of animals, plants and microorganisms, a correlation between the genotypically determined level of thermostability of the proteins and the average temperature for existence during an active period of the organisms' life has been established. Parallel studies on thermostability of proteins in living cells, cellular models and of isolated proteins, purified to various extents, leave no doubt that both modificational and genotypic changes in thermostability may be determined by essentially different biochemical mechanisms. "Fixation" of these shifts may be stronger or weaker. Accordingly, the differences in the heat stability of proteins in vivo, disclosed by cytophysiological approaches, may be either retained or lost during isolation of the proteins from the cell and during subsequent purification procedures. The varying degree of unspecificity of these shifts testifies in favor of a dissimilarity of the molecular mechanisms responsible for stabilization or labilization of proteins. A response of cells or of proteins can be simultaneously unidirectionally modified to a greater or lesser number of injurious agents, diverse in terms of their physico-chemical nature.

To achieve further development of the problem concerned, attempts must be focused on two unsolved questions: (1) what is the biological significance of the modification of the level of heat resistance of proteins during the life of an individual and in the process of phylogenesis? (2) what biochemical mechanisms are responsible for occurrence of these or those shifts in the level of protein thermostability? The next two chapters will show what can be said concerning these questions, while remaining firmly in touch with reality.

Chapter 4
Adaptive Modifications of Conformational Flexibility of Macromolecules as a Basis for Changes of the Protein Thermostability

What is the biological significance of modificational and genotypic variations in the heat stability of proteins? In order to provide answers to this question, one should primarily analyze the evidence concerning changes in the heat stability of proteins occurring in the evolution of species. Why, in the creation of species which have originated from a common root and have adapted to different temperature environments, has this process been accompanied by the appearance of corresponding differences in the heat stability of their proteins?

The easiest answer is that the warm-loving species must have more thermostable proteins in order to avoid thermal denaturation, since the latter would have interfered with the cell's life, and, eventually, resulted in the death of an organism. In many cases, such an explanation seems to be highly plausible. For instance, the grass *Aristida karelinii* that vegetates during the whole summer on the Kara-Kum sandhills and can survive temperatures of up to 42–46° C, would have been unable to live under those conditions if its proteins had been as stable against heat as those of the polar grass *Poa alpina*. An exposure to 42° C for 10 min arrests the protoplasmic streaming in the cells of this grass by thermally denaturing the cellular proteins (Alexandrov, 1963). The terrestrial mollusc *Helix lucorum* could not have withstood 40° C had its cells contained proteins similar in their thermostability to those of the mollusc *Pecten islandicus*, living in the sublittoral regions in the Barents Sea. At 30° C the ciliary beat in *P. islandicus* stops as early as within 5 min of heating (Alexandrov, 1952). Thermophilic bacteria can live and reproduce at 70–90° C because their proteins are far more stable against heat than analogous proteins in the mesophiles, which denature at such temperatures. Many organisms live in hot habitats at ambient temperatures close to the temperature of thermal death. As noted above, there is, for instance, a blue-green alga that lives at 65° C in a hot spring in Java. By elevating the temperature only 4° C the plant can be killed in as little as 5–10 min (Bünnning and Herdtle, 1946).

An attempt to apply the explanation given above to all cases in which a correlation between the cellular heat resistance and stability of proteins, on the one hand, and the ambient temperature, on the other, is observed, fails on account of various insurmountable difficulties.

4.1 The Correlation Between Thermostability of Proteins and the Environmental Temperature Conditions of a Species' Life Cannot be Explained by an Adaptive Significance of the Level of Thermostability

Strong arguments may be provided against the contention that this correspondence is the result of natural selection that acts on the resistance of cells and proteins to the denaturing action of heat. The objects of the selection are neither proteins, nor tissue cells, but organisms. Consequently, a selection for thermostability of organisms must bring into a correspondence with the environmental temperature those components of the organisms which are responsible for its thermostability. What, in fact, limits the life of organisms at maximal temperatures? It is absolutely clear that an unambiguous answer to this question is not at hand. In one group of organisms, an overheating would result in disturbances of one sort, whereas in others essentially different alterations would appear (Prosser and Brown, 1962). In many animal species, for instance, the site most susceptible to overheating proves to be the nervous system. The data in Tables 9 and 11 demonstrate that the temperature of a 30-min heating which produced thermonercosis in fish and frogs was rather low. Prolonging exposure of the organisms to heat necessitates the temperature of the thermonarcosis be even lower. The cause of thermonarcosis lies in the dysfunction of the nervous system. The lowest temperature for the upper temperature limit for life has been reported in Antarctic fishes. Somero and De Vries (1967) studied three species of these fish belonging to the *Trematomus* genus. The maximal lethal temperature was found to be around 6° C. Oxygen uptake by brain sections begins to decline already at 5° C. At the same time, the activity of some enzymes isolated from these fish increases linearly up to 30° C. The authors believe that there are especially thermolabile enzymes in the brain of Antarctic fishes, and the nervous system is of primary importance in determining the limit of thermostability of the entire organism. Analyzing the cause of thermal death of the crayfish, Bowler (1963) has concluded that it is due to a loss of the nerve-dependent coordination, which, in its turn, is the result of an alteration of the potassium-sodium balance.

The data referred to above show that thermal death of animals occurs as a consequence of an alteration produced in some particularly thermolabile cellular systems in an organism. This takes place at temperatures considerably lower than those which produce alteration of most tissue cells. Accordingly, an overwhelming majority of the cells are not involved in determining the level of thermostability of an animal organism[4].

Conversely, the level of thermostability of a given cell is certainly determined by a far lower number of the proteins found in it. As seen from Table 11, there are proteins in the cell which differ appreciably in their resistance to heat; e.g., in

[4] Unexpected findings have been obtained by B. Ushakov et al. (1972). These writers showed that between the heat resistance of tadpoles and that of their muscles even inverse dependence exists: the tadpole families exhibiting an increased heat stability of the organisms display a decreased thermostability of their muscles.

order for a 50% inactivation of the *Rana temporaria* muscle adenylate kinase to occur, a 30-min heating at 62.5° C is needed, whereas in the case of *R. ridibunda* this temperature is 73.5° C. In other animals too, adenylate kinase is conspicuous for its high thermostability. In this group of heat-stable proteins hemoglobin, serum albumin and collagen are also found. These proteins are denatured at temperatures which are in no way related to the temperature for life of these frogs. Rigby and his associates (Rigby, 1968c, 1971; Rigby and Hafey, 1972; Rigby and Prosser, 1975; Rigby and Robinson, 1975) believe that the temperature for denaturation (melting) of tropocollagen corresponds to the upper temperature limit for life of a given animal species. The data referred to above show that similar coincidence cannot be observed in other proteins.

In a series of papers, V. Ushakov (V. Ushakov, 1963, 1964, 1965a, 1965b, 1965c, 1966; V. Ushakov and Vasilyeva, 1965; V. Ushakov et al., 1971) has shown, in agreement with the earlier data of Inagaki (1906), that in muscles there occurs a stepwise alteration of separate protein fractions as the temperature is elevated. A fraction of the soluble sarcoplasmic protein related to the excitability of the muscles was found to be the most thermolabile. The contractile proteins displayed higher stability. Accordingly, various fractions of the muscle fibers were also gradually altered. The ability of tetanous fibers for responding to acetylcholine administration by contraction was the first to disappear. The bioelectrical component of an excitable muscle system was the more resistant, and still higher resistance was reported for the contractile function itself. It is quite obvious that thermostability of the muscle fiber (which, by the way, does not determine the resistance of organisms—their stability is considerably lower) is limited by one or more of its most thermolabile components, and decidedly not by any protein that an investigator happens to come across.

It should moreover be noted that it is frequently not the thermal injury itself that establishes the temperature limit for life of organisms, but other accompanying factors. In the case of marine organisms, this factor can be a decrease in the solubility of gases in water at elevated temperatures. Fry (1957) indicated that oxygen deficiency can account for the death of fish when they are placed in warm water.

Consequently, the temperature maximum for animal life is absolutely independent of the level of thermostability of the bulk of the proteins in an organisms. The natural selection of organisms for their thermostability cannot bring the level of thermostability of their proteins into correspondence with the habitat temperature.

It can be easily shown that this is the case in both plants and microorganisms. In contrast to the specific metabolic inhibitors that arrest this or that enzymatic reaction, heat is thought to be a non-specific injurious agent, though, in fact, by applying an appropriate dose of heat any enzymatic system can be destroyed. In plant cells, however, similar to the muscle fibers, various cellular components and functions appreciably differ in their thermostability (Figs. 6 and 7). In the *Tradescantia* leaf cells, for instance, a complete cessation of the streaming of protoplasm can be obtained in 5 min when the leaves are heated to ca. 44° C, a complete inhibition of photosynthesis at 48° C, an uncoupling of respiration from phosphorylation at temperatures of above 53° C, and an inhibition of respiration at above 60° C (Alexandrov and Alekseyeva, 1972). Feierabend and his co-workers (1969) grew rye seedlings at elevated temperatures (32–34° C). Unter these conditions, the process of chlorophyll synthesis and the initial stages of the carboxydismutase synthesis were almost entirely inhibited. At the same time, an overall protein synthesis in the cells as well as the synthesis of non-photosynthetic enzymes and leaf growth remained practically unaltered.

Only a single link in a metabolic chain may be affected during prolonged mild heatings. This, however, may suffice and a plant dies. The plant can be saved in this case by administering the metabolite which had become deficient as a result of the heat injury. A large number of such experiments have been performed in both higher plants and microorganisms (Kurtz, 1958; Langridge, 1963b; Farrell and Rose, 1967). One of the first publications relevant to this problem was written by Bonner (1957) and titled: *Chemical Cure of Climatic Lesions*. In this publication, the experiments of Galston are described (Galston and Hand, 1949; Galston, 1959), who showed that by supplying adenine, one can prevent thermal death of pea plants. Later, it was found that at elevated temperatures, adenine is destroyed at a higher rate than it is synthesized, so that the cell eventually runs out of this important metabolite. Duckweed dies at 30° C, but an administration of adenine enables the plant to grow well at this temperature.

Ketellapper (1963) carried out similar experiments with a number of plants (lupin, kidney bean, pea, etc.). Depending on the plant and the temperature, either vitamin B or ribosides were found to be effective in protecting the plants from thermal injury. At optimum temperatures, these preparations did not affect the plant growth. Langridge and Griffing (1959) exposed homozygous *Arabidopsis* races to supraoptimal temperatures. In two races, the alterations could be completely prevented by biotin administration; in the third race, cytidin was partially effective in diminishing the injurious effects, whereas the remaining two races showed no reaction when supplemented with these substances. Genetic experiments made these authors believe that in a wild population of *Arabidopsis thaliana*, there are genes, or sometimes only a single gene, the products of which are inactivated at an elevated temperature, thereby restricting the plant's growth. The work of Malcolm (1968b) has been cited above, in which he showed that the life of the psychrophile *Micrococcus cryophilus* at 25° C was limited by marked heat sensitivity of several aminoacyl-tRNA-synthetases. Transfer of *Escherichia coli* cells from 37° C to 40–45° C has been shown to inactivate homoserine transsuccinylase selectively. This extraordinarily thermosensitive enzyme catalyzes the initial reaction in methionine synthesis. The growth of bacteria inhibited with transfer to 45° C could be restored by addition of methionine to the medium (Ron and Davis, 1971a).

In his review paper, Langridge (1963b) gave a table compiled from the data available in the literature. The requirements of more than 40 species and strains of microorganisms in certain metabolites that appeared upon transfer of the cells to supraoptimal temperatures were listed. Frequently, life at a supraoptimal temperature becomes feasible with the addition of only one substance. Often this is one of the amino acids (tyrosine, cysteine, methionine) or a vitamin. As the temperature increases, the number of substances that are to be supplemented to maintain growth, as a rule also increases. The auxotrophy becomes wider, and favorable effects can then be observed by the administration of amino acid mixtures and of such complex additives as yeast extracts and peptone. Langridge indicated the following possible causes for the appearance of the requirements in supplementary metabolites during growth at supraoptimal temperatures: (a) a decrease in the concentration of gases; (b) an accelerated breakdown of metabolites; (c) a disturbance in the balance of synthetic and destructive processes; (d) a suppression of an adaptive production of enzymes; (e) a reversible and irreversible inactivation of enzymes.

The data discussed, as well as the evidence referred to by Levitt (1972), indicate that analysis of the causes of death of plant organisms produced by the action of supraoptimal temperatures must take two points into account: different heat stabilities of various cellular components on the one hand, and different temperature dependences of the rates of various cellular processes on the other. The latter results in a distortion of the normal relationships between the rates of important metabolic reactions. As a consequence, diverse alterations responsible for death of an organism may ensue: growth prevailing over reproduction, respiration over photosynthesis, protein degradation over its synthesis, etc. It is obvious that an imbalance of this kind is independent of the heat stability level of the bulk of cellular proteins. This, presumably, holds true also for animal organisms.

This evidence leaves no doubt that the natural selection of organisms can control the level of thermostability of only individual, strikingly thermolabile proteins in the cell. In this connection, an astonishing question arises as to how, in evolution, there could appear a correspondence between thermophily of a species and the heat resistance of thermostable proteins, that are explicitly not involved in endowing the organisms with an ability to survive at elevated temperatures. A difference in the resistance of thermostable proteins in species which are related, but are different in their thermophily, is no less pronounced than in the resistance of thermolabile proteins, and is sometimes even larger. Adenylate kinase, taken as an example, may illustrate the point. This enzyme displayed the largest difference (11° C) in the study in which analogous proteins were compared in *R.ridibunda* and *R.temporaria* (Table 11). Glushankova (1967a) has compared the resistance to a 30-min heating of actomyosin, alkaline phosphatase and adenylate kinase from related species of nereides *(Polychaeta)*, lamellibranchiate molluscs, cod fishes and lizards living in contrasting temperature environments. Actomyosin was the most thermolabile protein of these, whereas adenylate kinase was found to be the most resistant and alkaline phosphatase was in between. Glushankova concluded: "Interspecific differences in the heat resistance of the most thermostable protein—adenylate kinase—are manifested most distinctly and amount to 5.7–11-2° C, for actomyosin the difference is 1.2–4.4° C, whereas for alkaline phosphatase the difference is still smaller—1.4–2.2° C" (p. 138). Licht (1964a), however, studying the enzymes from eight lizard species, has observed a parallelism between the thermophily of the animals and thermostability of their muscle ATP-ase. The heat stability of the more thermostable alkaline phosphatase from the intestine was identical in four species. From this material, Licht concluded that alkaline phosphatase, being more thermostable, is of lesser significance for the determination of the resistance of lizards to heat, than is ATP-ase. Taking into consideration the accumulated observations concerning the specific differnces in the heat stability of proteins, this statement by Licht must be regarded as a fortuitous discovery.

Still more astonishing are the results of comparisons of the enzymes within the intraspecific groupings. If the level of thermostability per se were of any adaptive significance, then with the beginning of the evolutionary divergence, it might be expected that the differences would have been primarily revealed in the most thermolabile enzymes. Nothing of the sort has been found. Kusakina et al. (1971) have analyzed the intraspecific differences in the enzymes from six species and deduced that "the order in which the differences appear is not related to the degree of thermostability of the protein. In the rock lizards, such differences have been found only for two of the five proteins studied: alkaline phosphatase and hemoglobin (both are heat-resistant proteins). In the omuls, significant differences could also be observed only when highly thermostable proteins were taken for comparison: adenylate kinase, cholinesterase and hemoglobin" (p. 1000). Similar findings have also been reported for plants. McNaughton (1966) studied the resistance of enzymes which differed in their thermoresistance to heating at 50° C, from two populations of the reed mace platyphyllous, growing in different climatic zones. Malate dehydrogenase was found to exhibit higher stability against heat in the southern population than in the plants of the northern one. In contrast, aldolase, a less thermostable enzyme in both populations, showed no differences in resistance.

There is still another group of facts, which can not be understood in the context of the assumption that an organism's thermostability is determined by the heat stability of its tissue cells. As we have seen, it has been established by a large number of observations, made both under experimental conditions and in nature, that thermostability of an animal organism can vary appreciably without corresponding changes in the heat stability of its tissue cells or proteins. This has been shown in frogs (Arronet, 1959; Zhirmunsky and Shlyakhter, 1963), the meat fly (Amosova, 1963), molluscs (Arronet, 1959; Konev and Burtseva, 1970), leeches (Ushakov and Kusakina, 1960). Kirberger (1953) kept manure worms *(Eisenia foetida)* for 1–2 weeks at 3, 11, and 25° C. The higher the temperature, the higher was the heat stability of the organisms. No change, however, could be noticed in the heat resistance of respiration and that of the succinate dehydrogenase activity in the tissue sections.

Doubts concerning an adaptive significance of the level of thermostability of proteins have been expressed by a number of workers. Read (1967), among others, writes: "Heat stability of the enzymes ... is of no adaptive significance per se, since heat inactivation occurs only at temperatures well above those lethal to the entire organism" (p. 104). Braun and Fizhenko (1963) have expressed a justified surprise: "how could there appear in evolution those differences in the ability of the proteins to denature at temperatures which are of no vital importance?" (p. 252).

Still another point is to be considered: by admitting an adaptive significance of the level of thermostability of proteins, the answer might have been provided as to why the more warm-loving organisms possess more thermostable proteins. However, in this case no answer could be offered to this question formulated otherwise, but equally as valid: why the more cold-loving organisms possess more thermolabile proteins, i.e. what is bad in an excessive thermostability?

We are confronted with a dramatic situation. On the one hand a wealth of evidence testifies to the presence of a correspondence between the temperature environment and the level of thermoresistance of the proteins in nature. On the other hand, a multitude of facts contradict any adaptive significance of the level of the protein's thermal stability. Taking into consideration the modern concepts of evolution, it is difficult to agree with the idea that in the course of the phylogenetic development of animals, plants and microorganisms relationships, bearing no biological significance, have been infallibly reproduced and maintained.

An attempt to arrive at a solution of this problem has been made by B. Ushakov, who, together with his colleagues, has contributed much more than any other investigator to the accumulation of facts favoring a parallelism between the temperature ecology of species and thermostability of their proteins. Formerly, Ushakov (1959a) wrote, however, that "thermostability of the cells and protoplasmic proteins in poikilothermal animals is of no adaptive significance for a species" (p. 555). Later, Ushakov (1961) narrowed this thesis, stating, that "modification of thermostability of the cells, other than the nerve cells, cannot play any significant role in the regulation of the level of heat stability of an already formed multicellular organism" (p. 458).

This attitude makes Ushakov formulate an explanation of the nevertheless existing correspondence between thermophily of species and thermostability of the cells, other than the nerve cells. To this end he advances the following hypothesis. The level of thermostability plays a significant role in the sex cells and in the somatic cells at early stages in the embryo's development, when the supracellular levels of thermostability are still absent. At these stages, the heat stability of an individual is determined by thermostability of its cells. Thermostability of the tissue i.e., "*specialized cells of an adult organism is controlled by the natural selection at the earliest stages in ontogenesis and, in the majority of cold-blooded animals, represents a peculiar rudimentary feature of an early embryonic adaptation*" (B. Ushakov, 1963c; p. 38, italics by Ushakov). He continues: "In the process of differentiation, the majority of the cells in organisms acquire an elevated thermostability compared with that of the entire organism, and this property of the cells thereby loses its significance for the adaptation of animals to the injurious action of heat" (Ushakov, 1965b; p. 476).

This concept advocated by Ushakov is shared by his collaborator Andronikov (1965, 1967), who studied changes in thermostability of frog embryos at early stages of their ontogenesis and found that the heat stability of the embryos increased with development. Andronikov too believes that, at later stages in the ontogenesis, an adaptive significance of the level of heat stability of tissue cells is lost, and "species-specificity of the level of the adult organism's cellular thermostability should be regarded as a consequence of the early-embryonic specific divergence" (Andronikov, 1965; p. 136). Thus, the "rudimentary" hypothesis of Ushakov, in fact, discards the biological significance of the constantly revealed correspondence in closely related species between the heat stability of their tissue cells and proteins, on the one hand, and the thermophily of adult organisms, on the other. A number of facts are available, though, which can in no way be accommodated within the framework of this hypothesis.

It is certainly true that the embryonic stages of the development of poikilothermal organisms may be exceptionally heat-sensitive and that thermostability of the embryonic process can determine the boundaries for distribution of a species. This has already been shown as early as the classic works of Runnström (1927, 1930, 1936) and his forebears, and later confirmed in numerous investigations (Moore, 1939, 1942; Volpe, 1953, 1957; and others). Frequently the early developmental stages proceed at temperatures close to the maximal temperatures for life. This creates conditions favoring a more rigorous control by natural selection. The possible biological significance of a high thermolability observed in the embryonic cells will be discussed later in this book. It is worthy of note, however, that an enhancement of thermostability of cells in the course of development is not observed in all cases. Thus, Gorodilov (1969) has found that thermostability of the Neva salmon embryo did not increase, but grew less as the development progressed. It should be noted that this salmon spawns late in October and its development takes place against the background of progressively lowering temperatures. In this context, the data obtained by Payusova and Koreshkova (1972) concerning heat stability of the closely related fishes, *Leuciscus schmidti* and *L. bergi*, from the Issyk-Kul Lake are of considerable interest. The muscles from *L. schmidti* were found to be more stable against heat in spite of the fact that this fish usually spawns at lower temperatures than *L. bergi*. The former, however, after the embryonic stages, lives in warmer waters closer to the surface. *L. bergi*, on the other hand, tends to live deeper, in the zone of lower temperatures. The authors believe that the difference in the heat stability of the muscles of these two fish species is of adaptive nature and reflects the difference in the temperature environments of life not during the embryonic development, but typically for the entire period of the post-embryonic life of these fish.

Based on the "rudimentary" hypothesis, it is impossible to comprehend why, in a given adult organism, the cells and proteins that have to perform at different ambient temperatures exhibit a corresponding difference in their thermostability. A colleague of B. Ushakov, Skholl (1963b) showed that, in mammals, the level of the heat stability of the muscles located deep inside the body and those nearer to its surface varies in accord with the differences in the body temperature at these sites. The gigantic aquatic flea *Lethocerus americanus* is inactive at 0° C, at 2° C it starts to walk, at 4° C it swims, but it is impossible to make it fly at temperatures below 10° C. Accordingly, apirase in its leg muscles is inactivated at a lower temperature than the enzyme in the flying muscles. The temperature optimum of the enzyme from the flying muscles is shifted to a region of higher temperatures (Kenny and Richards, 1955). Such differences are apparently of importance only for adult organisms, and can therefore hardly be the rudiments of something useful only for the embryo.

Similar evidence is available for plant organisms as well, as was discussed earlier in the book (p. 86, 87). Among the pertinent findings, the higher thermostability of the cotton boll cells as opposed to that of the leaf cells should be recollected. Again, the higher thermostability of the adaxial epidermis of the leaf sheath compared to that of the epidermal leaf plate cells, and the higher thermostability of the cells in the summer leaves of lungwort compared to the spring, are examples to illustrate the point in question. And certainly these ecologically justified differences in the heat stability of various tissues of an adult organism cannot be regarded as rudiments preserved since the embryonic period.

Serum albumin is one of the proteins studied in *R.ridibunda* and *R.temporaria*, which revealed distinct interspecific differences. Thermostability of this protein from *R.ridibunda* is 7° C higher than that from *R.temporaria* (Alexandrov and Vitvitsky, 1970). This difference could not possibly have been created during evolution by selection in the embryonic period of the frog development, since this protein appears in the frog organism only following metamorphosis (Herner and Frieden, 1960). The pronounced differences in thermostability of collagen in species which differ in their thermophilies are also hardly explainable in terms of the "rudimentary" hypothesis. This protein too is absent at the early developmental stages of embryogenesis.

The principal opposing argument is, however, as follows. The selection of thermostability of embryos is expected to spread only to those proteins which are responsible for the heat stability of the embryos, i.e., the selection can bring into a correspondence with the ambient temperature the level of thermostability of only a few of the most thermolabile proteins. In fact, the "rudimentary" hypothesis leaves the principal problem of the correspondence of the level of heat resistance of thermostable proteins to the temperature of environment without any solution.

A wealth of evidence concerning a parallelism between ambient temperature and protein resistance to the denaturing action of heat, accumulated in the studies carried out in animals, plants and microorganisms, indicates that a biological regularity of utmost importance is hidden in this correlation. The next section is devoted to finding out the meaning and essence of this regularity.

4.2 A Hypothesis of the Adaptive Significance of a Correspondence of the Conformational Flexibility Level of the Protein Molecules to the Environmental Temperature Conditions of Species' Life

A large body of evidence demonstrates that in the process of the evolution of species a correspondence between the level of protein thermostability and the environmental temperature conditions of an organism's life during an active period is stubbornly maintained. On the other hand, no less trustworthy data make one admit that, in the overwhelming majority of cases, this correspondence per se is devoid of any adaptive significance. A natural resolution of this conflicting situation is the suggestion that the level of the resistance of proteins to intensive heat is quantitatively associated with some properties of the protein molecules, which are extremely important for the protein's functioning in full measure at the temperature that is normal for a given species, and, therefore, come under the vigilant control of natural selection. It is the shifts in these properties that we detect when testing the strength of proteins and cells by exposing them to the action of intensive heat treatments. The task is apparently reduced to the disclosure of these properties of protein molecules which are associated with the resistance of proteins to thermal denaturation. Before discussing the project designed for this task I shall give the causes which provoked the decision to devote myself to the solution of this problem.

During the development of the denaturation theory of injury and irritability by Nasonov and the author in 1940, a suggestion was made that the protoplasmic proteins in vivo are found in an unstable state. This admission was based on a number of indirect data and was also repeatedly expressed by other investigators (Lepeschkin, 1923; Bauer, 1935; and others). It was believed that the protoplasmic proteins are being continually denatured and spontaneously renaturated at a higher or lower rate, and was supposed that such an oscillation is necessary for normal living activity. Therefore, a suppression of the latter can be achieved not only with the aid of the denaturing agents, but also by creating conditions lowering the capacity for spontaneous physiological denaturation. The frequency and amplitude of such an oscillation of the protein status would have to be dependent on the temperature lability of the proteins and the power of the renaturating apparatus. Based on this dynamic concept, it was assumed at that time that there must exist a correlation between the temperature for life of organisms and the heat stability of their proteins. It followed that not only an insufficient protein thermostability might have adverse effects in hot climates, but their excessive stability in cold climates as well. An enhanced stability would have decreased the oscillation rate.

It is these considerations that prompted the author's decision in 1949 to initiate cytoecological research aimed at elucidating whether a correspondence exists between thermostability of the proteins and the temperature environment for life of a species. As we have seen, thanks to the considerable efforts of a large number of workers, the existence of this correspondence in nature may be regarded as proven. Should it be stated that this correspondence proves the validity of the original statement concerning the existence of a continual, spontaneous physiological denaturation and renativation of the cellular proteins? When assessing the affirmatory power of facts, one should remember the wise words of Einstein that most frequently facts say "no" to a theorist, sometimes "maybe", but they never say "yes".

Actually, the existence of a correspondence of thermostability of proteins to the temperature environment in which the organisms are found might be logically derived from the hypothesis of the permanent denaturation and renaturation of the protoplasmic proteins. This, however, is not the only possible explanation. At the same time, certain circumstances dictate the necessity to seek other approaches to solving the problem. The fact is that the principal argument in favor of the contention visualizing the process of denaturation–renaturation as a stationary mode of the protein's existence in the cell, was the observation that most proteins are rapidly denatured in vitro following their isolation from the cell, even at the normal temperatures for the organism's activity. Relevant findings were discussed in Section 3.2. It has been shown, however, that there are grounds to believe that in most, if not all cases, the high thermolability of proteins isolated from tissues is an artifact. The inability of an isolated protein to maintain its native conformation at a temperature that is optimum for the organism is accounted for by alteration of the bonds and by modification of the conditions which contributed to the stabilization of the protein in situ. This can be most explicitly demonstrated in an extracellular protein, namely collagen. Disruption of the intermolecular bonds in collagen upon its isolation from the collagen fibers results in reduction of its heat resistance by approximately 25° C.

4.2.1 Conformational Changes of the Functioning Protein Macromolecules

Another explanation can be offered for the correlation between thermostability of proteins and the temperature ecology of a species (Alexandrov, 1965c, 1967, 1969). This is due to the tremendous amount of facts accumulated in biochemistry over the last 20 years. References to the revolution in biochemistry and molecular biology that occurred in the second half of this century usually imply the insights gained into the principles underlying the functioning of the genetic apparatus and the protein-synthesizing machinery. The importance of these milestones to the progress of theory and practice is decidedly overwhelming. However, coincident with the vigorous attacks launched on these two battle-fields in the struggle with the unknown, there occurred in biochemistry some less spectacular though no less important events.

Biochemistry was satisfied, up to the fifties of this century with delineation of formulas of reactions which showed particular changes in the atomic composition of the reacting substances took place during the chemical interactions involved. The proteins were regarded, similarly to other substances, as reagents. It has now become obvious that biological macromolecules—proteins and nucleic acids—are no simple reagents, they are "mechanochemical aggregates" (Vorobyev 1966; Volkenshtein, 1967). Their biochemical performance amounts not only to changes in the atomic composition, but rather to modification of the spatial arrangement of the parts of a macromolecule, to changes of its own configuration. Nowadays thick volumes can frequently be seen dealing with the functioning of macromolecules and no chemical formula appears in a book. Authors are usually concerned with complicated translocations of parts of macromolecules in space, rather than with changes in their atomic composition; with an interplay of weak bonding forces which are responsible for the conformation of macromolecules being analyzed, rather than changes involving the covalent bonds.

In 1940, Nasonov and the author wrote a monograph "The reaction of the living substance to external stimuli", in which the findings obtained by the authors and by colleagues and researchers in other laboratories in both the Soviet Union and other countries were summarized. It was concluded that when the cells react to the action of injurious agents and stimulants, the protoplasmic proteins undergo reversible structural changes not involving changes in their atomic composition. Such modifications, it was believed, were similar to the modifications which occurred during denaturation of proteins in vitro. Later these changes were designated as "configurational" (Alexandrov, 1947). The ideas expressed in the book and in subsequent publications (Alexandrov, 1948; Nasonov, 1959; Alexandrov, 1959) were preludes to the subsequent development of biochemistry which has ascertained the primary role of the configurational, or, in agreement with current terminology, of the *conformational* changes of biological macromolecules, occurring in the realization of the living process. Wartime and language barriers precluded familiarity of biochemists outside the Soviet Union with the denaturation theory of injury and irritation. Progress in this field has been catalyzed by the works of Koshland.

The initial contributions by Koshland (1958, 1959, 1960, 1964b, and others) concerned the specificity of an interaction of enzymes with their substrates. As is known, most enzymes exhibit a strictly selective affinity toward their substrates. Often, a rather minor alteration of the substrate's chemical structure suffices for its interaction with the enzyme to become unfeasible. To explain this phenomenon, Fischer advanced in 1894 a theory of "lock and key", or "template". This theory accounted for the specificity of the binding of substrates to enzymes as due to the presence of a region in the enzyme molecule that is sterically complementary to the substrate molecule. According to Fischer's theory, which enjoyed universal recognition for over half a century, the steric correspondence of the enzyme to its substrate is a preformed one, i.e., it exists irrespective of the presence or absence of the substrate.

Koshland has pointed to a number of facts that do not fulfill the predictions of this theory. In terms of the "lock and key" theory, it is unexplainable why, in particular, the enzymes do not interact with those molecules which bear the same active groups as the substrate, but which are appreciably smaller in size. Again, no explanation could be provided as to why the binding of certain enzymes to their substrates is accompanied by changes in the state of some amino acid residues located at considerable distances from each other.

Findings of this sort led Koshland to advance a new theory designated by him the theory of "induced fit". According to Koshland, the steric correspondence of an enzyme to its substrate is not a pre-existing one, but is formed at that moment when an enzyme molecule interacts with the substrate. Instead of the image of "lock and key" that of "hand and glove" has been suggested. A glove takes the form of the hand only when a hand is inserted in it (Koshland, 1962; Koshland and Kirtley, 1966). The interaction of an enzyme and the substrate also induces a translocation of the atoms involved in a catalytic reaction, and as a result the active catalytic site is formed. This site may contain the atom groups, which prior to the interaction with the substrate, had been positioned at a distance from it. To recapitulate, the theory of Koshland is based on the notion that the substrate evokes significant conformational changes in the enzyme macromolecules.

The template theory aimed to explain only the selective affinity of the enzyme toward its substrate. The theory proposed by Koshland deals not only with the process of induction of a mechanism for the specific binding of substrate, but also for the induction of a new, specified arrangement of atom groups in the area of the catalytic site. In order for a substrate to be able to induce these changes, it must itself possess a definite structure. Small molecules, containing reactive groups similar to those of natural substrates, may fail in making an enzyme macromolecule undergo the necessary conformational rearrangements.

Koshland visualizes the conformational changes as changes in the mean positions of atomic nuclei which do not involve alteration of covalent bonds (Koshland and Neet, 1968). Such changes occur when groups of atoms turn around single bonds. The idea of conformational changes does not include, however, changes in the polarization of electrons occurring without accompanying changes in the positions of atomic nuclei. The essence of conformational changes of proteins should thus be regarded as a redistribution of the energy of bondings and interactions. This can result in a disruption of the existing interactions or in the establishment of new, weak interactions supporting the secondary, tertiary and quaternary structure of the protein macromolecule. Major shifts of amino acid residues can thus occur. According to the induced-fit theory, an enzyme must possess a certain degree of flexibility, i.e. the ability to undergo necessary conformational modifications under the influence of a substrate. "This flexibility is a key feature of the biological action of the enzyme" (Koshland, 1964a).

To date a large amount of facts have been accumulated, which, directly or indirectly, confirm the induced-fit theory. It is not intended to give here a list of relevant findings—in greater detail they can be found in a comprehensive review by Citri (1973); only a few of them will be cited for illustrative purposes. The most conclusive evidence is certainly that provided in X-ray structure analyses. Crystallographic examination, employing an X-ray diffraction, is performed on crystallized enzymes which are either in a free form or bound to analogs of the substrate. Comparison of the diffraction patterns reveals the conformational changes of the enzyme, which are responsible for the substrate binding. Investigations involving X-ray diffraction methods are extremely complicated and time-consuming with the result that only a limited number of enzymes have been studied so far in this regard. The most pronounced conformational changes have been detected in the bovine carboxypeptidase during the enzyme's interaction with glycyl-L-tyrosine as substrate. A 14 Å shift of the hydroxyl group of the tyrosine side chain and shifting of the guanidine group of arginine residue by 2 Å could be observed among others (Reeke et al., 1967). Less striking shifts have been detected in the egg white lysozyme when the latter was complexed with various analogs of the substrate: the tryptophan residue (Try 62) moves 0.75 Å toward the substrate: this is accompanied by translocations of other amino acid residues adjacent to the active site cleft (Blake et al., 1967). Similar studies made on α-chymotrypsin failed to detect any changes in the conformation of the enzyme produced by the interaction with substrate (Henderson et al., 1971).

Particularly interesting findings have been reported in the case of hemoglobin. The latter is not an enzyme, because it does not catalyze chemical transformations of its substrate, oxygen. Interaction of hemoglobin with oxygen so markedly resembles the enzyme-substrate binding that hemoglobin is called an "honorary enzyme". Hemoglobin is known to be a tetrameric protein. Its quaternary structure is composed of two α and two β-polypeptide chains. Initially, a structure analysis, using the X-ray diffraction at low resolution (5.5 Å), has revealed the changes occurring only in the quaternary structure during the oxygenation of hemoglobin: the subunits were found to come closer to each other. A minor translocation, no more than 1 Å, occurs at the sites of contact of the α_1-,β_1- and α_2-,β_2-chains, and a 7 Å shift at the contact sites of the α_1-,β_2- and α_2-,β_1-chains. A slight increase in the number of the van der Vaals interactions accompanied these translocations (Muirhead et al., 1967; Bolton et al., 1968).

Later investigations by the Perutz group (1970), using higher resolution (2.8 Å), showed that considerable modification of the tertiary structure of both the α- and, in particular, of the β-subunits account for the changes in the quaternary structure of hemoglobin occurring when it binds oxygen. There are, in the β-subunites, compared with the α-subunits, more steric obstacles for accommodating an oxygen molecule, which are removed at the moment the subunit interacts with an oxygen molecule. During the oxygenation reaction, the Fe atom goes from the state of high spin to that of low spin, its radius shortens slightly, and the Fe atom, moving 0.8 Å, attains the plane of the nitrogen atoms in the porphyrin ring. The heme group is constructed such that minor changes in the Fe atomic radius result in appreciable movement of the histidine associated with the heme. The F helix moves to the center of the β-subunit, tyrosine HC 2 (145) is pushed out of the "pocket" found in between the F and H helices and pulls along histidine HC 3 (146). The salt bridge between the latter and aspartate FG (94) is broken. The oxygenation of the α-subunits leads to similar, though somewhat different, changes of the tertiary structure of hemoglobin.

The X-ray diffraction studies clearly show the conformational changes in the enzyme induced by the substrate. Moreover, the extent of such changes can be evaluated with this approach. The data obtained in carboxypeptidase, lysozyme and hemoglobin have been used by the authors to develop concrete ideas about the functioning of these proteins. The X-ray diffraction method, however, has considerable inherent limitations. The state and activity of a crystallized enzyme are not completely equal to those of the enzyme in solution. Certain enzymes (papain, ribonuclease-S, chymotrypsin) retain their high catalytic activity when crystallized (Sluyterman and De Graaf, 1969), whereas that of other enzymes is drastically reduced. It has been shown, for example, that the catalytic activity of alcohol dehydrogenase and the activity of ferrihemoglobin is lower in the crystallized state than in solution. The former becomes 1000-fold, while the latter 21-fold less active. The crystallized carboxypeptidase shows 30–1000 times lower constants of the catalysis rate, depending upon the substrate, than the enzyme in solution (Spilburg et al., 1974). This decrease in activity is not related to impediment to diffusion of the substrate into the crystal. This is, apparently, referable to the restricted possibility of the enzymes performing the conformational transitions when being in a crystallized state (Theorell et al., 1966; Chance et al., 1966). Crystallographic studies yield information only about the events occurring during formation of the enzyme-substrate complex. The process undergone by an enzyme during catalysis and at the stage of liberation of the enzyme from the reaction products cannot be detected by this approach. Instead of the natural substrate, the crystallographer has to use, its analog, which irreversibly binds to the enzyme and is not catalytically transformed. One cannot be sure that analogs of substrates are as effective in inducing the conformational changes of enzymes, as are the natural substrates. These doubts are strengthened by the data communicated by O'Sullivan and Cohn (1966) to be discussed later (p. 137).

In summary, it can be said that if an X-ray diffraction analysis does not detect any changes in the enzyme's structure during its binding of the substrate, or reveals only minor translocations of the amino acid residues, as, for instance, in the case of ribonuclease, chymotrypsin, myoglobin and lysozyme, then this negative result should not be regarded as definitive. In spite of these considerations, Koshland (Koshland and Neet, 1968; Koshland, 1971) do not rule out the possibility that some enzymes behave in accord with Fischer's "template" theory.

More evidence concerning the conformational changes of the enzymes occurring during their interaction with the substrates and co-enzymes, has been obtained using various indirect approaches (Volkenshtein, 1967). Such investigations have shown the effects of the substrates and co-enzymes, or of their analogs on the hydrodynamic properties of the enzymes; on their resistance to the denaturing action of heat, urea and other agents as well as to the proteolytic digestion. Besides, the effects of these on the state and reactivity of various amino acid residues in the enzyme macromolecule and the capacity of peptide hydrogens for exchange with deuterium can be evaluated. The conformational changes occurring during the formation of the enzyme-substrate complex can be estimated by various optical methods, which detect changes in the absorption and fluorescence spectra, the degree of depolarization of fluorescence, the changes in the specific rotation of the plane-polarized light and the circular dichroism. Recently, the conformational changes of proteins have been observed with the aid of nuclear magnetic and electron paramagnetic resonance techniques. Frequently, researchers employ the method of "probes" or "labels", which amounts to fixation of a molecule or an ion to a specific site on the enzyme macromolecule under study. These probes, which can be easily detected with appropriate instrumentation, are extremely sensitive to changes in the atomic arrangements adjacent to the probes. A suitable fluorochrome may serve as a label for fluorescence analysis (Radda, 1971), whereas in electron paramagnetic resonance this might be a stable free radical i.e., a spin label (Ingram, 1969; Likhtenstein, 1971).

The data obtained by such methods cannot always be interpreted unequivocally. However, a "cross-examination" of a particular enzyme, performed with different methods, may often yield similar answers, and, in that case, the inferences suggesting that the conformational changes have taken place prove to much more reliable. The conformation of α-glycerine aldehyde-3-phosphate dehydrogenase molecule has been shown to become more compact and stable on binding the coenzymes NAD and NADH. Such a conclusion has been independently arrived at in studies employing essentially different methods: estimating the sedimentation velocity by ultracentrifugation and assessing the diffusion rate and the intrinsic viscosity. Complexing with NAD and NADH enhanced the resistance of the enzyme to both the denaturing action of heat and tryptic digestion. Changes in the optical rotatory dispersion reveal that the degree of orderedness of the enzyme's macromolecule increases when the latter combines with the co-enzymes. And, finally, this is accompanied by reduction of the exchange of some peptide hydrogens for deuterium. Calculations indicate that this decrease in the exchangeability cannot be ascribed to screening of the hydrogen atoms by the molecules of the co-enzyme. The reduction can be explained rather by a diminished accessibility of some peptide hydrogens for a solute that ensues as a result of a rearrangement of the enzyme's conformation (Elödi and Szabolcsi, 1959; Zavodsky et al., 1966; Bolotina et al., 1967).

Yagi and Ozawa (1962), studying the changes of the sedimentation coefficient, diffusion, specific volume, viscosity, and optical activity, detected conformational changes occurring in the amino acid oxidase macromolecules when they were complexing with the co-enzyme and substrate analogs.

The significance of the data obtained with indirect methods can be evaluated when such methods are used in conjunction with the X-ray diffraction analysis of the same protein. We have seen that crystallographic studies revealed considerable changes involving the tertiary and quaternary structures of hemoglobin during its binding of an oxygen molecule and the subsequent release of the latter from hemoglobin. The conformational changes occurring in the course of hemoglobin functioning had also been observed earlier, however, by various indirect methods. Thus, for instance, Benesch and Benesch (1962) have shown that the SH groups in oxyhemoglobin can react with iodoacetamide, whereas this reaction is infeasible in the case of deoxyhemoglobin. The latter binds higher amounts of bromethymol blue than does the oxygenated hemoglobin (Antonini et al., 1963). At the same time, oxyhemoglobin can be digested by carboxypeptidase (Zito et al., 1964) and pepsin (Ikonnikova, 1965) at a faster rate than deoxyhemoglobin. The conformational changes during the oxygenation of hemoglobin have been observed also by optical rotatory dispersion (Brunori et al., 1967).

The most profound conformational changes associated with the function of the protein have been found in the cytochrome c molecules during oxireduction. This process has been studied in great detail in the cytochrome c crystals by the group of Dickerson using the X-ray diffraction at 2.5 Å resolution (Takano et al., 1972; Dickerson, 1972). The authors described in detail the disruptions of the interatomic contacts and the translocations of many amino acid residues that resulted in a more significant alteration of the tertiary structure of a protein than had been observed for any of the proteins studied before. In the cytochrome c molecule there are, besides the mobile portions, also the rigid ones which are only slightly modified during oxireduction. The structure analysis demonstrated that the reduced form of cytochrome c is more compact and less accessible to the surrounding medium.

The data concerning cytochrome c, obtained by indirect methods, fully agree with the findings referred to above. The reduced cytochrome c is more resistant to denaturation by heat (Butt and Keilin, 1962), by surface tension (Jonxis, 1939), and is digested at a lower rate by trypsin (Nozaki et al., 1958). The exchange of peptide hydrogen atoms for deuterium is more impeded in the reduced form of cytochrome c as compared with its oxidized form. Thus, in the reduced protein, some 59% of the peptide hydrogens can be exchanged within 5 min, 20% are exchanged at a lower rate and 21% are not exchanged during 24 h. In the oxidized form, the proportion of peptide hydrogens exchanged during the respective time periods are: 68, 21 and 11%. Consequently, in accord with the more compact structure of the reduced cytochrome c demonstrated crystallographically, the peptide hydrogens in this protein are less accessible to water.

The work of O'Sullivan and Cohn (1966) cited earlier in this section is of considerable interest. These authors have studied conformational changes in creatine kinase during interaction of the enzyme with five different substrates (nucleoside diphosphates), the maximal rate of the enzymatic reaction being different for each substrate. Changes in conformation were observed by two essentially different methods: nuclear magnetic resonance and by estimating the number of the SH groups accessible to iodoacetate. These workers reported that the more reactive the substrate, the greater were the conformational changes detected by both methods.

The following is one more example taken from the work of Ryrie and Jagendorf (1971). The terminal reaction of phosphorylation in chloroplasts is catalyzed by the CF_1 enzyme. This high-molecular protein, built of subunits, is associated with the chloroplast membrane. During incubation of isolated spinach chloroplasts in a solution containing tritiated water in the dark, the hydrogens in the enzyme molecule do not exchange for tritium. Illumination of the chloroplasts results in an exchange of approximately 30% of the hydrogen atoms. If, following the exchange that takes place, the chloroplasts are returned to the dark in a medium prepared with ordinary water, no reverse exchange of the tritium for hydrogen occurs. The authors justly ascribe these results to a change in the conformation of the enzyme during its activity when exposed to light, and the hydrogen atoms, formerly hidden from the medium, come into contact with the latter. On transfer of the enzyme to the dark, its conformation returns to the original state and the tritium atoms are found in a trap.

The conformational changes of the enzyme during its interaction with the substrate are not restricted to the region of the active site. It has been already pointed out by Koshland in his studies on phosphoglucomutase (Koshland et al., 1962), that the substrate (glucose phosphate) modifies the reactivity of the tyrosine, tryptophan, lysine, cysteine and methionine residues, located in different regions on the phosphoglucomutase molecule. This can only be explained by the extensive conformational changes in the enzyme induced by the substrate, which involve an entire enzyme molecule. Koshland (1964a) compares the substrate to the spider which shakes its web when sitting in the center of it. Markus et al. (1968) have found the binding of a competitive inhibitor (2′-cytidylate) to the active site of ribonuclease protects the enzyme from the proteolytic digestion by three proteinases—subtilisin, trypsin and chymotrypsin—that disrupt the polypeptide chain in nine different loci. It is clear that the inhibitor molecule cannot itself screen all of these loci in the ribonuclease molecule. This conclusion can only be understood if one acknowledges that the presence of the inhibitor strongly modifies the conformation or the conformational flexibility of the entire ribonuclease macromolecule.

The number of reports confirming the occurrence of conformational changes of enzyme molecules during their interaction with substrates and co-enzymes is very large and is continually growing (Citri, 1973). With indirect methods such changes have been demonstrated in scores of enzymes, which so far have not been studied by the X-ray diffraction analysis. More often than not, the indirect methods, unlike the X-ray diffraction studies, do not reveal the existence of a concrete pattern of modifications in the atomic distribution during the conformational transitions. As a rule, one succeeds, with greater or lesser reliability, in stating only the very fact of the occurrence of conformational changes, by detecting an increase or decrease in the percentage of α-helical regions; in observing a transition of this or that amino acid residue from a hydrophobic environment into a polar one, or vice versa; etc. However, taken together, the numerous findings referred to in the literature testify that an overwhelming majority of the enzymes, if not all, perform in accord with the induced-fit theory and do not obey the predictions of the "template" hypothesis.

Every enzyme exhibits its own specific mode of conformational transition. By complexing with the substrate and co-enzymes, the enzyme can either be stabilized or made more labile (Grisolia, 1964). Many facts can be found in agreement with the concept advanced by Linderstrøm-Lang (Linderstrøm-Lang and Schellman, 1959), which states that a protein macromolecule is not found in a single definite conformation, but is in a state of continuous modification, going from one conformation to another. In addition, these conformations differ only slightly in their free energy. Different parts of a molecule may vary in their mobility. A substrate, a co-enzyme, or any other ligand may have affinity to the enzyme found in only one of the many possible conformations for this enzyme, and, having bound to it, stabilizes that conformation. This will result in accumulation of the protein molecules characteristic of a given conformation. The eventual outcome is a change in the ratio of conformers in a population of protein molecules. It is in this way that Markus explains the results of his experiments demonstrating the reduced digestibility of a protein upon its association with substrates or other ligands (Markus, 1965; Markus et al., 1967, 1968; and others).

Lumry and Biltonen (1969) fowarded the hypothesis of the role of conformational changes of enzyme molecules. In the process of interaction of the enzyme with the substrate, starting from the complex formation by these up to the liberation of the reaction products, changes in the free energy occur. The plot of the dependence of free energy upon the coordinate of the reaction displays a number of minima and maxima. This profile of chemical free energy is complementary to that of free energy of the conformational changes of the enzyme macromolecule. An increase in the chemical free energy elicits conformational changes which proceed to decrease the conformational free energy. A reduction of the chemical free energy is linked to an increase in the conformational energy. Consequently, the resulting profile of the free energy of the "enzyme-substrate" system versus the reaction coordinate becomes considerably smoothed. The authors pointed to a number of advantages such a balanced system must have both in terms of its thermodynamic and kinetic characteristics. This creates conditions in which natural selection would favor maintenance of those mutations leading to amino acid replacements, accompanied by an increased mutual complementarity of the changes in the chemical and conformational free energies.

Essentially similar reasoning can be found in a paper by Somero (1975a). This author states that enthalpy changes occurring in the formation of an enzyme substrate complex are "titrated" by enthalpy changes of the opposite sign determined by concurrently proceeding interactions of the enzyme with ligands.

The data reported by Ivanov and his colleagues (1973) for aspartate aminotransferase conform well to this concept. According to these authors, the complementary smoothing of the fluctuations in the energy level of the system is achieved through the inter-substitution between only two discrete states of the protein conformation. Modern views on the interaction of the chemical (involving electrons) and conformational movements in the enzyme-substrate complexes in the course of the enzymic catalysis have been discussed by Volkenshtein (1971).

Regardless of the differences in the interpretations of the vast amount of accumulated data, it is clearly obvious that an enzyme macromolecule must possess a certain degree of flexibility, plasticity, and lability in order to be able, under given physiological conditions, to perform the necessary conformational transitions.

The idea of flexibility of protein macromolecules and of the functional changes in their conformation has spread far beyond the limits of the problem of the enzymic catalysis and has been extensively employed, with greater or lesser appropriateness, for explaining almost all biochemical and physiological processes involving proteins.

This idea has been successfully applied to description of the processes of the allosteric regulation of enzyme activity. As known, the catalytic activity of very many enzymes is regulated by the stimulating or inhibiting effectors. The structure of an effector molecule is, as a rule, entirely different from the molecular structure of the substrate. The effector interacts not with the enzyme's active site, but with its remote part (relative to the active site) that is called the allosteric site. In order to understand the mechanism of allosteric regulation, it is important to realize in what way an effector, having associated with the allosteric site on an enzyme molecule, transfers a signal of stimulation or inhibition to the catalytic site on the same molecule.

Koshland (Koshland and Kirtley, 1966; Koshland and Neet, 1968; Koshland, 1969; Koshland, 1971) solves this problem based on the discussed ideas about the conformational flexibility of enzyme macromolecules. He believes that an effector molecule, bound to the enzyme's allosteric site, produces conformational changes in the enzyme molecule which extend as far as the area of the active site. In the case of a stimulating effector, these changes will facilitate interaction of the enzyme with the substrate, whereas in the case of an inhibitor they will preclude this interaction.

Another hypothesis has been advanced by Monod et al. (1965). There is no need, for the purposes of our discussion, to go into the details of their hypothesis. Briefly, Monod and his collaborators believe an enzyme can be found in two different conformations, and in only one of these can its affinity toward the effector be realized. Binding the effector drastically changes the equilibrium constant for the two forms of the enzyme and this leads to accumulation, in a population of enzyme molecules, of those forms, which are in the conformation favoring affinity to the effector. In the case of stimulation, this conformation promotes higher catalytic activity. The reverse holds true for the inhibitor.

Some authors (e.g. Mathias and Kemp, 1972) believe that the results they obtain can be explained in the context of both theories. Koshland, in his paper, refers to a number of examples which can hardly be explained by the views advocated by Monod and his co-workers, while they can readily be understood in the framework of his theory. For the discussion that follows it is not so important whether a ligand directly induces a change in the conformation of a protein macromolecule or whether it tips the established equilibrium between the preexisting conformers. What is important is that both concepts emphasize the existence of a certain level of conformational lability of enzyme macromolecules.

Spirin (1964) is right in saying that "... flexibility and mobility of the conformation of a protein molecule endows the latter with an ability to respond adequately to the presence of these or those factors in the medium, i.e. to be regulated by these factors" (p. 31). Abundant literary evidence has accumulated concerning changes in the secondary, tertiary and quaternary structure revealed by various techniques in many an enzyme during the interaction of the enzyme with stimulating and inhibiting effectors (Koshland and Neet, 1968; Koshland, 1969).

Transformation of zymogens into active enzymes, during which a disruption of one or several peptide bonds occurs, has been reported to be accompanied by conformational changes of the molecules. In the case of chymotrypsin this has been shown by the X-ray diffraction technique (Kraut et al., 1967; Henderson et al., 1971) and by the tryptophan residue fluorescence (Volotovsky and Konev, 1967). Several indirect methods have been used in studies on chymotrypsin (Neurath et al., 1956; Neurath, 1964).

There is little doubt left at present that a certain common principle, accounting for the interaction of the actomyosin type proteins with adenosine triphosphate, underlies the diverse forms of active movement of live cells and of their parts.

No one biochemical theory of contraction can ignore a change occurring in the conformation of the contractile proteins. All the theories proposed to date acknowledge the significance of disruption and reformation of the weak bonds supporting the macromolecular structure of contractile proteins (Bendall, 1969; Zaalishvili, 1971).

An idea of the conformational changes of the membrane protein components is widely accepted in various branches of membranology. This idea is most frequently discussed in studies on mitochondrial activity. Konev et al. (1970) give a number of references to works reporting conformational changes in mitochondrial proteins upon transition of mitochondria from the energized into the non-energized state and vice versa. Such changes have been detected by electron microscopy, by shifts of the maxima of fluorescence and changes in its intensity, by nuclear magnetic resonance, etc. Some of the hypotheses on the mechanism of phosphorylation, coupled to oxydation in the mitochondrial membranes, originate from the idea of conformational flexibility of proteins (Green et al., 1968; Green and Baum, 1970; Green, 1970). There are authors, however, who believe the participation of conformational transitions of mitochondrial enzymes in the process of formation of the primary energy-rich compound to be lacking firm corroborative evidence (Skulachev, 1972).

Here is an example from another field of membranology. A nerve impulse is mediated through the myonerval synapse by acetylcholine, which interacts with the receptor of acetylcholine—a lipoprotein—located in the subsynaptic membrane. As a result, permeability of the membrane to ions sharply increases, thereby generating a wave of excitation spreading along a muscle fiber. Vasquez et al. (1971) have studied the effects of acetylcholine on artificial membranes prepared from phospholipids, cholesterol and lipoproteins isolated from the myonerval synapses of the electrical organ of *Electrophorus electricus*. Acetylcholine enhanced the electroconductivity of this membrane, and, concurrently, structural changes visible by electron microscopy occurred and formations in the membrane appeared which were interpreted by the authors to be the channels of ion conductivity. The effect of acetylcholine can be accounted for by the conformational changes of the protein specifically binding acetylcholine.

Some investigators claim that a run of an excitation wave along the nerve is accompanied by conformational changes of the proteins incorporated in the axoplasm or in the nerve fiber membrane (Nasonov, 1959; Tasaki et al., 1968, 1969; Komissarchik et al., 1971). Tasaki and his colleagues arrived at this conclusion, based on the fact that a wave of electronegativity (a spike) in the nerve has been found to be accompanied by a change in birefringence and light scattering as well as by an enhancement in the fluorescence of the nerve prestained with a fluorochrome. Keynes (1970) also has observed a change in the light scattering and birefringence during a spike, but he was more reserved in making his statements.

The conformational changes of proteins are thought by numerous authors to occur during the reception of stimuli (Vinnikov, 1974). Most conclusively this has been proven in the case of the visual pigment rhodopsin, which is a receptor of photons and there by the first link in the act of vision. Rhodopsin is a chromoproteid: it consists of the protein component opsin and the chromophore retinen (an aldehyde of vitamin A). Absorption of a quantum of light leads to a modification of the state of the chromophore, followed by a chain of events proceeding in the absence of light. At these stages, consecutive transformations of the chromophore and conformational modifications of opsin take place. As a result the association of opsin with retinen breaks down. Dissociation of rhodopsin is accompanied by a change in the pigment color from red to yellow. All these processes, following as yet unknown routes, evoke a nerve impulse that creates a visual sensation in the brain. Etingof (1967) regards the fact of conformational re-arrangements of opsin as being established unequivocally.

As known, the central function of the blood serum albumin is the transportation and neutralization of a variety of biological compounds which this protein can bind. Among these are lipids, fats, fatty acids, vitamins, certain hormones, drugs, cations, anions, etc. Many of these ligands, which are bound by albumin in the intestine blood vessels or in other body parts, are released by albumin when it passes through the kidney blood vessels. The surprising capacity of this protein to form complexes with compounds of extremely diverse nature and to desorb them at specific locations in the body, is ascribed by many authors to the "conformational adaptation" of this protein (Karush, 1950).

Attempts are being made to explain such processes which are almost inaccessible to the researcher, such as the conjugation of homologous chromosomes in meiosis as related to the allosteric modification of protein conformation (Comings and Riggs, 1971).

Summary. The number of publications in various fields of biology in which conformational changes of proteins are discussed, is overwhelming (for a comprehensive review see Citri, 1973). Often the existence of such modifications has been conclusively proven, whereas in other cases, the authors, offering indirect arguments, lead the reader to acknowledge such changes to be highly probable. Occasionally, the expression "conformational changes of proteins" is used to create an impression of explanation of an as yet incompletely understood process. Researches employing progressively sophisticated approaches will provide further insights into whether the conformational changes of protein molecules in fact occur in each particular case and will facilitate quantitative evaluation of their type and extent. At any rate, the general agreement that conformational rearrangements are an indispensible component in the performance of the majority of proteins can hardly be questioned. Lumry and Biltonen (1969) consider as "very good" the assumption implying that conformational changes are operating in all those processes in which a protein interacts with other proteins or small molecules, and the specificity of the interactions can be shown to exist. Hence it follows that conformational flexibility or lability of a protein macromolecule, i.e. its ability to undergo transition into another structural state, is its key feature. Such a transition is accompanied by cleavage of some bonds supporting the secondary, tertiary and quaternary structure and establishment of new bonds and interactions between the atoms in a macromolecule.

4.2.2 Effects of Temperature on the Conformation of Protein Macromolecules

It is well known that the unique spatial packing of the polypeptide backbone of the protein molecule is supported by various forces (Linderstrøm-Lang and Schellman, 1959; Kauzmann, 1959; Scheraga, 1961; Ptitsyn, 1967; Polyanovsky, 1967; Brandts, 1967, 1969; Lumry and Biltonen, 1969). These are largely non-covalent, weak linkages. Hydrophobic interactions, among others, bring together non-polar amino acid residues. Since interactions of hydrophobic residues with water are unfavorable in terms of entropy, most of the residues are hidden inside a molecule, and, in globular proteins, they form a hydrophobic core located deep inside a globule (Bresler and Talmud, 1944). Ions bearing the same charge are electrostatically repulsed, while those with opposite charges are attracted. Hydrogen bridges are established primarily between the hydrogen of the carbonyl group in a peptide link and the imine nitrogen in the other link (N—H ··· O= =C); hydrogen bondings are thought to play a key role in supporting the secondary structure of the protein molecule (α-helix and β-structure of Pauling-Corey). In some proteins, their tertiary structure may be supported by the covalent disulfide bonds (–S–S–) and the salt bridges. Competition between the amino acid residues and molecules of water for the hydrogen bonds, as well as the influence of the hydrophilic and hydrophobic parts of the protein molecule on the water structure in close proximity to the residues, make the interactions between the molecule and water a cardinal factor, determining the stability of a protein molecule.

At various stages in the course of investigations directed to finding out the physical principles underlying the structure of protein molecules, attempts have been made to ascribe the chief responsibility for maintaining the protein spatial structure to this or that kind of interaction. At one time it was the hydrogen bonds, at another the hydrophobic interactions that were preferred. One common scheme can hardly be surmised that would be applicable to various proteins, much the same as it is impossible to ascribe an exclusive role to any single type of bond in establishing and preserving a unique structure of the protein macromolecule. This contention is strengthened by the facts known from the field of protein denaturation. There is in the literature a great deal of definition of the concept of "denaturation" (Neurath et al., 1944; Maurer, 1959; Kauzmann, 1959; Scheraga, 1961; Straub, 1964; Joly, 1965; Tanford, 1968; and others). As far as the essence of denaturation is concerned, however, it comes down to the fact that in the process of protein denaturation, more or less complete destruction of the secondary, tertiary and quaternary structure occurs without hydrolysis of peptide bonds. A protein molecule loses its unique native conformation and becomes a more or less random coil. This is accompanied by a loss of the intrinsic functional properties of the protein.

A characteristic feature of denaturation is that it can be produced by extremely diverse physical and chemical agents strikingly dissimilar in their nature: by high and low temperatures, various types of irradiation, high hydrostatic pressure, ultrasound, surface tension forces, acids and alkali, heavy metals, organic solubilizers, sulfhydryl active compounds, urea, etc. The mechanism of action of all these agents as well as the points of their application on a protein macromolecule are different but not fully understood in each particular case. There is no doubt that the sulfhydryl reagents cleave primarily the S–S bonds; organic solubilizers weaken the hydrophobic interactions, whereas the pH shifts affect the electrostatic interactions. In all cases though, irrespective of the first step in the process of destruction of the molecule by a denaturant, the result is essentially the same: the polypeptide chain loses its native packing and the higher order structures are destroyed. The products of denaturation of the same protein, resulting from the action of various agents, may be similar in their properties. This is, as a rule, a more or less disordered polypeptide coil. The process of destruction of a protein macromolecule by a denaturant is basically a cooperative one. Alteration of a given bond not only facilitates further cleavage of the remaining bonds of the same type, but leads also to weakening and disruption of other types of linkages. Eventually the entire spatial structure of a protein macromolecule collapses. The high unspecificity of the denaturation process, its cooperativity and the extremely diverse physico-chemical nature of denaturants testify to the fact that the spatial structure of the protein molecule is supported by the whole ensemble of bondings and interactions.

It is clearly obvious that the conformational flexibility of the protein macromolecule depends on the forces responsible for maintenance of the spatial arrangement of its atoms. These forces, in their turn, are determined by the primary structure of a polypeptide chain and by the state of the medium in which the protein macromolecule is found. In order for a protein macromolecule to undergo the conformational transitions required to perform its functions, the flexibility of the entire molecule or of its separate active parts must conform to the following two requirements. First, the *flexibility must be sufficient* for the necessary spatial translocations of the molecule's parts to be accomplished with the energy and time expenditures predetermined by the cell's organization. Second, this *flexibility should not be excessive* so that the macromolecule can be maintained in a required conformation. In other words, flexibility or rigidity of a protein macromolecule is to be preserved at a certain intermediate level. Such a state of the macromolecule might be designated as the *semilabile* or *semistable state.* For every protein, depending on its morphophysiological nature, an optimum would be this or that level of stability.

A measure of an overall stability of a protein macromolecule can be seen in the difference between the free energy of the macromolecule in its native ordered state and of the macromolecule found in the state of a denatured random coil: the larger the difference under given conditions, the more stable and rigid is the

Table 18. Effects of temperature on the entropy of protein denaturation. (From Tanford, 1968)

Protein and conditions of denaturation	ΔS value (cal/° C/mole) at different temperatures			
	0°	25°	50°	75°
Ribonuclease, temperature transition at pH 2.5	+ 31	+155	+340	—
Chymotrypsinogen, temperature transition at pH 2.5	− 80	+105	+360	+680
β-lactoglobulin, 5 M urea, pH 3.0	−260	− 72	+100	+260

macromolecule. A change in the free energy, ΔF, during a transition from a native state to a denatured one, is described by the trivial equation: $\Delta F = \Delta H - T\Delta S$, where ΔH and ΔS are the differences in enthalpy and entropy, respectively, between the native and denatured states, and T is the absolute temperature of the transition.

In terms of the development of the concept presented in this monograph, the principle problem is the elucidation of how the level of conformational flexibility of protein macromolecules alters with changes in the temperature environments. A priori answers to this question are difficult to find because the temperature dependence of the strengths of various forces supporting an ordered structure of the protein macromolecule varies with a particular force and is not always clear cut. Besides, difficulties arise also in evaluating the contributions of separate interaction forces to the enthalpy and entropy terms of the above equation.

From what is known about the structure of globular proteins it follows that the most significant events occurring during the transition are: an exposure of the hydrophobic groups, which had formerly been found in the interior of a native globule, outward into water; and an increase in the conformational entropy during the conversion of a rigid native structure with a limited freedom of atomic rotation, into a flexible random coil structure. An entropy change taking place during the process of denaturation is not only associated with a change in the conformational entropy caused by a transition of a protein macromolecule into a disordered state. A significant contribution in the opposite direction is being made by a change in the water structure in the vicinity of the exposed hydrophobic side chains. The ordered state of water near the hydrophobic groups is increased and this leads to a decrease in the entropy of the denaturation transition. As the temperature is elevated, the structure of water, induced by the hydrophobic chains, melts easily and, consequently, with elevation of the temperature, its contribution to the entropy of denaturation is reduced. As a result, the magnitude of ΔS of denaturation will increase as the temperature is elevated (Table 18).

When the temperature is increased within any region of the temperature scale, the entropy term ($T\Delta S$) steadily grows: this must weaken the strength of the spatial organization of the molecule. On the other hand, hydrophobic interactions, appreciably contributing to the maintenance of the structure, exhibit a complex dependence on the temperature (Scheraga et al., 1962). Concomitant

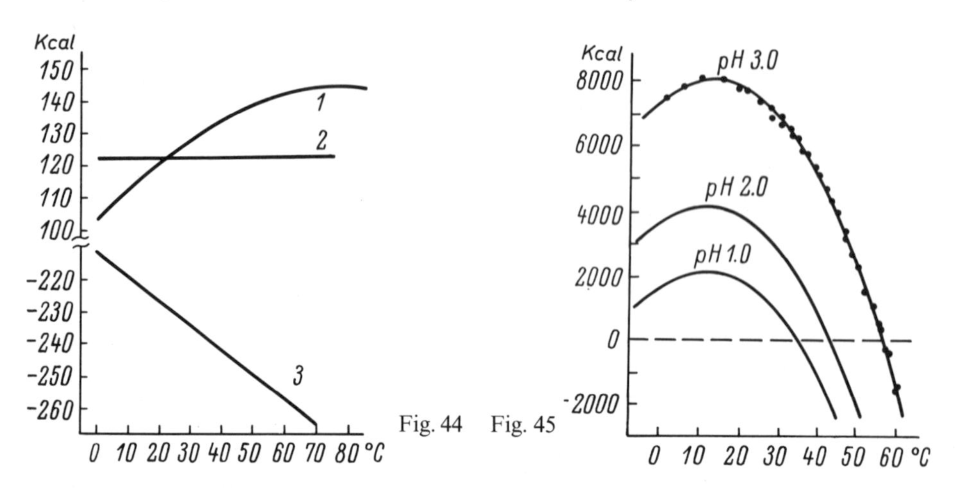

Fig. 44. Effect of temperature on contribution of free energy to stabilization of native structure of chymotrypsinogen by various factors (Brandts, 1967). *Abscissa:* temperature; *ordinate:* free energy. *1:* hydrophobic interactions, *2:* hydrogen bonds, *3:* the conformational entropy

Fig. 45. Effects of temperature on free energy of denaturation (F°) of chymotrypsinogen at different pH (After Brandts, 1967). *Abscissa:* temperature; *ordinate:* free energy of denaturation

with elevation of temperature over the whole range from 0 to 50–70° C, the force of expulsion of the hydrophobic amino acid residues from water increases, thereby the tendency for them to approach each other inside the globule is enhanced. Elevation of the temperature beyond 50–70° C tends to slow down the increase in the expulsion and the tendency of the hydrophobic residues toward mutual approaches becomes weaker. The temperature dependence of the strength of the hydrogen bonds inside a protein globule is difficult to determine; presumably it is not strong. The forces of the electrostatic interactions are not expected to be changed with variations in the temperature. Accordingly, the reaction of the protein macromolecule to a temperature fluctuation is determined essentially by the two temperature-dependent components comprising a system's entropy multiplied by the absolute temperature ($T\Delta S$): the conformational entropy and the entropy determined by the hydrophobic interactions. In the zone of physiological temperatures these components influence the protein stability, shifting it in the opposite direction, which fact, presumably, is of profound biological significance, since thereby the effects of temperature fluctuation on the state of a protein macromolecule are being buffered. The next question pertains to the relationships between these oppositely directed tendencies: which of these prevails and what actually happens to the stability of a protein macromolecule as the temperature is changed? Direct evidence that might shed light on this problem is lacking.

Brandts (1967, 1969) offers a scheme for the temperature dependence of the factors which stabilize the chymotrypsinogen molecule (Fig. 44). The experimental data obtained by him concerning the effects of temperature on the free energy of denaturation (ΔF) of chymotrypsinogen at different pH are depicted in Figure 45. A maximal free energy change occurring upon transition of chymotrypsinogen from a native state into a denatured one is seen at 10–13° C. Consequently, at this temperature the protein is most stable. As the temperature is elevated, the labilization of the protein grows, whereas lowering of the temperature also results in a slight decrease of its stability. The maximal stability of chymotrypsin, according to Brandts, is observed also at 10–13° C, and that of ribonuclease is around 0° C.

These data are only approximate since they have been derived from the constants of the equilibrium between the native and denatured forms of the protein found in non-physiological conditions. Pace and Tanford (1968) have reported the temperature maximum for β-lactoglobulin stability to be at 35° C, and, according to Hermans and Acampora (1967) myoglobin is most stable at below zero temperatures. Brandts and Hermans and Acampora alike worked at the extremes of pH, while Pace and Tanford used 5 M urea at pH 2.5–3.5. These workers had to create such conditions because under the physiological conditions by virtue of the intermolecular binding of denatured molecules, the system proves to be irreversible, and thus cannot be utilized for thermodynamic calculations. A similar temperature dependence of the stability of globular proteins, with the maximal stability at low temperatures, has been corroborated most recently by Privalov and Khechinashvili (1974). The authors have carried out more direct measurements of thermodynamic parameters by studying the calorimetric characteristic of the myoglobin, cytochrome c, chymotrypsin, lysozyme and ribonuclease preparations.

Another set of findings relevant to the temperature effects on the stability of the protein macromolecule should also be considered. There is an infinite amount of investigation in the literature dealing with the problem of thermal denaturation of proteins. Being a cooperative process, the transition from a regular native structure to a more or less disordered coil may occur in many proteins within a narrow temperature range. The denaturation zone corresponds to the temperature at which the forces stabilizing the proteins are being overcome. As shown, this zone in isolated proteins (see Sect. 3.2) is often found at a temperature below the physiological range of temperatures. Consequently, a protein in situ is stabilized not only by the forces of interaction between the components of its own polypeptide chain, but also by interactions with other macro and micromolecular substances. It follows from these experiments, however, that in the protein macromolecule itself, in the course of elevation of the temperature within the zone in which heating still has to augment the strength of the hydrophobic interactions, the destructive tendency associated with an increase in the $T\Delta S$ term, overcomes the opposing efforts of the hydrophobic amino acid residues. In many proteins, alteration of the protein conformation can be observed at temperatures below the denaturation zone and below the temperature optimum for life of the organisms from which these proteins had been isolated.

Privalov et al. (1971) studied the heat capacity of chymotrypsin, ribonuclease, and myoglobin solutions as a function of temperature that was gradually elevated starting from room temperature. As soon as the temperature was only slightly elevated they detected an enhancement of the heat capacity of the proteins. The authors believed this to be the result of progressing labilization of a globular protein molecule. The denaturation transitions, on the other hand, occur at appreciably higher temperatures. Studying the temperature dependence of the circular dichroism of ribonuclease A, Simons et al. (1969) found that as the temperature was increased in the range of 15–50° C (pH 6.46), the ribonuclease molecule was affected by a noncooperative alteration that was followed by a cooperative transition occurring at a more intensive heating. The second, cooperative stage of the modification of the ribonuclease molecule was halfway through at 60° C.

The temperature optimum for life of the thermophilic bacterium *B. stearothermophilus* corresponds to 65° C. At the same time, in a number of enzymes isolated from this thermophile, conformational changes unrelated to a loss of their activity have been detected at temperatures around 50° C: in α-amylase a conformational shift is observed at 50° C (Pfueller and Elliott, 1969), in ATP-ase—at about 55° C (Hachimori et al., 1970), and in fructose-1,6,-diphosphate aldolase—at 50–53° C (Sugimoto and Nosoh, 1971).

Stepanov and Voronina (1972) reported that an RNA-binding protein from rat liver and loach embryo reversibly binds RNA at temperatures of 0–4° C, in contrast to natural informosomes. The loach protein shows stable binding with RNA at 21° C and the rat protein only at 37° C. This finding might be tentatively explained as due to the conformational changes occurring in these proteins at the temperatures corresponding to the physiological temperature optima for the loach and rat.

In many proteins, the SH and SS groups are hidden in native molecules inside the globular unit. Most, if not all, of these groups are unable to react with appropriate reagents until a more or less complete disordering or alteration of a native molecule has taken place. There are 17 SS groups in the serum albumin molecule. Davidson and Hird (1967) have demonstrated that the disulfide groups in the bovine serum albumin cannot be reduced by glutathione at room temperature, although beginning from 35° C, along with elevation of the temperature, an increasing number of SS groups can be reduced and at 55° C as much as 50% of the bonds react with glutathione. Consequently, molecules of this protein are labilized by heat already in the zone of physiological temperatures. Still more impressive data have been reported for the human serum albumin obtained with the hydrogen-deuterium exchange technique (Hvidt and Wallevik, 1972). It proved that at pH 6.2 the conformational stability progressively decreases as the temperature is raised from 5 to 35° C, which fact accounts for an enhancement of the H-D exchange.

The number of non-polar groups exposed to the surrounding medium increases in a molecule of the rabbit myosin as the temperature is elevated from 20 to 50° C. This was established by Lim and Botts (1967) who recorded an enhancement of the 8-anilino-1-naphthalene sulfonate fluorescence. This compound fluoresces when bound to non-polar groups of the protein macromolecule. The myosin molecule was found to be in the most compact conformation at 20–25° C.

Klee (1967) investigated the effects of temperature upon the resistance of the bovine pancreatic ribonuclease to chymotrypsin, trypsin, carboxypeptidase and aminopeptidase digestion. Below 35° C the molecules of this enzyme were tightly packed and the peptide bonds were inaccessible to the action of these four peptidases. An increase in temperature induces labilization of the enzyme globular structure so that its sensitivity to proteases grows. Complete accessibility of the bonds to all of the proteases tested is achieved by heating the enzyme preparation above 55° C. Studies concerning the temperature effects on the optical rotatory dispersion revealed that local alterations in this protein conformation can be detected beginning at 10° C.

Troitsky and his co-workers (1971), employing the optical rotatory dispersion technique found that in a number of proteins conformational transitions occur at physiological temperatures. Konev and his colleagues (Konev et al., 1970a; Mazhul et al., 1970; Konev et al., 1970b; and others) estimated in 14 proteins the temperature dependence of the degree of depolarization of fluorescence and that of the position of the maxima of fluorescence spectra at the exciting wavelength of 297 nm. The fluorescence pattern was largely determined by the state of the tryptophan residue. Conformational transitions were detected in all of the proteins. The temperature at which semi-transitions occurred varied with the protein from 6° C (hexokinase) to 22° C (human serum albumin). The authors concluded that: "... along with temperature elevation (within the physiological range—V. A), a consecutive replacement of discrete protein forms A–B–C occurs eventually resulting in a saltatory increase in the intramolecular mobility" (Mazhul et al., 1970, p. 8). The authors further indicate: "... the temperature-related augmentation of the radical mobility is associated, to a somewhat greater extent, with the hydrophilic tryptophanyls, located at the surface of the globule, rather than with the hydrophobic ones, found in proximity to the protein core". With the aid of the methods already mentioned, they have detected conformational changes of proteins in live cells of the frog eosophagus ciliary epithelium, which occurred within the range of 3–11° C. Further development of investigations relevant to the temperature-dependent transitions in cells and cellular membranes is discussed at length in a monograph by Chernitsky (1972).

The Arrhenius plot of temperature dependence of the maximal reaction rate allows the activation energy of a reaction to be evaluated. If the activation energy is not changed over a certain range of temperatures, the dependence is expressed by a straight line. The higher the angle of the curve's slope in reference to abscissa, the higher is the value of activation energy. The Arrhenius plot of enzymic reactions frequently displays a break at a certain temperature. In this case, the plot consists of two rectilinear branches corresponding to different activation energies. If a discontinuity in the Arrhenius plot is obtained for an enzymic reaction carried out in a simple in vitro system, this might be most plausibly explained as due to a conformational modification of the enzyme occurring at a certain temperature (Massey et al., 1966; Kumamoto et al., 1971). Figure 46 depicts two Arrhenius plots of the temperature dependence of the transformation rate of isocitrate into α-ketoglutarate (Palm and Katzendobler, 1972). This reaction is catalyzed by NAD-dependent isocitrate dehydrogenase. The temperature at which a conformational transition of isocitrate dehydrogenase occurs is a function of NAD concentration. In the presence of 0.15 mM NAD, a discontinuity in the plot is seen at 21° C, at a higher NAD concentration (1.0 mM) the break is shifted to 9.5° C. In its conformation specific for lower temperatures, isocitrate dehydrogenase is functionally less active.

Sultzer (1961) reported a discontinuity in the Arrhenius plot of oxidation of sodium octanoate from two species of mesophilic bacteria occurring at 20 and 25° C, and for a psychrophilic bacterium at 15° C. A number of examples of enzymatic reactions and physiological processes, which can be described by the

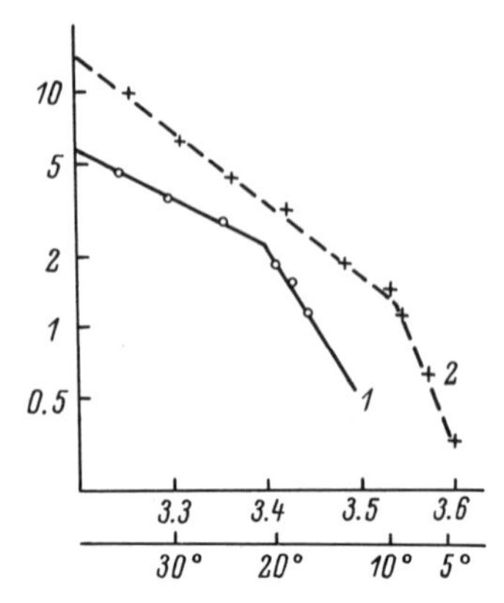

Fig. 46. Arrhenius plots of yeast isocitrate dehydrogenase activity in the presence of 0.15 *(1)* and 1 *(2)* mM NAD (Palm and Katzendobler, 1972). *Abscissa:* reciprocal of absolute temperature ($1/T° K \cdot 10^3$); *ordinate:* logarithm of maximal reaction rate

Arrhenius plots with a break in the region of physiological temperatures, are referred to in a monograph by Konev et al. (1970). A discontinuity is thought by the authors to be accounted for by temperature-dependent conformational modifications of enzyme macromolecules. It deserves pointing out, however, that in those cases when a plot with a break describes a reaction catalyzed by an enzyme associated with the membrane, the primary cause of the appearance of a discontinuity may be related to temperature-dependent changes in the physical state of the membrane lipids (Wilson and Fox, 1971; Raison et al., 1971a; Towers et al., 1972; Konev et al., 1972; Inesi et al., 1973). This topic will be dealt with in more detail later (p. 244).

A multitude of data may be cited in addition to those already discussed, which demonstrate conclusively that temperature variations within the physiological limits affect the spatial structure of protein molecules. In spite of the constraints imposed by the hydrophobic interactions—and this is probably one of their principal roles—an elevation of temperature tends to labilize of protein macromolecules. Although a protein molecule can be regarded as an integral system, one can hardly agree that its integrity and the cooperativity of its destruction are absolute. It might be acknowledged, therefore, that a temperature increase should, to a greater extent, labilize the peripheral, most reactive parts of the molecule, which are found at a distance from the internal hydrophobic core of the globule.

At a lower temperature, around 0–5° C, many proteins are inactivated. Among such proteins one finds glucose-6-phosphate dehydrogenase (Kirkman and Hendrickson, 1962), glutamate dehydrogenase (Fincham, 1957), α-β-hydroxybutyrate (Shuster and Doudoroff, 1962), lactate dehydrogenase (Zondag, 1963), 17β-hydrosteroid hydrogenase (Jarabak et al., 1966), pyruvate carboxylase (Scrutton and Utter, 1965), glutamate decarboxylase (Shukuya and Schwert, 1960), α-amylase (Pfueller and Elliott, 1969), the nitrogen fixating enzyme (Dua and Burris, 1963), pyruvate kinase (Kuczenski and Suelter, 1970) l-threonine deaminase (Feldberg and Datta, 1971) and others. If exposure to a low temperature is not too prolonged, the activity of an enzyme upon its return to room temperature is usually restored. It has been established that cold inactivation is accompanied by dissociation of an oligomeric protein into subunits (Frieden, 1971). Kawashima et al. (1971) observed a reversible cold inactivation of the R and DP-carboxylases from tobacco that was not accompanied by a change in the constant of association. This protein consists of eight large and six small subunits. The authors believe a partial dissociation occurs in the cold—the compact quaternary structure of the protein is loosened. The cause of this dissociation is thought by most workers to lie in the weakening of the hydrophobic interactions supporting the quaternary structure of oligomeric proteins in the cold. Sometimes, however, an aggregation of molecules takes place during cold inactivation (Jarabak et al., 1966). It might, however, be noted, that in order to preserve the activity of most proteins, the biochemist keeps them in the cold and not at room temperature. Proteins inactivated in the cold are apparently only those whose conformation is determined primarily by the forces of hydrophobic interactions.

Summary. An analysis of the evidence concerning the effects of temperature on protein conformation warrants belief that as the temperature is elevated within the range of physiological temperatures, either general or local labilization of macromolecules occurs, in spite of the restricting influence of the hydrophobic interactions. The counter-balancing tendency of hydrophobic interactions is more effective in smoothing an increase of the entropy term value when the temperature is raised over the range of low temperatures—somewhere around 0–10° C. Lowering of temperature in this zone promotes weakening of the hydrophobic interactions in some oligomeric proteins to such a degree that the latter are eventually dissociated and inactivated.

4.2.3 Adaptive Changes of the Conformational Flexibility of Protein Molecules with Changes in the Environmental Temperature

Earlier (p. 144) it was indicated that the level of conformational flexibility of proteins, the performance of which is associated with re-arrangement in their conformation, must be *sufficient, but not excessive*. The realization of this prerequisite is feasible for a given macromolecule only over a limited range of temperatures, which must lie within the limits of the physiological optimum for life of the tissues of an organism. For physical reasons this temperature span cannot be too large. Considerable changes of habitat temperature can alter an optimum balance between lability and rigidity, and thereby inflict damage on the biochemical efficiency of a protein. Cells and organisms alike dispose of various regulatory mechanisms, which, to a certain extent, can compensate for an unbalanced correspondence. Involvement of the compensatory mechanisms, however, requires additional energy and material expenditures. It seems that in terms of economy, it is more profitable to restore by some means the optimum correspondence between the conformational flexibility of protein molecules and the temperature of environment.

The facts presented in the preceding section make one admit that lowering the temperature below the zone of a temperature optimum for life will reduce the lability of protein molecules and augment their rigidity. Conversely, elevation of the temperature above the optimum will enhance their lability. In the former case, the lability may prove to be insufficient, whereas in the latter it may be excessive. In both instances, however, this may interfere with the realization of the function for which a given protein is responsible. Adaptation of protein molecules to new temperature environments, as the temperature is lowered, must be achieved through labilization of the molecules. Again, when the temperature is elevated, the rigidity of the protein molecules will be increased. There are three apparently feasible ways of altering the conformational flexibility of protein macromolecules: a modification of the primary structure, an association with, or release of ligands, and a modification of the composition of ambient medium in which a protein functions (changing the concentration of apolar substances, ionic strength, pH, etc.). Later the extent to which each of these possibilities is realized will be discussed in more detail. At this point, however, an answer should be provided as to how information can be obtained on the occurrence of a shift of the conformational flexibility of a protein during a change in temperature, and if it occurred, in which direction.

4.2.4 Thermostability of the Protein as an Indirect Indicator of the Conformational Flexibility of Protein Macromolecules

There is no method available at present that might serve to estimate quantitatively the degree of conformational flexibility of protein macromolecules. We have, however, at our disposal an indirect indicator suitable for revealing a modification of flexibility and detecting a direction of such a shift. In fact, a decrease in conformational flexibility must be achieved through strengthening of bonds and interactions supporting the secondary, tertiary and quaternary structure of proteins. This should affect the response of protein molecules to the denaturing agents. A decrease in flexibility must enhance the resistance of a protein to denaturants and, primarily, to the denaturing action of heat. An augmentation of flexibility, on the other hand, is expected to diminish thermostability of the protein. Riegel (1965) is right in saying that usually testing of the interaction forces for their "strength" is performed through assessment of their resistance to a "heat attack". In sum, thermostability may be used as an indirect measure for comparative evaluation of the level of conformational flexibility of protein molecules, although it cannot be employed for estimating the absolute value of protein flexibility. Some of the restrictions inherent in this method will be discussed later in this monograph. The thesis of the necessity of maintaining a correspondence between the temperature of environment and the level of the conformational flexibility of protein molecules will be considered now, as well as the validity of employing thermostability as an indirect criterion of flexibility in the context of the whole body of evidence referred to in Chapters 1–3.

4.2.5 Conformational Flexibility of Protein Molecules and Adaptation of Organisms to the Environmental Temperature in Phylogenesis

It was shown in Chapter 2 that in evolution a correspondence has been established between the primary thermostability of the cells and thermophily of species. Later (Sect. 3.2.2) it was demonstrated that the primary cellular thermostability reflects the heat resistance of the protoplasmic proteins. Hence, it was concluded that a correlation between the level of protein thermostability and the environmental temperature for a species' life is maintained in nature. This regularity holds true on the general biological scale since it can be detected in animals, plants and microorganisms, and in both the protoplasmic and extracellular proteins. It was clearly shown in Section 4.1 that, in the majority of cases, a correspondence of the protein heat resistance to the environmental temperature bears no adaptive significance per se. The only possibility to explain the infallibly maintained correlation between protein thermostability and ambient temperature in evolution, is that the level of thermostability of proteins is quantitatively related to some other characteristic of proteins, which is most important biologically, and which is temperature-dependent within the physiological range of temperatures. If, based on the above evidence, one admits that this characteristic is the conformational flexibility of protein macromolecules, it will become clear why, in the process of evolution, the frequently rather useless parallelism between ambient temperature and thermostability of proteins has been so carefully preserved. This statement being accepted, the biological significance of the data referred to in Chapter 2 and Section 3.2.2 may be formulated as follows. *There has been maintained in nature, in the process of evolution of species, a correlation between the level of conformational flexibility of protein macromolecules and the mean environmental temperature during an active period of life of organisms of a given species.*

The evolutionary changes of the conformational flexibility level are revealed when comparing the heat stability of analogous proteins and—more indirectly—comparing the primary thermostability of analogous cells in closely related species or within intraspecific groupings, living in dissimilar temperature environments. In Section 3.2, the shortcomings and advantages of both methods—both protein- and the cell-oriented—were demonstrated. Putting aside the language of thermostability and adopting that of conformational flexibility, we are able to provide answers to a number of puzzling questions formulated in Section 4.1. In that section it was shown that in agreement with an adaptive significance of protein thermostability, attempts can still be made to solve the problem of why an adaptation to elevated temperatures is accompanied by an increase in the heat stability of proteins. It would be impossible, however, to explain why an excessive thermostability interferes with adaptations of the organisms to low temperatures. And the second question is no less justified than the first one.

If, however, an adaptive significance is seen in the flexibility, but not in the resistance, then it becomes apparent that it is not only a decrease in the conformational flexibility of protein molecules during adaptation of organisms to elevated temperatures that is important; but also an enhancement of the flexibility, namely, a reduction of the rigidity of the molecules occurring in adaptation of organisms to low temperatures is no less important. When the temperature is lowered, the original flexibility may prove to be insufficient, parts of the molecular machine may become less movable, like engine parts in which the lubricant solidifies in the cold. This thesis about the harm that an excessive rigidity of protein molecules can do to the adaptation to life at a lowered temperature should be taken into consideration in solving the problems of acclimation and selection of cold-loving breeds and varieties of animals and plants.

Another question posed in Section 4.1 cannot be answered if an adaptive significance is ascribed to the heat stability of proteins per se. It was unexplainable why the level of protein thermostability, that definitely plays no role in the determination of the resistance to heat of an entire organism, correlates with the temperature ecology of the species. Why can this regularity be traced in the thermostable proteins, which denature at temperatures exceeding by scores of degrees the zone of physiological temperatures? If an adaptive significance is visualized in the flexibility, but not in the heat stability of molecules, then this situation becomes fully explainable. So long as a change in flexibility unidirectionally affects the level of thermostability of macromolecules, a correspondence of their thermostability to the temperature of a species' existence must be detectable in any protein, the activity of which is associated with conformational changes, i.e. in any protein, for which maintenance of its conformational flexibility at a certain level is important. Moreover, this must be the case for any protein irrespective of whether a given protein is a thermolabile or a thermostable one [5].

At the same time, the existence of an adequate interspecific difference in the primary thermoresistance of the cells injured by heating to a temperature far above that which an organism can tolerate, also becomes understandable. The cells reflect species-specific differences in the heat resistance of the protoplasmic proteins and, indirectly, demonstrate differences in their conformational lability.

It was shown earlier that in related species which differ in their thermophily, some proteins display no difference in thermostability. The reasons for such negative results may be as follows. The activity of a given protein is not associated with conformational changes and maintenance of its flexibility at a semi-stable level is not important for this protein. (The author of the induced-fit theory himself believes that not all the enzymes perform in accord with this principle.) The capacity for interacting specifically with substrate by means of conformational flexibility has been acquired later in evolution in the process of improvement of enzyme functioning. The primary form had been the interactions of a preformed template type (Koshland and Neet, 1968). The second possibility is that antagonism between the contributions of the hydrophobic interactions to the free energy of a molecule and those of the conformational entropy is well balanced. Therefore, no appreciable difference appears in the conformational flexibility of molecules in transition from the temperature of existence of a given species to the temperature characteristic of life of another species. In such a situation, there is also no reason for the appearance of interspecific differences in the heat stability of a given protein (for more about this possibility see p. 262). The third reason is that the activity of a given protein may be associated with a specific region of its macromolecules, and during evolution a correspondence has been maintained between the environmental temperature and the conformational flexibility of only that specific site, but not of the entire molecule. Variations in local flexibility may not affect that part of the molecule which is responsible for the criterion chosen to assess the protein stability. And, finally, isolation of proteins

[5] Tonomura et al. (1964) proposed designating the proteins which change their conformation in the process of realization of their biochemical functions as "transconformers". There is hardly a need to introduce this term, since most, if not all, proteins belong to this category.

from cells may affect the mechanism which maintains an interspecific difference in the conformational flexibility and also, at the same time, a difference in the heat stability specific for the living status. The latter cause is the most plausible and examples have been given above (see Sect. 3.2) illustrating how an interspecific difference in protein thermostability found in vivo is lost during isolation of proteins from tissues.

Other explanations might as well be offered regarding the absence of any differences in heat stability (or in conformational flexibility) of proteins in related species that differ in their thermophily. One might rather be surprised by the fact that this difference can be detected in so large a number of proteins than that it is in some cases not readily observed. The finding that many proteins undergo conformational changes in the process of their biochemical functions, led to a conclusion about the existence of a positive correlation between the level of conformational flexibility of these proteins and the temperature ecology of a species. If this is true, the opposite conclusion may be stated: if comparative molecular-ecological investigations, concentrated on a given protein, can reveal a correlation existing between the protein flexibility and the temperature of a species' life, then it might be inferred with greater probability that conformational mobility plays an important role in the activity of this protein. Thus, as far as is knwon, there is no evidence that would allow a relationship between the function of collagen and its conformational lability. Considering, however, the fact that thermostability of collagen responds most sensitively to a difference in thermophily of a species, it might be admitted that collagen activity is intimately associated with the conformational mobility of its molecule. The essence of this association is as yet unknown but its elucidation is certainly worth the efforts involved.

Conformational flexibility of protein molecules depends not only on temperature: it is sensitive to the action of all the factors, affecting stability of the spatial structure of the protein molecule. Therefore, if an organism adapts to an environment with modified salinity levels, different pH, hydrostatic pressure, etc., and if the supramolecular mechanisms of adaptation prove to be inefficient in protecting the proteins of that organism against the adverse effects of such shifts, then, in order to maintain a normal balance between the flexibility and rigidity of protein macromolecules under new conditions, these molecules must be appropriately restructured. Many halophilic organisms, for instance, live in media containing 25% salt. The interior of these organisms is not protected against salt (Christian and Waltho, 1962; Larsen, 1967); accordingly, the enzymes isolated from these halophiles function only if placed in high salt concentration (Maksimov, 1973). In low salt (0.25% NaCl), halophilic enzymes are inactivated; 25% solution of NaCl and KCl restore their activity (Holmes and Halvorson, 1963, 1965; Bayley and Griffiths, 1968; Hubbard and Miller, 1969; Lanyi and Stevenson, 1970; Lieberman and Lanyi, 1972; Keradjopoulos and Wulff, 1974; and others). Salts protect enzymes of the halophiles from thermal denaturation. As the salt concentration increases, the temperature of the maximal activity of an enzyme increases too. Nitrate reductase from *Halobacterium* sp., for example, displays the highest activity in the presence of NaCl in concentrations of 0.17, 0.85, 2.56, and 5.31 M at temperatures 65, 75, 85, 85–93° C, respectively (Marquez and Brodie, 1973). Some

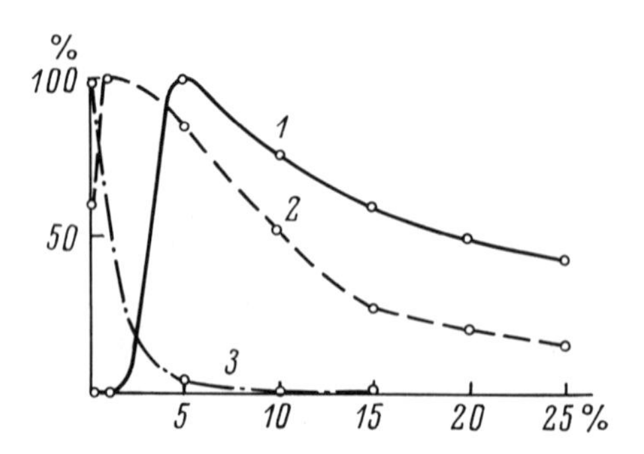

Fig. 47. Effect of NaCl concentration of activity of malate dehydrogenase isolated from obligate halophilic bacterium *Halobacterium salinarium (1)*, facultative halophile WR-1 *(2)*, and pig liver *(3)* (Holmes and Halvorson, 1965). *Abscissa:* NaCl concentration; *ordinate:* enzyme activity (% of maximal activity)

enzymes from halophilic bacteria (e.g., ornithine carbamoyl transferase from *Halobacterium salinarium*) is reversibly inactivated in the absence of salt (Dundas, 1970).

Thus, halophilic proteins, having acquired the ability to retain a functional conformation in concentrated salt solutions around 4 M, have lost the ability to preserve it at normal salt levels (Fig. 47). In this case, apparently similar to the situation with temperature, the general rule is proved, namely: *the protein can maintain the status of sufficient, but not excessive, conformational flexibility within only a limited span of intensities of an agent which affects its stability.* With a decrease in salt concentration, halophilic proteins undergo conformational changes resembling those taking place during transition into a state of a random coil (Hsia et al., 1971). It is highly likely that such behavior of these proteins has been established during evolution by virtue of an appropriate modification of their primary structure. According to Bayley (1966), *Halobacterium cultirubrum* ribosomes differ markedly from those of other organisms in their higher content of acidic proteins, having an isoelectric point around 3.9. This was reported to be the case also for a total amino acid composition of the envelope proteins in the halophiles (Reistand, 1970).

It is intriguing to study the peculiarities of the protein from the abyssal organisms, which live at hydrostatic pressures of up to 1000 atm. What would be the specific features of their performance at 1 atm?

This section emphasized the significance of thermal and other environmental factors for the establishment, in the process of evolution, of a certain level of conformational flexibility of protein macromolecules. We are concerned primarily with those factors which affect proteins in their natural localization. Some bacteria living in extremely acidic or alkaline media are still capable of maintaining a pH of their protoplasm about 6–7. The enzymes isolated from *Thiobacillus thiooxidans*, living at pH 0.5, are found to perform most efficiently in vitro at neutral pH (Skinner, 1968). This, according to Levitt (1956a), is a good example of adaptation by avoidance at the cellular level.

The rigidity of protein macromolecules is certainly determined not only by environmental factors, but also by the milieu of that intracellular "ecological niche", which a given protein occupies. An enzyme, operating in vivo in a solution, may have an environment which differs considerably in a number of parameters from that surrounding the enzyme incorporated into a membrane. A number of enzymes are known which require for their stability or activity an addition of organic solvents—ethanol, methanol, acetone—if they were isolated from the cells and transferred to a solution (Nozaki et al., 1963; Takemori et al., 1967; George et al., 1969; Ono et al., 1970; Nyns, 1970; Kornberg and Malcolm, 1972). These enzymes are apparently found in vivo in a hydrophobic environment and the conformation of their macromolecules is adapted to the conditions characterized by a lower dielectric constant.

Consequently, the level of stability of proteins reflects not only the factors common for all the proteins of a given cell, but also those local specific conditions in which a particular protein variety is found. That is why when studying the significance of different environmental conditions for the stability of proteins of different species or individuals, it is imperative to compare homologous proteins. If comparisons are made on heterologous proteins, the individual peculiarities and specificities of their intracellular milieu, their "intracellular ecology", would exert profound influence on the comparisons.

4.2.6 Conformational Flexibility of Protein Molecules and Reactive Changes in the Primary Thermostability of Cells

The foregoing sections dealt with adaptation of the level of conformational flexibility to the environmental temperature in the course of phylogenesis. This concept may also be applicable to modificational adaptations discussed in Chapter 1 and Section 3.2.1. It was shown that thermostability of protoplasmic proteins, and, consequently, the primary thermostability of cells, can be modified in response to changes of the temperature as well as by the action of other agents. Both the heat stability of proteins and primary thermostability of the cells vary in the course of the development of an organism. It might be admitted that the biological significance of these changes lies not in the acquisition by the proteins of a new level of resistance to the denaturing action of heat, but, rather, in the transition of proteins to a new level of conformational flexibility. Based on this viewpoint, one can provide an explanation for a number of facts hardly explainable if an adaptive significance is recognized in the protein heat stability per se.

It was shown in the preceding section that in phylogenesis it is not only *an increase* in the protein stability achieved in the process of adaptation of a species to elevated temperature that bears an adaptive significance. No less important, however, is *a decrease* of the stability with adaptation to cold. Based on this point of view, it is easier to comprehend the phenomenon of temperature adjustments in *Protozoa* and lower plants, in which a shift in their primary thermostability can occur, depending on the temperature, in either direction. These organisms convincingly demonstrate the discomfort of maintaining an excessive stability at low temperatures.

In the higher animal and plant tissue cells, the capacity for adjustment has been lost. It probably made way for more perfect, but so far undetected regulatory mechanisms. However, judging from the data concerning the primary heat resistance of cells, cellular models and isolated proteins, the level of protein conformational flexibility in these organisms is to be regarded as far from being constant. In the higher plants in particular, the primary thermoresistance of their cells can be reactively enhanced under the action of supraoptimal temperatures and of some other agents. In heat hardening of plants, the heat stability of all the functions studied so far is increased, which is the case for both thermolabile functions (protoplasmic streaming, photosynthesis) and rather thermostable ones (e.g., respiration). An adaptive significance of such an overall stabilization becomes readily understood if the latter is considered as resulting in a general decrease in the conformational mobility of cellular proteins. With return to normal temperature, the plants hardened by an exposure to heat lose the increased stability of their proteins.

In Sections 1.2.2.3 and 3.2.1.1, findings were discussed concerning an increase in the heat stability of cells and of some proteins in a number of plants, elicited by the action of low temperatures effective in producing the state of cold hardiness. According to the terminology of Precht (1964a) this phenomenon should be called the "paradoxical reaction" as opposed to a "reasonable" one. It is hardly plausible that the severe natural selection would have tolerated a species' fancy for paradoxes. It would be more correct to admit that beyond the "paradoxical" appearance of a phenomenon is hidden its "reasonable" destination. In fact, Hazel and Prosser (1970, 1974) refer to a number of examples, illustrating an adaptive significance of the "paradoxical reaction". In our case, the biological significance of an increase of protein stability produced by low temperatures may, presumably, amount to protection of the proteins against cold denaturation. A number of workers, following Gorke (1907), have regarded denaturation of cellular proteins as one of the causes of cold injury in plants. The existence of such cellular proteins, which are being denatured during cooling of plants has been demonstrated by Heber and his collaborators (Ullrich and Heber, 1958, 1961; Heber, 1958a, 1958b, 1959). With the return of a warm season, however, the plants which had elevated stability of their proteins in the winter, lower it once again. This can be detected by a decrease in the primary thermostability of the cells. Apparently with an increased stability it is difficult to commence active growth and development in the cool springtime.

It has been seen that plant cells are able to increase their primary thermoresistance not only in response to a change in temperature—a water loss and a wound injury may yield similar effects. An impression is created that plant cells possess an unknown mechanism (s), which regulates the level of conformational flexibility of their proteins. These mechanisms may be triggered by various extreme influences. That a plant cell responds to a heat treatment by increasing the resistance of its proteins, should in no way, however, be regarded as indicative of the fact that the responding apparatus has been created during evolution solely to react to overheating. The fact is that similar reactions to overheating can be obtained not only with plants found in habitats where overheating is frequent, but also in those plants which in nature have never been exposed to overheating. With equal ease we have been able to obtain the effect of heat hardening in the terrestrial plants on the Franz-Joseph Land and in the brown algae living there offshore (unpublished experiments of Alexandrov and Shukhtina). Marine algae have been living there for millenia under "thermostated" conditions at a temperature of around 0° C. If heat hardening in Middle Asian plants serves to adapt to heat and an elevated temperature is a normal trigger for provocation of hardening, then, in polar plants, this reaction should be regarded as a response of the cells to a non-adequate stimulus. In the polar algae, this mechanism reacts to heat apparently in much the same way as a nerve-muscle preparation reacts to heat and mechanical stimuli by contracting its muscle. What, aside from heat, may be a natural stimulus for the plant cells and what constitutes such a stimulus for the polar plants in particular is difficult to decide. It might be a shift in the concentration of some metabolites which affect stability of the cellular proteins.

Are the plant cells capable of increasing the original level of the conformational flexibility of their proteins? The experiments on the action of heavy water (D_2O) on the higher plant cells testify in favor of such a possibility. D_2O has been shown to stabilize protein macromolecules and make them more resistant to various denaturing agents (Maybury and Katz, 1956; Hermans and Scheraga, 1959; Rigby, 1962; Appel et al., 1967; Lobyshev et al., 1974). It has been established further that D_2O is effective in stabilizing both plant and animal cells, which prove to become less sensitive to heat and high hydrostatic pressure when placed in heavy water. This stabilization is accompanied by a decreased activity of a number of cellular functions (Alexandrov et al., 1964a; Alexandrov et al., 1965, 1966). Microorganisms are known to be able to adapt to 100% heavy water. The cells of higher plants can also adapt to heavy water, though to a lesser extent. Such adaptation can be observed if *Campanula persicifolia* or *Tradescantia fluminensis* leaf pieces are immersed in 50% D_2O for several days. As time goes by, the initial alteration (a decrease in the protoplasmic flow rate, an increased viscosity) disappears in spite of the continuing action of D_2O. If the cells adapted to D_2O are transferred to ordinary water and their primary thermostability is estimated, the latter, as a rule, is found below the norm (Fig. 48; Denko, 1967; Alexandrov and Denko, 1967). These findings indicate that adaptation to a stabilizing agent (D_2O) is accompanied by a decrease in the level of resistance of protoplasmic proteins. In this case, shifts in the stability of a protein occur, which are apparently opposite to those taking place during hardening of plant cells.

Along with the reactive changes of the primary thermoresistance of plant cells evoked by diverse factors, certain modifications of stability occur, associated with growth processes. They probably reflect the changes in the conformational lability of cellular proteins, which correlate with variations in the metabolic activity level in the process of growth.

In animals also (Sect. 1.2.1.3, 1.2.2.2, 1.3.4, and 3.2.1.2) changes of primary thermostability of cells and of heat resistance of proteins can be observed; they may be associated with the action of external agents, with stages of embryogenesis, and with sex cycles. In the latter case, hormonal factors are decisive. It is absolutely clear that a reduction of the heat resistance, notable in many animal species during the reproductive periods, should not be associated with an adaptive significance of the level of cellular and protein thermostability per se.

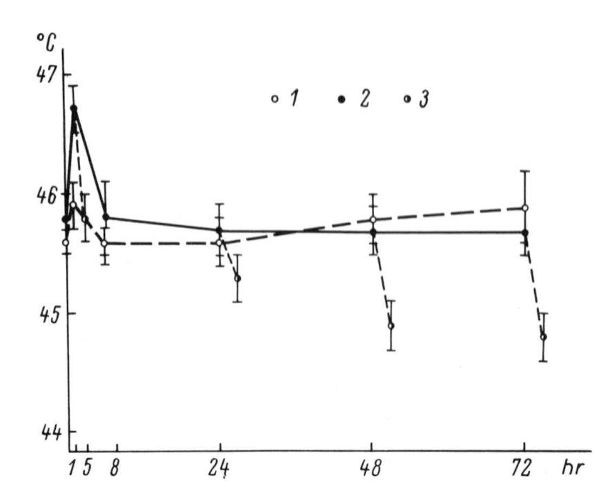

Fig.48. Effect of 50% D_2O on heat stability of *Tradescantia* leaf epidermal cells (Alexandrov and Denko, 1967). *Abscissa:* time period elapsed after infiltration of leaves with heavy water; *ordinate:* temperature of 5-min heating that stops protoplasmic streaming. *1:* in H_2O, *2:* in D_2O, *3:* after transfer from D_2O into H_2O

It would be only natural, in this case too, to suspect that a shift in the conformational flexibility of protoplasmic proteins is important to a change in metabolism. Modification of the cellular and protein thermostability, however, should be regarded primarily as an external manifestation of that shift.

Summary. A starting point for our discussion is the contention that of primary importance to life of the organisms is a correspondence between the level of conformational flexibility of protein macromolecules and the intensity of the environmental factors affecting the protein spatial structure. Among these factors, the environmental temperature is a central one. On the other hand, there is a correlation between the degree of conformational flexibility of protein macromolecules and their resistance to the denaturing action of heat. This allows the use of protein thermostability and the primary thermostability of cells related to it as indirect criteria for assessing the level of the conformational flexibility of proteins. Naturally, this correlation cannot be expressed at present in definite quantitative terms applicable to various proteins. It might be inferred that in one group of proteins a significant increase in their conformational flexibility would appreciably decrease their thermostability, whereas in other proteins this decrease would be slight. This concept so far is a qualitative one; it emphasizes the unidirectional pattern of the changes of flexibility and thermolability, and its validity is largely demonstrated in comparative studies carried out in taxonomically related objects. It should not be expected, therefore, that a difference in thermostability of proteins and cells in two species which differ in their thermophily, expressed in degrees centigrade, would be equal to a difference in average habitat temperatures for these two species.

Despite this limited significance, an evaluation of the primary cellular thermostability and assessment of the resistance of proteins has offered the possibility of revealing charges of major biological importance in the level of protein conformational flexibility for the genotypic and modificational adaptations of organisms. Biebl (1962c) wrote that short-term experiments employing thermal treatments which disclose the differences in the temperature of coagulation of the protoplasm in different species, are "unsuitable for substantiating any conclusions concerning thermostability required for life in habitats that differ in temperature regimes. Ecologically significant boundaries for life can be seen only in those temperatures at which a plant is injured during a prolonged exposure" (p. 143). The same thought is expressed in the work of Kappen and Lange (1968). If, however, the data gathered in short-term exposures of the cells are regarded not as indicative of the heat resistance of an organism, but as a reflection of a certain level of conformational flexibility of the protein molecules, then the results gained in the experiments with pulse heatings should be regarded as full of profound ecological meaning, because they demonstrate the relationship between this biochemically important variable and the temperature conditions of a species' existence.

The idea of an ecological significance of the level of conformational flexibility of the protein was formulated by the author already in 1965. Similar thoughts can be found in publications by Brock (1967, 1969), in which he assumes that the existence of thermolabile proteins in microbes living at low temperatures is related to higher flexibility of their proteins, and, consequently, to higher efficiency of these at low temperatures. According to Brock, growth at low temperatures requires thermolabile structures. Babel et al. (1972) believed that the ability of microbes to live at low temperatures is limited by the inability of the proteins to undergo transitions from one conformation into another. Similar ideas were also expressed by Rose and Evison in 1965. According to their findings, the transport of sugar into the cells of mesophilic microorganisms is inhibited in the cold, unlike the situation with the psychrophiles, because a transport protein becomes hyperfolded under these conditions.

Johnston et al. (1975b) compared the temperature dependence of the myofibrillar ATP-ase activity for Antarctic and tropical fishes and accessibility of the SH groups in the myosin preparations from them. Based on the data obtained, these authors write: "We suggest that the higher activity of the myofibrillar ATP-ase of Antarctic fish at low environmental temperatures is associated with weaker bonding, in other words, with a more open molecular structure". Weakening of the bonds makes the structures of higher order in poikilotherms adapted to cold more susceptible to thermal inactivation at higher temperatures.

Low and Somero (1974) compared free energy (ΔG^{+}), enthalpy (ΔH^{+}) and entropy (ΔS^{+}) of several enzymatic reactions in poikilothermal and homeothermal animals. The values of all three parameters were found to be higher for homeotherms. The authors explain this finding as due to a lower number of weak bonds that are formed during formation of a activated enzyme-substrate complex in the case of homeotherms. This, in its turn, is determined by higher conformational rigidity of homeothermic enzymes. The higher rigidity, according to the authors, in this case is required to provide enhanced resistance of the higher order structures of the enzyme at body temperatures of around 40° C.

The significance of the level of conformational flexibility of proteins for adaptation of organisms to environmental temperature has been fully acknowledged by Somero in his later publication of 1975 (Somero, 1975a).

The hypothesis advanced by the author has been criticized by Levitt (1969). Admitting the prevailing importance of the hydrophobic interactions in maintaining the protein macromolecule's conformation, on the one hand, and an enhancement of these interactions with a temperature increase, on the other, Levitt believes that a protein macromolecule tends to be stabilized, but not labilized, as the temperature is elevated. This conclusion made by Levitt is inconsistent. The stability of a protein macromolecule is determined by the magnitude of the free energy (ΔF) during a transition of the protein into a disordered denatured state. Levitt does not take into account the fact that the temperature dependence of ΔF is determined not only by changes in thermodynamic parameters associated with an alteration of the hydrophobic interactions supporting the protein macromolecule, but also by the conformational entropy, which steadily increases as the temperature grows. This topic was taken up in Section 4.2.2. Levitt refers to Brandts as supporting his arguments, but, in fact, it was Brandts (1967) and a number of other workers, who showed that the temperature maxima for the stability of a large number of proteins lies appreciably lower than an optimum temperature for life. Consequently, an elevation of the temperature in reference to the optimum should reduce the stability of protein macromolecules.

4.3 The Level of Conformational Flexibility of Protein Molecules and Their Resistance to Non-Thermal Agents

4.3.1 Non-Thermal Denaturants

We have seen that an increase or a decrease in the conformational flexibility of protein molecules is adequately reflected in their thermolability. With that change the resistance to other denaturing agents is also expected to be altered. A concordant modification of the resistance to all denaturants and in all cases is, however, hardly a realistic expectation. Everything depends on what particular fixative interactions have contributed to a modification of the protein conformational flexibility and where the primary point of application of a given denaturant lies. This conclusion might have been deduced based on the findings referred to in Section 3.1, where it was shown that modificational and genotypic differences in the primary cellular thermoresistance are often, but not always, accompanied by similar differences in the resistance of cells to a number of other injurious agents. In order to acknowledge the validity of such a comparison, one should agree that the agents in question inflict an injury to cells by denaturing certain target protoplasmic proteins in them.

More direct evidence concerning degree of non-specificity of stabilization of a protein with accompanying reduction of its conformational lability, might be found in studies performed on proteins themselves. Such investigations, however, are scarce. In addition all of them deal with genotypic differences, and, as might be expected, are rather variegated. For instance, according to Braun and Fizhenko (1963) *Rana temporaria* hemoglobin exhibits higher thermostability than that from *R. ridibunda*, but in regard to high hydrostatic pressure the sensitivity of hemoglobins from both frogs is identical. Actomyosin from *R. ridibunda* was more resistant to heat than that from *R. temporaria*, whereas the resistance of these actomyosins to ethanol was the same (Braun et al., 1959). On the other hand, myocardium actomyosin from *R. temporaria* was more resistant than actomyosin from the skeletal muscles of these frogs to heat, alcohol and urea. In rabbits, myocardium actomyosin is more resistant to heat and alcohol than homologous protein from skeletal muscles (Braun et al., 1963). Both more intensive heating and higher hydrostatic pressure are required in order to suppress functioning of the glycerol models of the palate ciliary epithelium from *R. ridibunda* as opposed to analogous preparations from *R. temporaria* (Arronet and Konstantinova, 1964; Arronet, 1966). Mutants of *Neurospora* have been obtained which differed in thermal stability of tyrosinase. Sussman (1961) has shown that thermostable tyrosinase is also more stable to urea and formamide than its thermolabile counterpart.

The resistance of homologous proteins from thermophilic and mesophilic microorganisms against various denaturing agents has been investigated by a number of workers. In this case too, rather diverse results have appeared. Koffler et al. (1957) have compared the resistance of flagellar proteins from four thermophilic and five mesophilic bacteria species. They found the more thermostable proteins from the thermophiles display also a greater resistance to urea, acetamide and sodium dodecylsulfate, 6-phosphogluconate dehydrogenase from *B. stearothermophilus* is more resistant than homologous enzyme from *E. coli*, not only to heat, but also to urea (Veronese et al., 1975). According to Marre et al. (1962), TPN-H-cytochrome reductase in thermophilic algae combines an elevated resistance to heat, urea and acetamide. Catalase, however, in these same algae, is more resistant to heat and acetamide than the corresponding protein in the mesophiles, but no difference in the sensitivity to urea could be detected. It follows from these data, that in one and the same organism, different proteins have acquired during evolution a higher rigidity by increasing the strength of different bondings. Ferredoxin from a thermophilic bacteria of the *Clostridia* genus is denatured by heat at a lower rate than that from the mesophilic representatives of this genus; however, no difference could be observed in its response to the action of high guanidine concentrations (Devanathan et al., 1969). Enolase from the extreme thermophile *Thermus aquaticus* VT-1, in contrast to enolase from yeasts and rabbit muscles, is much more stable against heat, ethanol and guanidine chloride (Stellwagen et al., 1973).

Amelunxen and his collaborators (Amelunxen, 1967; Sauvan et al., 1972) compared glyceraldehyde-3-phosphate dehydrogenase from *B. stearothermophilus* and from rabbit muscles, and detected higher resistance of the thermophilic enzyme to heat, urea and shifts of pH toward either acidic or alkaline extremes. Dehydrogenase of the thermophile is considerably more resistant to urea than that from yeast. The findings reported by Amelunxen concerning the effects of temperature on the stability of this enzyme to urea are extremely interesting (Amelunxen et al., 1970). This author showed that at 30° C glyceraldehyde-3-phosphate dehydrogenase from the thermophile was appreciably more resistant than the rabbit enzyme to 8M urea. However, at temperatures corresponding to the optimum for life of *B. stearothermophilus* (55–60° C) thermophilic dehydrogenase is as sensitive to urea as the rabbit enzyme at 30° C. This fact indicates that glyceraldehyde-3-phosphate dehydrogenase molecules are equally labile in both the thermophile and mesophile at temperatures corresponding to those encountered in their normal environments. Such a conclusion fully agrees with the concept being developed in this book concerning a correspondence between the level of conformational flexibility of protein macromolecules and the temperature conditions of organisms' life.

From the scanty relevant data it may, nevertheless, be concluded that a change in the protein conformational flexibility may affect not only the protein response to heat, but also to other denaturants. This, among other things, may account for the relative non-specificity of shifts in the resistance of cells during the temperature adaptations of the adjustment and/or hardening type; during adaptation to the action of nonthermal factors as well as during changes of the level of cellular lability related to growth processes and developmental cycles. Consequently, apart from heat resistance, stability against the action of other denaturants might, to a certain degree, serve as an indirect measure of the level of conformational flexibility of protein molecules. However, the more selective the action of a denaturant with respect to various forces responsible for maintenance of the spatial structure of protein macromolecules, the more limited the possibility for employing resistance to this agent as a measure of overall flexibility of macromolecules. In comparison with other denaturing agents, heat is apparently a more universal destroyer of a macromolecule's structure and, therefore, thermostability in a more reliable indicator of the conformational lability.

4.3.2 Proteinases

Proteinases are another class of destroyer of protein molecules, but, in contrast to denaturants, the destruction they produce is not limited to the structures of higher order. Proteinases cleave peptide bonds, thereby breaking down the primary structure of a protein. Different proteolytic enzymes offer different, sometimes extremely specific, requirements to atomic groups neighboring the peptide bond being hydrolyzed (Laidler, 1958). The probability of hydrolysis depends also on the steric accessibility of the linkages being disrupted. Therefore, the rate of hydrolysis depends, to a considerable degree, upon the spatial structure of the protein molecule being attacked. Changes in the conformation of a protein molecule, often very slight indeed, may appreciably affect the rate of its hydrolysis.

It has long been recognized that native proteins are considerably more resistant than denatured ones to digestion by various proteinases. Hemoglobin denatured by urea is digested by papain 100 times as fast as the native protein (Lineweaver and Hoover, 1941). There are proteins, e.g., bactericin from *Pseudomonas aeruginosa* and the bacterial α-amylase, which cannot be digested at all, at least by some proteases, if they are in native state (Okunuki et al., 1956; Kageyama, 1964). In such cases, digestibility may be used as a measure of the degree of protein denaturation under given conditions (Hagihara et al., 1956a, 1956b; Okunuki et al., 1956; Okunuki, 1961). When exposing egg albumin to heat, Strachitsky and Kologrivova (1946) observed a parallel increase in the levo-rotation and in the rate of tryptic digestion. Thermal denaturation increases the digestibility of serum albumin by trypsin even before a disruption of the S-S bonds has occurred (Gorini and Andrian, 1956). In studies on live infusoria it has been shown that intracellular proteinases—catepsins—similarly digest the denatured proteins more readily than native ones (Viswanatha and Liener, 1955). Thus, the catepsins found in rat liver lysosomes hydrolyze proteins that had been denatured (Coffey and de Duve, 1968). Linderstrøm-Lang believed that there are many proteinases which act exclusively on denatured proteins, and they themselves catalyze the transition of a globular protein form a native from into a denatured one (Linderstrøm-Lang, 1950; see also: Lundgren, 1941; Korsgaard, 1949).

Digestibility is an indicator that is extremely sensitive to changes in conformation of a protein macromolecule. Putnam et al. (1943), compared employing several criteria, native serum albumin with an albumin preparation renaturated after denaturation produced by urea and guanidine. Both protein preparations showed no difference in terms of their hydrodynamic and electrophoretic properties. The renaturated protein, however, similarly to the denatured one, exhibited appreciably higher digestibility by trypsin. Bernheim et al. (1942) reported similar results when they compared native serum albumin and pseudoglobulin preparations with those renaturated after denaturation with urea. Ram and Maurer (1958) have treated serum albumin with trichloroacetic acid and ethanol followed by dialysis and lyophilization of the product. The resultant protein preparation did not differ from the original native preparation in the following parameters: crystallization, sedimentation, electrophoretic mobility, thermostability, and interaction with an appropriate antiserum. The preparation, however, was digested by trypsin at a higher rate than the native protein.

Fukushi et al. (1968) denatured α-amylase from *B. subtilis* with 8M urea. After removal of urea, an absorption spectrum and the Moffit parameters of the optical activity a_0 and b_0 immediately returned to normal values. Enzymatic activity, on the other hand, was restored only gradually and over the subsequent 6 h was found within 60–90% of the normal value. An enhancement of the resistance to proteolytic digestion in the course of α-amylase renativation was seen to be recovered at the same rate as the restoration of its enzymic activity. Consequently, changes of the proteolytic resistance were reflecting the course of restoration of the original state of the enzyme more truly than were the changes in the absorption spectrum and the optical rotatory dispersion.

The resistance to proteolytic digestion is changed also during conformational changes of proteins accompanying a normal biochemical performance of proteins. Yeast cytochrome c in an oxidized form is more easily digested by trypsin and bacterial proteinase than in a reduced state (Nozaki et al., 1958; Mizushima et al., 1958; Margoliash and Schejter, 1966). Similar data are available concerning cytochrome c from horse and pig heart (Yamanaka et al., 1959). Oxyhemoglobin and oxymyoglobin are digested by carboxypeptidase at a higher rate than are hemoglobin and myoglobin respectively (Zito et al., 1964). The rate of digestion of enzymes by various proteinases can be modified by substrates and allosteric effectors (Grisolia, 1964; McClintock and Markus, 1968; Schimke, 1969).

In the overwhelming majority of cases those influences which modify the resistance of proteins to the action of proteolytic enzymes also modify in the same direction their heat resistance and stability against other denaturing agents. Thus, Ca and Mg ions elevate the resistance of serum albumin to both heat and tryptic digestion (Gorini and Andrian, 1956). Sodium caprylate and related compounds stabilize serum albumin toward heat, urea and papain digestion (Rice et al., 1945). Detergents and fatty acids diminish the rate of digestion of serum albumin by trypsin and α-trypsin (Kondo, 1959; Epstein and Possick, 1961), and increase its resistance to heat and urea (Boyer et al., 1946a; Boyer et al., 1946b).

Glucose stabilizes the yeast hexokinase to heat and trypsin (Berger et al., 1946; Kunitz and McDonald, 1946). Binding of a single methyl orange molecule suffices to increase the resistance of a serum albumin molecule to the digestive action of five proteolytic enzymes, which hydrolyze peptide bonds in absolutely different parts of a polypeptide chain, namely, to chymotrypsin, trypsin, papain, pronase, and subtilisin. A maximal protective effect is achieved with the dye to albumin molar ratio 5:1 (Markus, 1965; Markus et al., 1967). Concurrently, thermostability of serum albumin is increased too (Vitvitsky, 1969). The binding of 2-cytidylate to the ribonuclease active site stabilizes the enzyme against subtilisin, trypsin, chymotrypsin, and heat (Markus et al., 1968).

Ca^{2+} enhances the resistance of *B. stearothermophilus* α-amylase to heat, to decrease in pH, and to the digestive action of papain, trypsin, chymotrypsin, and pronase (Pfueller and Elliott, 1969). A staphylococcal nuclease is stabilized by deoxythymidine-3,5-diphosphate and Ca^{2+} to heat, trypsin subtilisin, α-chymotrypsin, and thermolysin. This is accompanied by a decrease in the ability of the enzyme to exchange its hydrogens for tritium. This overall stabilization of the macromolecule, however, leaves unprotected the first five peptide groups at the NH_2-end, which are hydrolyzed in both the presence and absence of the ligands that do not affect flexibility of this part of the nuclease molecule (Taniuchi et al., 1967, 1969; Schechter et al., 1968). Kynurenine transaminase can be stabilized by diethylstilbestrol against heat and chymotrypsin (Mason and Gullekson, 1960). Various tanning agents concordantly increase the resistance of fish collagen to both thermal contraction and tryptic digestion (Gustavson, 1942). As mentioned earlier (p. 159), substituting of H_2O for D_2O increases the resistance of solubilized proteins to various denaturing agents. At the same time, it was shown that procollagen becomes more resistant in D_2O not only to thermal denaturation, but also to the proteolytic action of collagenases (Alexandrov et al. 1975). In contrast a completely deuterized protein phycocyanin, obtained from the algae grown for a long time in a medium containing D_2O, was digested by trypsin and α-chymotrypsin 1.5 times as fast as the ordinary phycocyanin; the former, however, proved to be more sensitive than the latter to the denaturing action of heat (Hattori et al., 1965). A molecule of glyceraldehyde-3-phosphate dehydrogenase becomes more rigid in the presence of a co-factor (NAD): its α-helical content increases, the number of hydrogen atoms exchangeable for deuterium decreases. This is accompanied by a reduction of the rate of digestion of the enzyme by trypsin and its thermal denaturation proceeds more slowly (Szabolcsi et al. 1959; Elödi and Szabolcsi, 1959; Polgar, 1964; Zavodsky et al., 1966; Bolotina et al., 1967).

In the presence of NAD-H_2, glutamate dehydrogenase becomes less resistant to both heat and a bacterial proteinase digestion (Inagaki, 1959a, 1959b), and addition of NADP-H_2 decreases the enzyme resistance to heat, urea and trypsin (Grisolia et al., 1962). Binding of an *E. coli* RNA polymerase to DNA increases the resistance of the enzyme to heat, NH_4Cl and tryptic digestion. Addition of nucleoside triphosphates and other compounds supporting RNA synthesis in vitro results in further enhancement of the RNA polymerase resistance to salts, urea, guanidine, and trypsin (Khesin et al., 1967). The mitochondrial systems: NAD-H_2-oxidase (I, II, V-th complexes according to Green) and succinate oxidase (II, III, VI) in a functional state (in the presence of substrates and co-factors) acquire an extremely stable conformation; as a result, their resistance to heat, trypsin and cobra venom phospholipase increases (Luzikov et al., 1967; Luzikov et al. 1967a, 1967b; Luzikov, 1973). The conformation of isoleucyl-RNA-synthetase is stabilized in the presence of substrates under conditions which promote formation of adenylate complex. With that the enzyme resistance to heat and trypsin increases, and concurrently, alkylation of the SH-groups and their titration proceed at a lower rate (Baldwin and Berg, 1966). The difference in the resistance of various forms of aspartate aminotransferase (apoenzyme, holoenzyme with natural and modified co-enzymes, complexes of the enzyme with inhibitors) to α-chymotrypsin was found to correspond to the difference in their resistance to the denaturing action of urea and heat (Makarov et al., 1974). A correlation between the resistance to heat and proteolysis has been found also when comparing the response to these same treatments of myoglobin, hemoglobin, their cyanoderivatives and tropomyosin (Saburova, 1973; Saburova and Markovich, 1973). Glucose elevates the resistance of rat hexokinase to both heat and proteolytic digestion (Grossbard and Schimke, 1966).

A parallelism in the sensitivity to heat and proteolysis is frequently observed when different isozymes are being compared. Thus, lactate dehydrogenase isozyme H_4 from human spermatozoa displays a higher resistance to heat, trypsin, chymotrypsin and papain than does the isozyme M_4 (Clausen and Øvlisen, 1965). The three isozymes of rat hexokinase may be arranged in the same sequence with respect to their sensitivity to heat, trypsin and pancreatic protease (Grossbard and Schimke, 1966).

In the light of the material presented above, it is justified to admit that changes in the level of conformational flexibility of protein macromolecules—both genotypic and modificational ones—are likely to be reflected not only in a modification of the protein resistance to the denaturing action of heat, but also to the digestive action of proteases. This problem, unfortunately, has received only limited attention; nonetheless, there is evidence to support this contention.

Connell (1961) found that fish myosin is denaturead by urea at a faster rate than the myosin from warm-blooded animals, and contains more titratable SH-groups though their amount is the same; in addition, the fish myosin is appreciably easier digested by trypsin. A comparison of the data reported by Vinogradova (1963a) and Glushankova (1967a) with those obtained by Skholl (1965) reveals that the fish contractile proteins are denaturead far more easily by heat than corresponding proteins of warm-blooded animals.

Gustavson (1942, 1955) demonstrated that fish collagen is much more sensitive to heat than is the mammalian collagen, and, in addition, is considerably faster digested by trypsin. In regard to their thermostability, collagens from various sources can be arranged in the following order: rat > carp > cod. They appear in the same sequence if arranged according to their resistance to proteolytic digestion by collagenase (Kazakova et al., 1958; Kunina et al., 1962). Cyto-

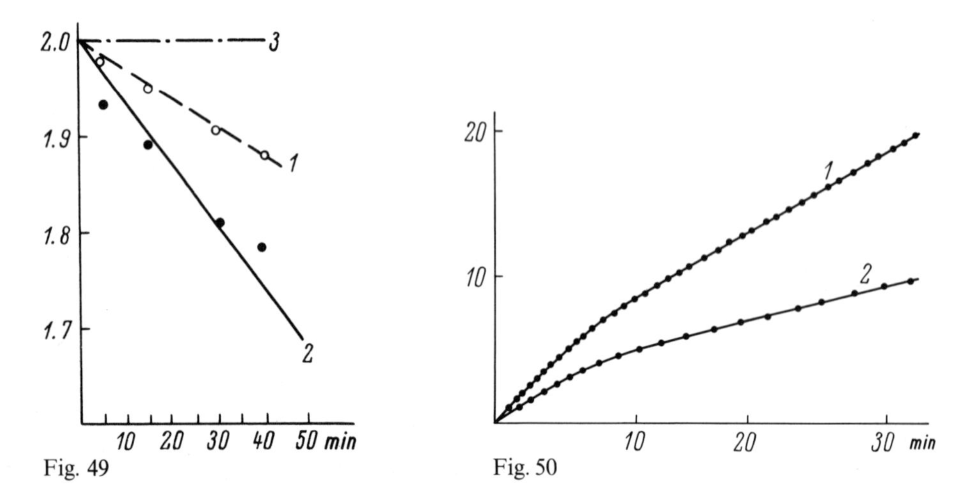

Fig. 49. Rate of digestion of tropocollagen with collagenase at 15° C (Alexandrov and Andreyeva, 1967). *1:* tropocollagen from *Rana ridibunda* skin, *2:* tropocollagen from *Rana temporaria* skin. *Abscissa:* duration of digestion; *ordinate:* logarithm of relative viscosity (% of initial value), *3:* control (no enzyme)

Fig. 50. Stability of serum albumin from northern frog *R. temporaria (1)* and southern *R. ridibunda (2)* against tryptic digestion at 25° C (Vitvitsky, 1970). *Abscissa:* duration of digestion; *ordinate:* extent of digestion, expressed in mmoles of NaOH required for neutralization of reaction mixture

chrome c from birds is more resistant to heat and proteinases than it is in other vertebrates (Margoliash and Schejter, 1966).

In the works cited, comparisons have been made between proteins of phylogenetically distant species. Ecologically justified differences in the digestibility by proteinases can be found also for the species belonging to one genus. As illustrated in Table 11, a number of proteins display higher thermostability in the more warm-loving *Rana ridibunda* as compared with *R. temporaria.* Their resistance to proteases has been estimated for five proteins. It was found that *R. ridibunda* tropocollagen is digested by collagenase at a lower rate than the same protein isolated from *R. temporaria* (Alexandrov and Andreyeva, 1967). The same holds true for digestion of other proteins of these two frogs: serum albumin by trypsin (Vitvitsky, 1970); adenylate kinase by trypsin, myosin by trypsin, chymotrypsin and pronase (Konstantinova, 1972, 1974) (Figs. 49–51). Interesting results were obtained by comparing glutamate dehydrogenase from these frogs. No difference in the thermostability of this enzyme from the two frog species could be detected either by heating pieces of liver tissue, or heating isolated and purified glutamate dehydrogenase preparations. Tryptic digestion, however, revealed that *R. ridibunda* possesses glutamate dehydrogenase characterized by higher proteolytic resistance than the enzyme isolated from *R. temporaria* (Andreyeva et al., 1975). It should be remembered that the resistance to proteinases frequently proves to be a more sensitive test for the stability of the conformation of a protein macromolecule, than resistance to the denaturing action of heat (p. 165).

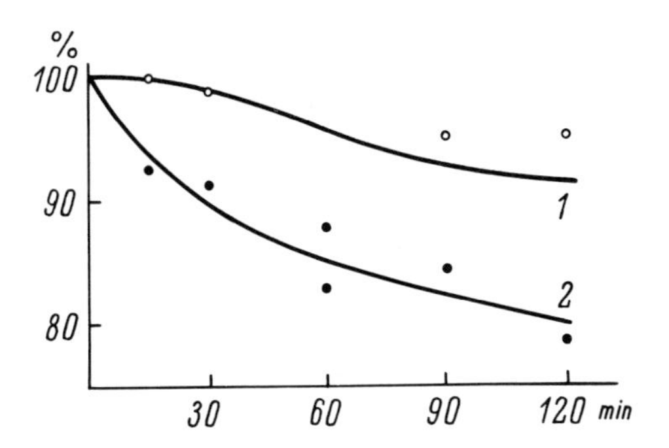

Fig. 51. Stability of adenylate kinase of southern frog *R. ridibunda (1)* and northern *R. temporaria (2)* against tryptic digestion at 25° C (Konstantinova, 1972). *Abscissa:* duration of digestion; *ordinate:* residual activity (% of initial activity)

Jollès et al. (1966; Dianoux and Jollès, 1967) found that lysozyme isolated from the white of goose eggs was considerably less thermoresistant than analogous protein from the white of the hen eggs. The goose lysozyme was, in addition, more rapidly digested by trypsin and displayed higher specific activity. These differences the authors ascribe to a lower cystine and tryptophan content in the goose lysozyme. It is difficult to decide if those differences are of any ecological significance.

The examples discussed above indicate that the differences in the resistance of proteins to heat, referable to those in the thermophily of the species concerned, are accompanied by like differences in the resistance of these proteins to proteolytic digestion. Similarly, this parallelism between thermostability of proteins and their digestibility has been also observed in ontogenesis. According to Stewart and Halvorson (1954), the *Bacillus terminalis* spore alanine racemase is strikingly more resistant to heat and acetone treatment than the enzyme isolated from the vegetative cells. Moreover, the spore alanine racemase proved to be more resistant to papain digestion.

As indicated (p. 103), heat hardening of pea leaves enhances the resistance of glucose-6-phosphate dehydrogenase and ferredoxin. Concurrently, both proteins increase their stability against tryptic digestion too (Feldman et al., 1975b).

In spite of the great similarity in a number of principal properties, the dog myocardium myosin is digested by trypsin at a lower rate than is the skeletal muscle myosin (Brahms and Kay, 1963; Mueller et al., 1964). In conformity with these findings are the data reported by Braun and his colleagues (1963) concerning higher resistance of the rabbit myocardium myosin compared with that of the skeletal muscles to the denaturing action of heat and ethanol. In like manner, the frog myocardium myosin exhibits higher resistance than the protein from skeletal muscles to heat, ethanol and urea. Rumyantsev (1960) and Lukyanenko and Nikolayev (1967) have shown that thermostability of the myocardium tissues of frogs and fishes is higher than that of the skeletal muscles. It is of interest to note in passing, that the more stable myocardium myosin displays lower ATP-ase activity (Brahms and Kay, 1963).

Summary. All the facts discussed in this section support the belief that the stability of proteins against the digestive action of proteinases, as thermoresistance, may serve as an indirect criterion of the level of conformational flexibility of protein macromolecules. In this regard, more adequate results should apparently be expected when endopeptidases are employed, rather than exopeptidases, the latter hydrolyzing peptide bonds only at the ends of a polypeptide chain.

The above warrants the contention that in nature there exists a correlation between thermophily of species and the resistance of their proteins to proteolytic digestion.

Apart from the resistance to the denaturing action of heat and to the proteolytic digestion by proteinases, comparison of the levels of conformational flexibility of proteins may use a more direct method, i.e., the deuterium or tritium exchange technique. In the protein macromolecule there are, aside from the non-exchangeable hydrogen atoms linked to carbon, the hydrogens associated with oxygen, nitrogen and sulfur. These hydrogens can be exchanged, at a higher or lower rate, for deuterium. The rate of exchange is largely determined by the steric accessibility of hydrogens and reflects thereby the conformation of a protein macromolecule as well as lability of this conformation. Unfortunately, this method has not yet been applied to solving ecological problems. In this context, the works conducted by Privalov at the Institute for Protein Research, the USSR Academy of Sciences are most promising. In a preliminary communication, Privalov (in press) has reported his findings concerning hydrogen/deuterium exchange in cod, pike and rat tropocollagens carried out at different temperatureas. At 25° C the exchange rate is lower, the higher the thermophily of the species. Identical exchange rates were observed at temperatureas corresponding to the environmental temperatureas characteristic of a given species. The exchange rate was the same for tropocollagens from cod at 5° C, for that from pike at 10° C, and for the rat—at 37° C. Again, Lazarev and his co-workers (1975) also found identical rates of deuterium exchange in tropocollagen preparations obtained from rat, frog, pike and cod at temperatures corresponding to respective habitat temperatureas typical of those organisms. In this way, with the aid of deuterium-exchange technique, a correspondence between the level of conformational flexibility of tropocollagen and the temperature for life of a species has been directly demonstrated.

Chapter 5
The Plausible Points of Application of the Natural Selection During Alteration of a Correspondence Between the Level of Conformational Flexibility of Protein Molecules and the Temperature Ecology of a Species

The facts and considerations referred to in the preceding chapter lead to the conclusion that for proteins to function optimally it is important to possess a definite level of conformational flexibility at a given mean temperature of a species' life. The resistance of proteins to the denaturing action of heat is causatively related to this level which should be neither too high nor too low. Every protein has its own measure of what is a high or a low level of flexibility. Moreover, the relationship between flexibility and thermoresistance is a specific feature of every protein. This phenomenon is responsible for the existence of large differences in the heat stability of different proteins in the cell.

Comparative studies on the heat stability of analogous proteins in related forms, differing in their thermophily, suggest that in the evolution of organisms, a correlation has been maintained between the level of protein conformational mobility and the temperature conditions of life. This means that during evolution changes in the temperature ecology of a species elicit, in the appropriate direction, a change in the conformational flexibility of protein macromolecules. If, in agreement with Darwin's theory, a shift in conformational flexibility is to be regarded as a result of a combination of undirected hereditary variability and of the directed pressure exerted by natural selection and, if one assumes that the organisms, not the proteins, are being selected, then the question should be asked: by what means does selection distinguish the organisms in which the level of protein conformational flexibility by chance fits the temperature of a species' existence, from those organisms in which this correspondence has been distorted. In other words, it is important to elucidate what actual deprivations and/or discomfort are evoked by altering a required correspondence. One of the approaches to this problem might be to compare the performance of homologous enzymes taken either from the species living under contrasting temperature conditions, or form the individuals of a given species adapted to different temperatures. The relevant evidence accumulated in the literature allows comparisons between the temperature ecology of a species and the following parameters: activation energy, affinity of enzymes to substrates, and temperature dependence of enzyme activity.

5.1 Activation Energy

The activation energy, E, is that energy which is to be conferred upon the reacting molecules for the reaction to be feasible. The value E is, as a rule, expressed in calories per mole of reactant. This value is calculated by measuring the dependence of a reaction rate upon the absolute temperature and can be estimated from an Arrhenius plot (p. 149). Another measure of activation energy is the temperature coefficient Q_{10}, which is given by the ratio of a reaction rate at temperature t to that at t-10°. The Q_{10} of most chemical reactions is around 2–3. A simple relationship exists between the activation energy and Q_{10}:

$$E = \frac{RT^2 \ln Q_{10}}{10},$$

where R is the gas constant, and T is the absolute temperature. The principal reason for an enzyme to interfere in a biochemical reaction as a catalyst is to reduce the activation energy of the reaction. As a result, the reaction rate at a given temperature is sometimes increased by up to nine orders of magnitude.

Attempts have long been made to correlate the value of activation energy or that of Q_{10} with the temperature ecology of a species and with modificational adaptation of organisms to temperature (Bělehrádek, 1935; Blagoveshchensky, 1950, 1966). Arguments in favor of an adaptive significance of the changes in Q_{10} or E have sometimes been extremely controversial. If, for the organisms adapted to cold, these parameters were found to decrease, a favorable result was interpreted as due to a decrease in the temperature dependence of the process concerned. Conversely, if an adaptation to cold increased the activation energy, that could also be regarded as having its positive aspects, i.e. only a slight increase in temperature suffices for the activity of an object adapted to cold to be enhanced considerably. But what actually occurs?

There is a wealth of data in the literature concerning measurements of activation energy of various processes in entire organisms, cells and enzyme systems of various degrees of purity. Such systems have been isolated from the organisms belonging to the species and races characterized by different temperature conditions of their existence, or from the individuals kept at different temperatures. Unfortunately, in reviewing this literature, one fails to detect a general regularity. In the works of one author, species or individuals adapted to cooler environments exhibit lower activation energies for some processes as compared with those in organisms living in warmer conditions; other workers report contrary observations, and still others find no difference whatever. Sometimes, in one publication a description of different processes is given and different results are reported for each process. Here are some illustrative examples. Precht (1949) kept *Amphipods* and *Daphnia* at different temperatures for five to seven days. The activation energy for the pleopodia strokes in the *Amphipods* was not changed, while in the *Daphnia* kept in the cold, the activation energy for the heartbeat increased. According to Hunter (1964), in *Drosophila* kept at 15° C, the Q_{10} of respiration was higher than that in the organisms exposed to 25 and 30° C. Moore (1939) and Volpe (1957) studied the Q_{10} for growth of a number of *Amphibia* from various climatic zones, featuring markedly different temperatures for development of the organisms. According to these writers, the northern species showed lower Q_{10} values. Scholander and his co-workers (1952) found no difference in the Q_{10} for respiration of Arctic and tropical lichens. Semikhatova (1959) gives the Q_{10} values for respiration of 42 plant species—high mountain, Arctic, temperate zone and tropical plants—both those she had obtained in her experiments and taken from the reports of other investigators. The measurements have been made over a wide range of temperatures. No correlation could be detected between the Q_{10} values and the climatic zone the plants had been taken from. Having exposed some aquatic plants to cold (4–8° C) for three months, Harder (1925) detected a decrease in Q_{10} for the carbon dioxide uptake compared with that in the plants kept at room temperature. The activation energy for growth of the thermophilic blue-green alga *Synechococcus lividus* is of the same magnitude as that for growth of the mesophilic and psychrophilic yeasts and bacteria (Meeks and Castenholz, 1971).

In the works cited above, either E or Q_{10} values have been estimated for the physiological processes, which are based on a sequence of biochemical reactions. It is difficult to decide what phenomena are being revealed in a change of temperature dependence of such a process. This may result from modification of the enzyme properties in that reaction, which is the rate-limiting one for the entire process. This might be also a consequence of a substitution of one rate-limiting reaction for another (Sorokin, 1960). It appears that more unambiguous results can be obtained in studies on separate enzymic reactions. The number of such observations, however, is also insufficient for drawing any positive conclusions concerning ecological significance of changes of E or Q_{10}.

Numerous contradictory data are available relevant to the effects of the cultivation temperature on the E or Q_{10} of enzymic reactions. Roberts (1967), for instance, grew one winter and two spring varieties of wheat at 6 and 20° C. The E values for invertase were lower in the winter variety grown in the cold than were those for the enzyme from the "warm" wheat. In contrast, the spring variety showed no difference related to the cultivation temperature. Hochachka and his collaborators (Baldwin and Hochachka, 1970; Moon and Hochachka, 1971) kept trout at 2 and 17° C, but could not find any difference in E for either brain acetylcholinesterase, or isocitrate dehydrogenase from liver. The activation energy for brain acetylcholinesterase isolated from goldfish acclimated to 6 and 30° C was the same in both cases (Hebb et al., 1972). Acclimation of goldfish to 5 and 25° C for a month was also ineffective in modifying the activation energy for muscle succinate dehydrogenase (Hazel, 1972a, 1972b). On the other hand, acclimation of goldfish to 1° C resulted in considerable reduction of the activation energy for the myofibrillar ATP-ase activity as compared with that in the fish acclimated to 26° C: E in the former was 14.3 Kcal/mole, in the latter $E = 21.9$ Kcal/mole (Johnston et al., 1975a). Finally, the Mg-dependent ATP-ase from the neural cord of the earthworm acclimated to 14° C showed no differences when compared with the enzyme isolated from the worms kept at 25° C; at the same time, (Na^+, K^+)-dependent ATP-ase from the "cold" worms displayed appreciably lower *E*, i.e. 7 Kcal/mole instead of 15 Kcal/mole (Lagerpetz et al., 1973).

Unequivocal results also could not be obtained by comparison of the enzymes isolated from the species and races living at different temperatures. Sultzer (1961) reported that the E for oxidation of sodium octanoate was higher for a mesophilic bacterium than for a psychrophilic one. Mutchmor and his co-workers (Mutchmor and Richards, 1961; Mutchmor, 1967) found ATP-ase from more warm-loving insects to exhibit a lower E value. The activation energy for the frog succinate dehydrogenase is almost three times as low as that for this enzyme isolated from warm-blooded animals (Vroman and Brown, 1963). Read (1964a, 1964b), working with various species of cold-blooded animals, obtained results that varied with the nature of the enzyme under study, namely: the activation energies for ATP-ase and succinate dehydrogenase were lower in the more warm-loving species. This regularity was less pronounced for the mollusc ribonuclease. The activation energies for invertase and proteinase from two *Arthropoda* species, characterized by different thermophilies, proved to be the same. Identical *E* values have been reported for fructose diphosphate aldolase in the rabbit, trout and two Antarctic fishes. According to Somero (1969b), the same activation energy value has been found for lactate dehydrogenase from the tuna living at 20° C and for that from the Antarctic fish *Trematomus*, living at −2° C: the *E* for the rainbow trout pyruvate kinase (the temperature for life is 10° C) was higher than that for the analogous enzyme in the more warm-loving tuna, and higher than the *E* value for the enzyme from the cold-loving Antarctic fish. The activation energy for fructose diphosphate aldolase from rainbow trout and Antarctic fish was the same in both cases. This holds true also for the enzyme from Antarctic fish and rabbit (Komatsu and Feeney, 1970). Olsson (1975) was unable to detect any interspecific distinctions in *E* of lactic and malic dehydrogenase isolated from guinea pig hedgehog, bat, frog and cod.

Scores of publications may be additionally referred to, in which the *E* and Q_{10} values have been compared with the temperature for life of species or individuals, but the overall picture of an extreme diversity and controversy in the data reported will not be changed. Comparison of the findings reported by different workers is complicated also by the fact that the Q_{10} value, as shown by several writers (Somero, 1969b; Behrisch and Hochachka, 1969b; Hochachka and Lewis, 1971; and others), depends on the substrate and co-factors' concentrations as well as on pH. It frequently proves to be significantly higher for high substrate concentrations, as opposed to low ones, which approach the physiological concentrations. Moreover, as we saw earlier (p. 150), an Arrhenius plot is frequently interrupted by breaks, and consequently, the *E* and Q_{10} values for different temperature ranges will be different. All this taken together provided grounds for a number of writers to reject the idea that a change in activation energy for biochemical reactions is involved in adaptation of the organisms to the temperature of environment (Scholander et al., 1953; Rao and Bullock, 1954; Hochachka and Somero, 1969; Behrisch, 1969; McNaughton, 1972; and others). It is worthy of note, however, that in a later publication, Hochachka and Somero (1971) have taken a less negative attitude in this regard. They rule out any relevance of changes in activation energy of enzymes as being involved in the realization of the organisms' adaptation to rapid changes in environmental temperature. As for the adaptation of species to different temperatures in the course of evolution, despite the existing contradictory evidence in the literature. Somero admits, regarding some enzymes, an adaptive significance of a decrease in activation energy for life at low temperatures. Thus, Somero and his collaborators (1968) found an extremely low E value (3780 cal/mole) for succinate dehydrogenase from the Antarctic fish *Trematomus bernacchii*. At the same time, in his recent publications, Somero (Low et al., 1973; Somero, 1975a) agrees that enzymes of ectothermic animals are more effective in lowering the energy barrier of reaction, than are homologous enzymes of endothermic organisms (birds, mammals).

Newell and his co-workers (Newell and Northcroft, 1967; Newell and Pye, 1968) believe that, within the range of temperatures for life of littoral invertebrates and algae, the Q_{10} of respiration of the animals in the state of rest, is low (less than 1.2); this provides for a relative independence of the process on temperature. The plot of temperature dependence of the respiration rate varies over a season in such a manner that these changes favor a correspondence of a minimal Q_{10} value to ambient temperature at any given time. Quite opposite data can be found in a work by Artsikhovskaya and Rubin (1954). They too observed seasonal variations in the Q_{10} of respiration of plants, but, according to their observations, the maximal Q_{10} values were observed at an environmental temperature typical of a given season. It is in this that the authors visualize an adaptive significance.

From all the material presented in this section one can hardly draw any conclusions of a general nature concerning the significance of differences and variations in the activation energy or in the temperature coefficient for adaptation of organisms to the temperature of environment. An uncertainty in this problem stems, apparently, from the fact that, as shown by Blumenfeld (1974), the activation energy value derived from the Arrhenius equation may have no physical sense at all in the case of reactions involving biopolymers. At a change in temperature, the first event to take place is a structural change in the protein itself, and, consequently, at temperatures T_1 and T_2, in fact essentially different compounds partake in the reaction concerned. The situation is still more complicated when activation energy is being evaluated for a certain physiological process based on a complex of many different reactions.

5.2 Affinity of Enzymes to Substrates, the Michaelis Constant

The rate of an enzymic reaction is markedly affected by the affinity of the enzyme to its substrate. The relationship between these parameters is taken up in the equation:

$$v = \frac{V}{1 + \frac{K_m}{S}},$$

where v is the rate of the enzymatic reaction at a substrate concentration S, V is the maximal reaction rate achieved at full saturation of the enzyme by the substrate, and K_m is the Michaelis constant. This constant is numerically equal to that concentration of the substrate at which the rate of the reaction is half that of the maximal rate attained at saturation of the enzyme with the substrate. The higher the affinity of the enzyme the substrate, the lower the concentration of substrate required to attain the maximal reaction rate, and consequently the less will be the numerical value of K_m. Thus, the K_m value is inversely proportional to the strength of the enzyme's affinity to the substrate; provided an excess of the substrate is available, a change in K_m will only slightly affect the reaction rate. If, however, its concentration is considerably lower than the saturating one, a decrease in the K_m will enhance the reaction rate, and an increase in K_m will produce an opposite effect. Concentrations of substrates in the live cell are, as a rule, rather low; therefore, a change in the K_m a priori may serve as a regulator of the rate of enzymic reactions. Consequently, if a change in temperature had evoked a corresponding change in the K_m, this might have compensated for a deviation of the reaction rate from the norm, which might be expected to occur under the influence of a temperature shift.

For the first time probably, the existence of such a mechanism of temperature adaptation has been shown for AMP-deaminase from the muscles of poikilothermal animals (frog, carp and trout) by Zydowo et al. (1965), although the very fact of an icrease in the Michaelis constant at an elevation of temperature of an enzymic reaction had been established much earlier (Hakala et al., 1956). Later this idea was developed in a series of extremely important studies performed by Hochachka, Somero and their collaborators (Hochachka and Somero, 1968, 1971, 1973; Somero and Hochachka, 1968, 1969; Somero, 1969a, 1969b; Behrisch, 1969; Somero and Johansen, 1970; Baldwin and Hochachka, 1970; Moon and Hochachka, 1971; and others). The principal conclusions drawn from these investigations are as follows. In the majority of the enzyme from poikilothermal animals studied to date, the K_m values depend on the temperature at which a reaction runs. A minimal K_m value is, as a rule, found at a lower temperature within the temperature range for life in a given habitat. The temperature dependence plot of the K_m of glucose-6-phosphate dehydrogenase from the king crab, the mean temperature for life of which is around 7° C, may serve as an example (Fig. 52). An increase of the environmental temperature that accelerates the enzymic reaction should simultaneously increase the K_m; this reduction in the affinity of the enzyme to its substrate affects the rate of the enzymic reaction, tipping it over in the opposite direction and compensating for the consequences of a temperature shift. Lowering the temperature increases the affinity of the enzyme to substrate, thereby acting as a positive effector, and thus neutralizes the retardant action of cooling. This mechanism of compensation is put into action immediately a change in the environmental temperature occurs.

In the species which live under different temperature conditions, corresponding differences can be detected in the positions of the temperature minima of the K_m (Fig. 53). This allows for regulation of the reaction rates with temperature variations found within the zone of optimum which covers different portions of the temperature scale for different animal species.

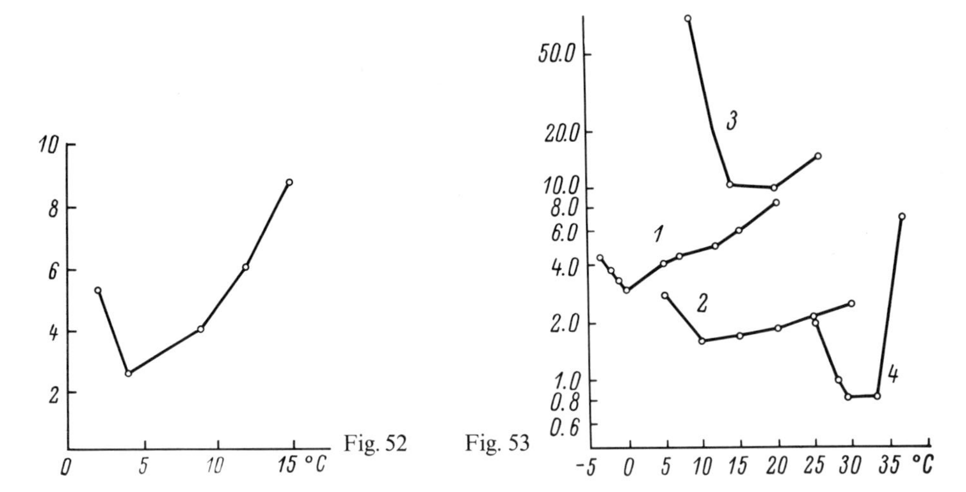

Fig. 52. Effect of temperature on Michaelis constant of glucose-6-phosphate dehydrogenase from Alaskan king crab in a reaction with glucose-6-phosphate. (Somero, 1969b). *Abscissa:* temperature; *ordinate:* K_m (M, $\times 10^5$)

Fig. 53. Effect of temperature on Michaelis constant of lactate dehydrogenase from different poikilothermal animals in a reaction with pyruvate (Somero, 1969b). *Abscissa:* temperature; *ordinate:* K_m (M, $\times 10^4$). *1:* Antarctic fish *Trematomus borchgrevinki*, *2:* trout adapted to 5° C, *3:* trout adapted to 15° C, *4:* lungfish *Lepidosirien paradoxa*

With transitions of individuals of a given species to a new temperature environment, the K_m temperature dependence plot is accordingly modified. In trout acclimated to 17° C for four weeks, a minimum for the K_m of the muscle isocitrate dehydrogenase was 15–17° C, whereas in trout kept at 2° C, K_m was found to decrease continually with a drop in temperature down to zero. At the same time, no change in activation energy could be detected (Moon and Hochachka, 1971). A modification of K_m, provoked by acclimation, was accompanied by changes in the isocitrate dehydrogenase isozyme pattern. Baldwin and Hochachka (1970) found different temperature dependence of the K_m for two isozymes of the trout brain acetylcholinesterase. The authors believe the presence of these isozymes is important for adaptation of the organisms to environmental temperature. According to Somero (1969a), pyruvate kinase exists in two different forms in the king crab muscles: the "cold" one has a minimum for K_m at 5° C, whereas the "warm" one, at 12° C. The cold enzyme is only active at temperatures of below 10° C, and the warm enzyme, only above 9° C. Such a dual protection makes the rate of the reaction catalyzed by pyruvate kinase to a considerable degree independent of temperature over the entire range of optimum temperatures for the life of this crab (4–12° C). Somero believes that it is not the two isozymes that perform this function. According to electrophoresis and isoelectrophocusing, there is only one enzyme which is capable of going from one form into another depending upon ambient temperature.

At a later date, the "summer" and "winter" varieties featuring different temperature maxima of K_m have been found to exist in five rainbow trout enzymes:

pyruvate kinase, liver NADP-isocitrate dehydrogenase, fructose-6-phosphate kinase, brain acetylcholinesterase and liver citrate synthetase (Somero and Hochachka, 1971). Still later, "cold" and "warm" glucose-6-phosphate dehydrogenase isozymes were found, which could be distinguished electrophoretically in the mullet liver (Hochachka and Clayton-Hochachka, 1973). In this enzyme, an increase in its activity, produced by a temperature rise, was compensated for not only by reducing the affinity to substrate, but also via a decrease in the affinity to a coenzyme (NADP) and through the enhancement of the activity of the inhibitor (NADP-H). The K_m for the frog *(Rana pipiens)* liver and kidney arginase at 2° C is lower than that at 25° C. In the winter this value is lower than in summer (Boernke, 1973).

Nickerson (1973) has developed an idea of multistability of proteins. He maintains that there are proteins, which exist in several stable conformations. A transition from one form into another is associated with overcoming the energy barrier and can be provoked by changes in the environment, in particular, by temperature fluctuations. A substitution of one conformer for another with different K_m, at a change in the environmental temperature, may have an adaptive significance.

The results reported by Newell and Pye (1971) are in line with the data obtained by Somero and Hochachka discussed above. They showed that oxygen consumption by isolated mitochondria of the mollusc *Littorina littorea* at relatively high substrate concentrations (pyruvate) was strongly dependent upon temperature and the temperature dependence plots displayed distinct maxima. A maximum for the mitochondria isolated from the molluscs acclimated to a warm environment, was found to be located in the region of higher temperatures than that for the mitochondria from the organisms kept in the cold. If, however, the mitochondria were incubated with 10 times lower substrate concentrations, at which oxygen uptake could be regulated through the affinity of enzymes to substrates, then the oxygen consumption ceased to be dependent on temperature, and remained at a constant level over the range of temperatures from 2 to 16° C. Hazel (1972a) reported a linear increase in K_m for goldfish muscle succinate dehydrogenase toward succinate as the reaction temperature was elevated from 5 to 25° C. This fact was used by the author to account for the mild effects of temperature on reaction rates at low, physiological concentrations of substrates. However, the K_m values for goldfish acclimated to 5 and 25° C were identical. This enzyme exists in only one form and has no isozymes. The temperature-dependence plot of K_m of NADP-isocitrate dehydrogenase from the ide *Idus idus* shows a W-shape. The reason for the existence of two minima is not clear. In the case of the enzyme from the fish adapted to a warm habitat, both minima are shifted toward the region of higher temperatures (Künnemann and Passia, 1973).

The data discussed clearly show the significance of K_m changes in both the instantaneous and gradual adaptations of poikilothermal organisms to variations in the environmental temperature. Similar dependence could not be found for some enzymes of the homeotherms. The K_m for the rat pyruvate kinase was found to be temperature-independent. Lactate dehydrogenase and pyruvate kinase isolated from the fin and kidney of the sperm whale displayed identical temperature dependence of K_m, in spite of the fact that the temperature of the tissues in these two organs is essentially different (Somero and Johansen, 1970). Gerez de Burgos et al. (1973) found that K_m for lactate dehydrogenase increased as the reaction temperature was raised. This increase was detected for both the snake and bovine enzymes, however, in the latter case an increase was less pronounced. K_m for the bovine arginase, in contrast to that for the frog enzyme, was the same at 2 and 25° C (Boernke, 1973). On the other hand, Cowey (1967) reported a dramatic increase in the K_m for the muscle glyceraldehyde-3-phosphate dehydrogenase provoked by a temperature increase; this was the case not only for the enzyme from cod and lobster, but also for that from rabbit. The pig heart NADP-isocitrate dehydrogenase shows a sharp minimum of K_m at 37° C, for both the substrate (isocitrate) and the co-substrate (NADP) (Künnemann and Passia, 1973).

In *Monotremata* (platypus and echidna), characterized by imperfect thermoregulation and variable body temperature, the affinity of lactate and malate dehydrogenases to their respective substrates is decreased as the temperature is elevated, as also occurs in poikilothermal animals. This provides for the temperature independence of reaction rates over the temperature range from 25° C to 35° C (Baldwin and Aleksiuk, 1973).

However, as frequently happens in biology, the results of subsequent investigations carried out by Hochachka, Somero and other workers on enzymes from poikilothermal animals somewhat deranged the formerly perfect concept, since the worst that can happen to a hypothesis is its further experimental development. Thus, Behrisch and Hochachka (1969a, 1969b) showed that the K_m for fructose-1,6-diphosphatase from the lungfish is temperature-dependent, whereas K_m for this enzyme from the rainbow trout is not changed with temperature fluctuations. On the other hand, the affinity of the trout enzyme to the inhibiting effector adenine monophosphate is increased as the temperature is elevated. Consequently, stabilization of the catalysis rate in the trout with elevation of temperature is achieved by augmentation of the enzyme-inhibitor affinity, and not by reducing the affinity to the substrate, as would be expected to occur in the lungfish. It was found out later that the temperature effects on the Michaelis constant may strongly depend on pH. The affinity of lactate dehydrogenase to the substrate—pyruvate—at pH above 7.5 falls abruptly as the temperature is elevated, whereas at pH below 7.0 the affinity becomes independent of temperature. The K_m lactate dehydrogenase isolated from the tuna fish adapted to 20° C exhibits a mild temperature dependence (Hochachka and Lewis, 1971). Isocitrate dehydrogenase from trout liver cells is represented by a number of isozymes so that in different individuals the isozyme patterns differ. If the cells contain isozymes A_2, B_2, and C_2 (the enzyme is considered to be a dimer), then the affinity to isocitrate is drastically reduced as the temperature is increased. If, however, there is only one isozyme A_2 present in the liver cells (and such individuals constitute approximately 27% of a population), the affinity of isocitrate dehydrogenase to the substrate remains unchanged from 0 to 25° C (Moon and Hochachka, 1972).

Komatsu and Feeney (1970) compared the temperature dependence of the K_m for the muscle fructose diphosphate aldolase in two Antarctic fish species (*Trematomus borchgrewinki* and *Dissostichus mawsoni*) with that for the analogous enzyme isolated from rabbit muscles. No difference could be found in the temperature dependence of the K_m for the enzymes in these organisms and the authors therefore disagree with the significance of compensatory changes in the affinity of enzyme to substrate as being involved in adaptation to life at low temperatures. Doubts of this kind have been expressed also by Baldwin (1971), using the following arguments. The K_m for tuna fish brain acetylcholinesterase remains constant over a wide range of temperatures. On the other hand, the rate of reaction of the enzyme with acetylcholine within this range of temperatures is also changed only slightly (Q_{10} = 1.1–1.2). Consequently, stabilization of the reaction rate in this case has not been achieved by a compensatory temperature-dependent modification of the enzyme-substrate affinity, but some other events have been responsible for the phenomenon observed. Furthermore, the catalysis rate is reduced approximately by half for the Antarctic fish acetylcholinesterase by increasing the temperature from 2 to 10° C, whereas the enzyme–substrate affinity in 35 times lower under these conditions. Baldwin believes that if temperature-dependent variations of K_m are of any adaptive significance, the latter is not implicated in the compensatory regulation of the rate of catalysis during temperature changes.

Hebb and his co-workers (1972) studied the temperature dependence of the affinity of goldfish brain acetylcholinesterase from the organisms acclimated to 6 and 30° C, toward both substrates of the enzyme—choline and acetyl-CoA. It was found that the maximal affinity to these substrates roughly corresponds to the environmental temperature for each of the two groups of the acclimated fish. According to the authors, these modifications of temperature dependence of the enzyme–substrate affinity during acclimation are not referable to the presence of different isozymes. Moreover, the authors are not inclined to regard the facts obtained as indicative of a mechanism responsible for the temperature independence of the rate of the enzymic reaction. They believe that an enhanced enzyme-substrate affinity at acclimation temperatures provides for stabilization of an enzyme in given conditions. As reported from the laboratory of Prosser (Wilson et al., 1973), no changes in isoenzyme patterns could be observed for lactate dehydrogenase and malate dehydrogenase isolated from brain, muscles, heart and liver tissues of the goldfish acclimated to 5, 15, and 25° C for four weeks. Again, according to the findings communicated from the laboratory of Precht, no difference could be found in the isoenzyme patterns of lactate dehydrogenase isolated from brain, gills, intestine and muscles of the ides acclimated to 10 and 20° C (Künnemann, 1973).

The same workers sometimes report very contrasting results obtained in studies on not too distant forms. Thus, Wernick and Künnemann (1973) acclimated two fish species to 10 and 20° C. In agreement with the findings of Moon and Hochachka for trout isocitrate dehydrogenase, the K_m for the "cold" ides was minimal at 10° C, and in the "warm" fish, at 20° C. On the other hand, the acclimation temperature had no effect on the temperature dependence of the K_m for the *Rhodeus amarus* glutamate dehydrogenase.

Relevant data in plants are extremely scanty. In the author's laboratory, Feldman (1973) could not detect any temperature dependence of K_m for acid phosphatase from the snowflake leaves. McNaughton (1972) has estimated the K_m for glycolate oxidase in two populations of *Tyha latifolia* (reed mace) from warm and cold climates, and obtained results which are not consistent with the concept advocated by Hochachka and Somero.

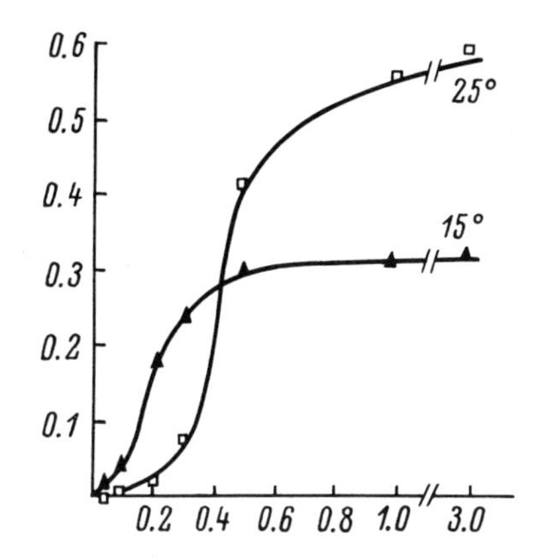

Fig. 54. Effect of substrate concentration on activity of phosphofructokinase at 15 and 25° C (Freed, 1971). *Abscissa:* concentration of fructose-6-phosphate (M, × 10^3); *ordinate:* reaction rate

Thus, to evaluate the role given to the changes of K_m in temperature adaptations of organisms, further investigations are needed. They will reveal to what extent the compensatory effects of temperature dependent changes of the Michaelis constant are resorted to in nature, and in what enzymes and organisms this regularity may be operative.

An interesting mechanism, effective in restricting an acceleration of an enzymic reaction elicited by a temperature increase, has been reported by Freed (1971) for the glycolytic system from the king crab muscles. According to his observations, the plot of the dependence of the reaction rate upon the substrate (fructose-6-phosphate) concentration changes its shape with an increase of the reaction temperature. At low temperatures it is a hyperbola, as the temperature is elevated the curve becomes progressively S-shaped (Fig. 54). In other words, the higher the temperature, the more cooperative is a catalytic process. As a result, in the region of low substrate concentrations (up to $3 \cdot 10^{-4}$ M), an increase in the reaction temperature may lead not to an enhancement, but to a reduction of the rate of a precess. This finding of Freed that, at low concentrations of fructose-6-phosphate, Q_{10} of the process is less than unity, agrees with this contention.

5.3 A Temperature Optimum for Enzyme Activity

In studies concerning the mechanisms of adaptation of animals, plants and microorganisms to environmental temperature, major efforts are being spent in attempts to discover the dependence of various physiological precesses and biochemical reaction on temperature. Plots describing such a dependence are usually of a dome-like shape, in which zones of a temperature optimum, minimum and maximum can be recognized (Fig. 55). As a rule, the descending branch of the curve, from an optimum to a maximum, is more steep than that from an optimum to a minimum. Such plots have been reported for entire organisms as well as for isolated tissues and purified enzymes. Relevant investigations are focused on two

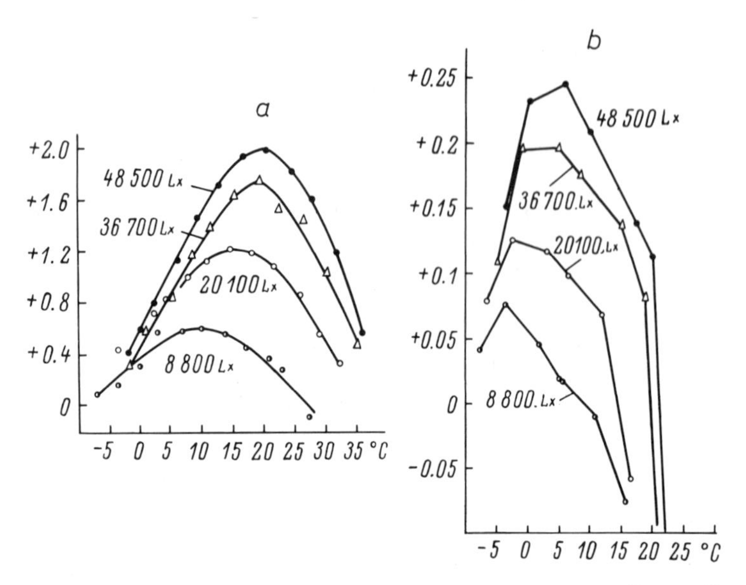

Fig. 55 a and b. Temperature dependence of CO_2 uptake by lichens at different illumination intensities. (Lange, 1969). *Abscissa:* temperature; *ordinate:* mg CO_2 (per gram per hour). (a) lichen *Ramalina maciformis* from the Negev desert; (b) Antarctic lichen *Neuropogon* sp.

major aspects: modificational and genotypic. On the one hand, in the case of modification-oriented studies, attempts are being made to discover the effects of adaptation of individuals to this or that temperature environment on the temperature dependence of the rate of a given process. On the other hand, genotypic studies deal with comparisons of the temperature-dependence plots of the rates of different processes in species, races and strains, that differ in their thermophily. It should be remembered, however, that such plots are largely only relative and their positions in reference to a particular portion of the temperature scale depend, to a considerable degree, upon the duration of the reaction concerned. Prolongation of a reaction, as a rule, shifts an optimum toward lower temperatures. The position of a plot is affected by a number of factors and, primarily, by the composition of the medium, by the substrate, co-factor concentration, etc. The curves of temperature dependence of photosynthesis (Fig. 55), for instance, are usually shifted to the right, toward higher temperatures as the light intensity is elevated up to saturation (Pisek et al., 1969; Lange, 1969; and others). By increasing the CO_2 concentration, similar changes can be produced (Costes et al., 1967). All this should be taken into account when analyzing the data obtained by different workers.

The steepness of an upward branch of a curve depicting the rate of a process as a function of increasing temperatures is determined, within the temperature zone from a minimum to an optimum, by the activation energy which is usually assumed to be characteristic of and constant for a given process. If a series of biochemical reactions underlies the process concerned, then the shape of the curve reflects the temperature dependence of that reaction which proceeds at the lowest rate, this reaction being the rate-limiting reaction for the entire process.

One cannot assume that throughout a given portion of the temperature scale the limiting role will be given to one and the same reaction. A decline in the process rate with elevation of the temperature in the zone between an optimum and a maximum, is thought to be accounted for by the denaturing effect of heat on the enzyme operating in the rate-limiting reaction. It is on these ground that the temperature effective in producing a decrease in the rate of a process is frequently used for evaluating the heat stability of an enzyme. It should be stressed, however, that such evaluation of thermostability may be at variance with estimations made on an enzyme solution exposed to heat in the absence of the substrate and cofactors. The latter components may significantly affect the stability of the enzyme, making it either higher or lower. Besides, according to Hochachka (1968b), in some cases, a reduction of the rate of a process at elevated temperatures may be related not to denaturation of the proteins, but to formation of an inhibitor of the reaction.

A peak of a curve in the zone of optimum is sometimes rather sharp, whereas in other cases it is extended and flattened considerably. As follows from the preceding section, the mild temperature dependence of a reaction rate in the zone of optimum may be due to the fact that with an increase in temperature the acceleration of the reaction is diminished by a lowering of the affinity of enzymes to substrates. This mechanism of compensation, however, is effective only if the substrate concentration is far below the saturation level. This has been most explicitly demonstrated in the muscle AMP-deaminase by Zydowo et al. (1965). At substrate concentrations much below saturation level, they obtained for the enzymes from homeothermal animals (rabbit, rat, pigeon) curves of the temperature dependence of the reaction rate, exhibiting a well-pronounced sharp optimum located at 35° C. The curve for the enzyme from poikilothermal animals had extremely flattened peaks. Reaction rates showed almost no variation in frogs at 30–40° C, in carp between 25 and 35° C, and in trout within the range of 20 to 30° C. (By way of a reminder it can be said that Somero and Johansen (1970) could not find a compensatory dependence of K_m upon temperature for mammalian enzymes too.) At high saturation concentrations of substrate, the differences in the temperature dependence of AMP deamination rate in homeothermal and poikilothermal animals disappeared, i.e., in both cases a sharp optimum could be seen in the temperature-dependence curves, the location of which was notably shifted toward higher temperatures.

In spite of the numerous complicating aspects and reservations to be made, the results of a large number of works leave no doubt as to the fact that in the process of adaptation of organisms to new temperature environments, changes occur in the temperature dependence of enzyme activity, both of genotypic and modificational nature. From the standpoint of the concept formulated in Section 4.2.5, a semistable state of the enzyme necessary for its effective operation can be maintained only within a more or less limited temperature range. It might be expected, therefore, that adaptation to new temperature conditions would be expressed primarily in a shift of the curve as a whole toward lower temperatures during adaptation to cold, and, conversely, in an opposite direction in adaptation to warmer conditions of life rather than in an extension, in either direction, of the temperature-dependence curve of the rate of enzyme activity. Indeed, one can find

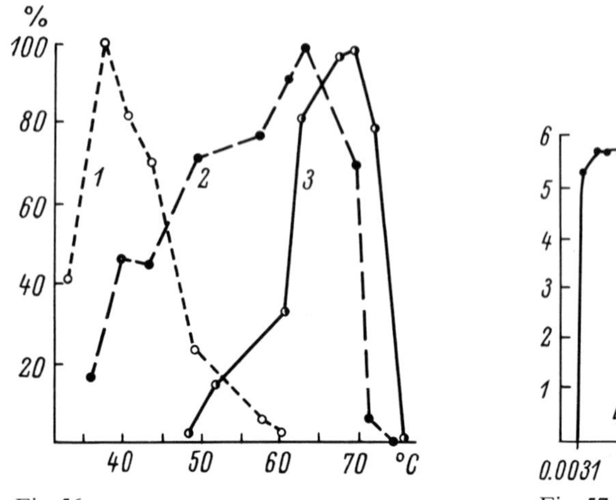

Fig. 56

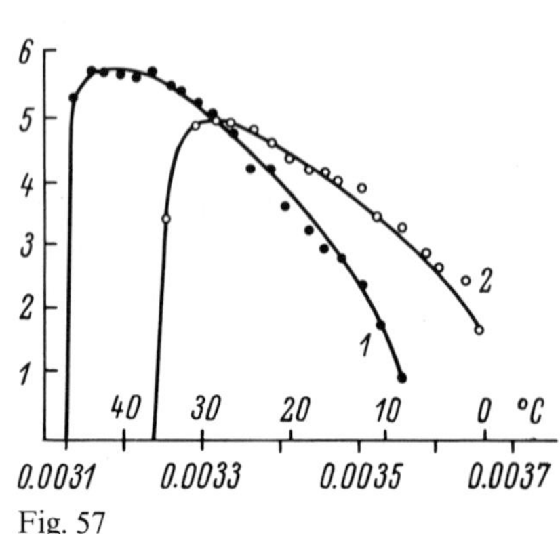

Fig. 57

Fig. 56. Temperature dependence of $^{14}CO_2$ uptake in the dark by bacteria living at different temperatures (Brock and Brock, 1968). *Abscissa:* temperature at which $^{14}CO_2$ incorporation took place; *ordinate:* rate of uptake (% of maximal rate). *1, 2, 3:* bacteria living at 37, 58.5, and 68.1° C, respectively

Fig. 57. Arrhenius plots of temperature dependence of growth of mesophilic *Escherichia coli (1)* and psychrophilic strain 21-3c *(2)* (Ingraham, 1962). *Abscissa:* logarithm of reciprocal of absolute temperature; *ordinate:* logarithm of growth rate

numerous examples in the literature of such lateral shifts of curves during genotypic and, to a lesser extent, in modificational adaptations. As a consequence of a lateral shift of the entire curve for objects adapted to cold, the rates of processes and reactions at temperatures below an optimum would be higher than in organisms adapted to heat, provided measurements are made in both cases at the same temperature. Opposite results will be obtained if the rate of a process has been estimated at a temperature above that of an optimum for an organism adapted to cold. With adequate and sufficient shifts of temperature-dependence curves of a process rate, full compensation may be achieved; in this case, the rate of a process in "cold" and "warm" objects will be identical at temperatures to which these objects have adapted. According to the classification of Precht (1958, 1968), such a mode of adaptation corresponds to the second type of adaptation. If compensation is incomplete, it might be included in the third type of adaptation.

Following are some examples to illustrate the point. Brock and Brock (1968) studied the temperature dependence of $^{14}CO_2$ incorporation in bacterial populations taken from three different areas of a hot spring which differed markedly in temperature (37.0, 58.5, and 68.1° C). The plots reported by these authors (Fig. 56) demonstrate that the position of an optimum on the temperature scale is directly dependent on the temperature of species' life. These workers reported the same results for the assimilation of glucose. Allen and Brock (1968) placed tubes with nutrient media at different temperatures ranging from 25 to 75° C and transferred heterotrophic microorganisms taken from various sources onto these media. Depending on the temperature, this or that species composition has eventually been

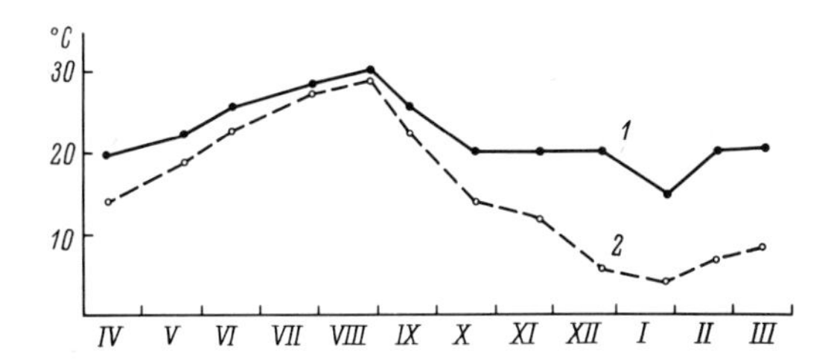

Fig. 58. Seasonal variations of temperature optimum for photosynthesis of phytoplankton *(1)* and water temperature *(2)* in a pond in Tokyo University (Aruga, 1965). *Abscissa:* months; *ordinate:* temperature

established in the tubes. The temperature optimum for the labeled glucose assimilation was then estimated for the populations in each tube. Positions of the temperature optima corresponded to those temperatures at which the microorganisms had been adapted in a given tube. Figure 57 shows a lateral shift of Arrhenius plots of the temperature dependence of growth rates of the mesophilic bacterium *E. coli* and the psychrophilic one, 21-3c (Ingraham, 1962).

Aruga (1965) studied the temperature dependence of photosynthesis in a phytoplankton from a pond, the species composition of which varied over the year. The curves for seasonal changes in the water temperature and the temperature optimum of photosynthesis are reproduced in Figure 58. From April to October a parallelism between these parameters is very close, however, in the cold season it becomes distorted.

In the majority of the works referred to in this Section, probes characterized by a mixed species composition have been covered. Studies on temperature dependence of various indicators and in some species which differed in their temperature ecologies, reported very similar results. A correspondence of the positions of minima, optima and maxima to a habitat temperature has been detected in a number of investigations on photosynthesis: in 19 species of tree and grass plants, growing at different altitudes (Pisek et al., 1969)., in three grass species taken from different altitudes (Loginov and Akhmedov, 1970), in eight *Astragalum* species (Loginov and Nasyrov, 1970), in 12 grass species from different families (Murata and Iyama, 1963). Lateral shifts have been noticed in the curves of temperature dependence of apparent photosynthesis in the grass *Agrostis palustris* adapted to a cold climate, and for that in *Cynadon dactilon*, a grass found in hot regions. The plots were laterally shifted with respect to each other by more than 10° C (Miller, 1960). The temperature optimum of photosynthesis in an Antarctic diatomic alga is around 7° C (Bunt et al., 1966). Photosynthesis in three algal species displayed differences in its temperature dependence which neatly corresponded to the difference in respective thermophilies of these algae (Felföldy, 1961). This is the case also in the plots reported by Lange (1969) for the CO_2 exchange in lichens growing in the Negev desert and in Antarctic lichens (Fig. 55). Given equal illumination conditions the positions of temperature maxima in the former species are shifted 15° C to the right. Nevertheless, the position of a maximum for the lichens from a hot desert, corresponding to about 20° C, appears to be unjustifiably low if taken at face value. As pointed out above (p. 68), this is, however, readily explainable as being due to the fact that the Negev desert lichens photosynthesize actively only during the cool morning hours, whereas in the heat of the day they are in a passive, desiccated state in which the resistance of proteins to heat becomes considerably higher. It is obvious that a combination of the capability to be active during midday heat with the ability of function during the cool hours is unattainable. In order to meet the first requirement, such a high stability of protein macromolecules is needed that it is incompatible with the conformational flexibility required for performing at temperatures of around 10–20° C.

Mooney and his colleagues (1964) grew in a greenhouse 20 plant species taken from different altitudes. The temperature optimum of photosynthesis in the species growing at 9600–12950 feet above sea level was around 20° C, and for the plants living at 4500–8700 feet it was 30° C. Similar data have been reported for the three populations of the European goldenrod *Solidago virga-aurea* growing in three different climatic zones (Björkman et al., 1960), and for the geographically isolated populations of the mountain sorrel *Oxyria digyna* taken from the biotopes featuring contrasting temperature conditions (Mooney and Billings, 1961; Billings and Mooney, 1968). West and Mooney (1972) studied the temperature dependence of photosynthesis in three sage-brush species growing at different altitudes. They found genetically determined differences in the positions of the temperature optima for photosynthesis; in *Artemisia arbuscula*, growing at high altitudes, the optimum was somewhere between 10–20° C; in *A. nova*, which grows at lower horizons—around 25° C, and in *A. tridentata*, found in between these two habitat altitudes, the optimum was around 20° C. At 10° C the intensity of photosynthesis was greater, the higher the altitude of the plant under study. These authors showed that two species, beside the genetically fixed dissimilarities, also exhibited modificational plasticity in the response of photosynthesis to temperature. In contrast to these plants, *Tidestromia oblongifolia* (*Amarantaceae* family), growing in Death Valley in California where the temperature frequently reaches 50° C, has its temperature optimum for photosynthesis at around 47° C (Björkman et al., 1972).

Tieszen and Helgager (1968) evaluated the temperature dependence of the Hill reaction in two ecological races of tussock grass *Deschampsia caespitosa*. The temperature-dependence plot of the more warm-loving Alpine race was completely transposed to the right of the temperature scale as compared with the plot for the Arctic race. Additional evidence of this kind may be found in a review by Hiesey and Milner (1965).

Genotypic adaptations of plants to the environmental temperature, achieved by a shift of temperature dependence of their functions, have been reported not only for photosynthesis, but also for respiration (Scholander et al., 1952; Mooney and Billings, 1961; Klikoff, 1966; Chatterton et al., 1970). Similar evidence is available concerning growth rate. An optimum for growth of five grasses in the temperate zone is around 20–25° C; for subtropical and tropical grasses the maximum for growth is found at 30–35° C. In the first group of species, growth could still be detected at 5° C, in the second group chlorosis developed at temperatures below 15° C and the plants eventually died (Cooper and Tainton, 1968).

The above data concerned genotypic adaptations of plants. A wealth of evidence has also been accumulated for animals, likewise demonstrating the wide

Table 19. Comparison of reproduction and development of the frog species belonging to *Rana* genus which differ in their thermophily. (From Moore, 1939)

Species	*R. sylvatica*	*R. pipiens*	*R. palustris*	*R. clamitans*	*R. catesbeiana*
Water temperature at which development starts (° C)	10	12	15	25	—
Northern boundary of distribution (° North)	67°, 30′	60°	51–55°	50°	47°
Lower temperature limit for embryo development (° C)	2.5	6	7	11	—
Upper temperature limit for embryo development (° C)	24	28	30	35	—
Time lapse between stages 2 and 20, development at 18.5° (h)	68	116	126	138	170

occurrence of lateral transpositions of curves for the dependence of a process rate on temperature, obtained in studies on genotypic adaptations of species and races to the environmental temperature (for a reviews see: Bullock, 1955; Vernberg, 1962). In this regard, the works of Moore (1939) and Volpe (1957) should be mentioned as being extremely illustrative. Moore studied five frog species characterized by different distribution areas and reproduction periods. Table 19 reproduces the data reported by Moore. The more southern a species, the higher are both minimal and maximal temperatures at which its development is still feasible. In addition, the wideness of the temperature span within which development of the species occurs varies only slightly for different species. As a result of such a lateral shift of temperature dependence of growth, the length of the time period required for the development of the species, characterized by different thermophilies, to take place at one and the same temperature (18.5° C), is greater the more warm-loving the species proves to be. Similar results have been reported by Volpe for six toad species, which differed in their temperature ecologies. The findings by Volpe have been fully confirmed by Ballinger and McKinney (1966) who studied the temperature dependence in other toad specis of the same genus *Bufo*.

More or less complete compensation of respiration intensity attained by shifting the temperature optimum for respiration, has been found in comparisons of molluscs, fish and crustaceans living in the Arctic and tropics (Scholander et al., 1953). These findings have been corroborated by Read (1962) in his studies concerning the effect of temperature on respiration in two marine molluscs with different thermophilies. Fox (1939) studied the heart-beat rate, pulsation of vessels, and beating of epipodites in a large number of poikilothermal animals, and found that at a given temperature, the activity of these functions in most cases was higher in the more cold-loving organisms. The growth rate of embryo molluscs of one species at a given temperature was found to be appreciably higher in those molluscs taken at the Alaska littoral as opposed to the organisms picked up in California (Dehnel, 1955).

Strikingly illustrative examples of genotypic adaptation of the temperature dependence of intensity of various cellular functions are provided by microorgan-

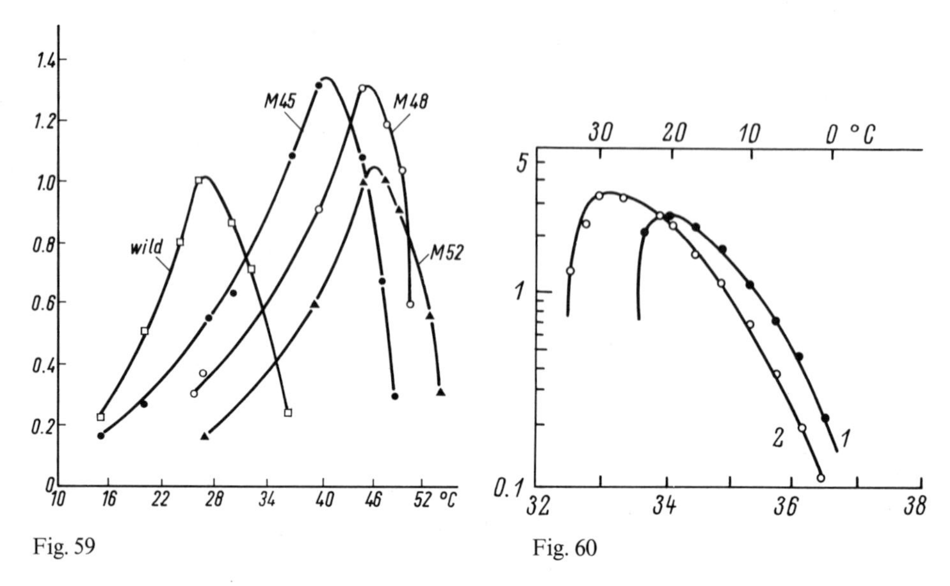

Fig. 59

Fig. 60

Fig. 59. Effect of incubation temperature on growth rates of wild and mutant strains (M45, M48, and M52) of *Pseudomonas fluorescens* (DeCicco and Noon, 1973). *Abscissa:* temperature; *ordinate:* growth rate (doublings per hour)

Fig. 60. Arrhenius plots of temperature dependence of growth of psychrophile *Micrococcus cryophilus*. *1:* wild type, *2:* a thermostable mutant (Malcolm, 1969a). *Abscissa:* log of reciprocal of absolute temperature, $\times 10^4$; *ordinate:* log of growth rate

isms. It is widely known that psychrophilic microorganisms are unable to live at a temperature which is optimum for the mesophiles: it is too high for them. On the other hand, most thermophiles show extremely poor growth, if any, at an optimum temperature for the mesophiles, since it is too low for them (Farrell and Rose, 1967). Clones of the blue-green alga *Synechococcus* sp. were isolated from a hot spring. When grown in culture, the clones displayed the following optima for growth: 45, 50, 55, and 65° C. A clone with an optimum at 65° C never reproduced at temperatures below 50° C (Peary and Castenholz, 1964). DeCicco and Noon (1973) selected several spontaneous mutants, which exhibited various degrees of thermophily, from a *Pseudomonas fluorescens* culture. The higher the temperature maximum for a mutant, the higher was also its minimum. As a result, the extent of the temperature range, within which growth could occur, proved to be identical for the wild type and for all of the mutants studied (Fig. 59).

Micrococcus cryophilus is an obligate psychrophile. It shows no growth above 25° C. Malcolm (1969a) isolated from a culture of this organism a mutant that could grow at temperatures almost 10° C higher than the wild type. Growth rate of the mutant, however, was appreciably lower than that of the wild type at temperatures from 0 to 20° C (Fig. 60). Malcolm found that, in this mutant, thermal stability of two enzymes involved in protein synthesis, was increased, namely, that of glutamyl-tRNA-synthetase and prolyl-tRNA-synthetase. The findings of Malcolm (1969a) show that an enhancement of thermostability—from the present author's point of view, an increase in the conformational rigidity of protein macromolecules—imposes constraints on the metabolic rate at low temperatures. The observations reported by Brock and Brock (1971) are in agreement with those of Malcolm. These authors showed that in thermophilic bacteria living at 90° C, an optimum for incorporation of some amino acids into protein is around 90° C, so that at 60–70° C the rate of amino acid uptake is practically nil.

Pertinent examples might certainly be listed, but even those already referred to above explicitly testify to the fact that in plants, animals and microorganisms, the temperature dependence of the intensity of various functions (photosynthesis, respiration, motile activity, growth, reproduction, etc.) is adequately changed during genotypic adaptations of the organisms to the environmental temperature. This change results in the individuals of species and races which are adapted to a somewhat lower temperature, displaying much higher metabolic and physiological activity at those temperatures which are found in the interval between their respective temperature minima and optima, then those adapted to a higher temperature. Their activity is lower at temperatures found between the temperature optima and maxima. In other words, adaptation to low temperatures is achieved at the expense of the ability for normal functioning at a higher temperature. Adaptation to an elevated temperature has an opposite effect. This scheme is fully consistent with the assumption that the basis for such adaptations is provided by changes in the level of conformational flexibility of protein molecules; of enzymes, in particular.

In the foregoing the observations cited dealt with complex physiological processes. From the standpoint of the concept being developed here, it is important to know whether, in fact, a change of temperature dependence of a physiological process during compensated genotypic adaptations is related to a corresponding change in the temperature dependence of enzyme functioning. Evidence in the literature concerning this problem is less abundant; however, a number of facts emerge which support this view. Only a few observations may be cited which report lateral transpositions of temperature dependence curves of the activities of plant enzymes. Thus, McNaughton (1965) assayed the glycolytic oxidase activity at 17, 27 and 37° C. The enzyme was taken from four reed mace populations growing in two climatic zones. The data he obtained indicate that in the populations growing in cooler environments, the enzyme activity is maximal at 17° C, whereas for the populations from a warmer habitat, the maxima are found at 27 and 37° C. Treharne and Cooper (1969) studied the temperature dependence of ribulose diphosphate carboxylase and phosphopyruvate carboxylase from two typical temperate-climate grasses and two tropical grasses. All the grasses were grown under identical conditions. The plots for oat and *Zea mays* enzymes were laterally transposed relative to each other. In other cases, a difference in the temperature dependence could not be found. A subsequent publication from that laboratory (Treharne and Eaglis, 1970) described curves of the temperature dependence of the activity of ribulose diphosphate carboxylase isolated from *Dactylis glomerata* grown from seeds of Portuguese and Norwegian populations. The enzyme from the southern population was more active at elevated temperatures, whereas at lower temperatures the activity of the northern population enzyme proved to be higher. Negative results heve been obtained by Feldman (1973) in

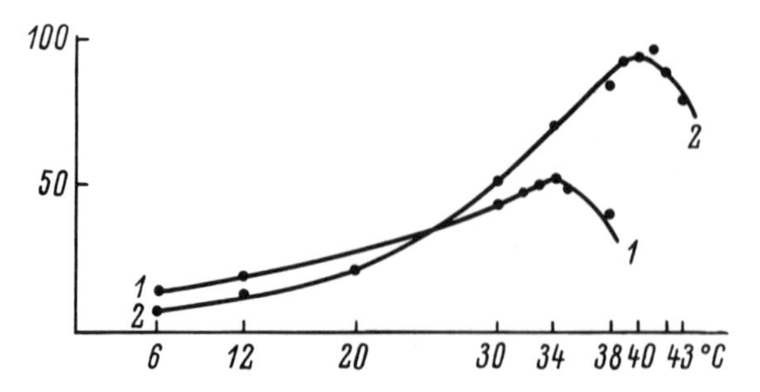

Fig. 61. Temperature optima for ATP-ase activity of actomyosin from *Rana temporaria (1)* and the more warm-loving *R. ridibunda (2)* (Vinogradova, 1965b). *Abscissa:* assay temperature; *ordinate:* enzyme activity

two snowflake species which differed in their thermophilies: the curves of the temperature dependence of acid phosphatase activity in the spring and summer snowflakes practically coincided.

Findings of this sort are more numerous in animal enzymes. Mutchmor and his collaborators (Mutchmor and Richards, 1961; Mutchmor, 1967) have studied the temperature dependence of the muscle ATP-ase activity in a number of insects, showing different preference to cold. These differences were found to be in good correlation with the difference in the positions (on the temperature scale) of the temperature dependence curves of the ATP-ase activity.

The temperature corresponding to an optimum activity of the Mg^{2+}-ATP-ase and (Na^+, K^+)-ATP-ase in brain tissue homogenates of the cold-water fish *Leuciscus rutilus* was lover than that for the warm-water fish *Helosoma temmincki* (Kohonen et al., 1973). Vinogradova (1963a, b, c) compared the temperature-dependence plots of actomyosin activity of two species of crabs, skate, molluscs and frogs, characterized by different thermophily. In all cases an optimum for the activity of actomyosin isolated from a more warm-loving species was found at a higher temperature than that for the enzyme taken from a closely related less thermophilic counterpart. The curves for the frog actomyosin preparations showed a distinct intersect (Fig. 61). According to Tartakovsky and his co-workers (1973), a highly purified myosin preparation obtained from *Rana ridibunda* displays 1.5–2° C higher stability against heat than the protein from *R. temporaria*. In agreement with this, a temperature optimum for the *R. ridibunda* myosin activity also is 3.5–4° C higher. The temperature-dependence curves of the pancreatic amylase activity obtained for the enzymes isolated from two lizard species which differed in their preference to heat, are laterally transposed with respect to each other (Abrahamson and Maher, 1967). Very impressive data have been collected by Licht (1967) on the muscle ATP-ase from four lizard species (Fig. 62). A complete correspondence was found to exist between the positions of the plot for a given species and the preferred temperature characteristic of that species: the greater the preference to warmth in a given species, the less the enzyme's activity was inhibited as the temperature was elevated within the supraoptimal zone, and the more pronounced was the suppression of the enzyme's activity as the temperature was lowered within the zone of suboptimal temperatures.

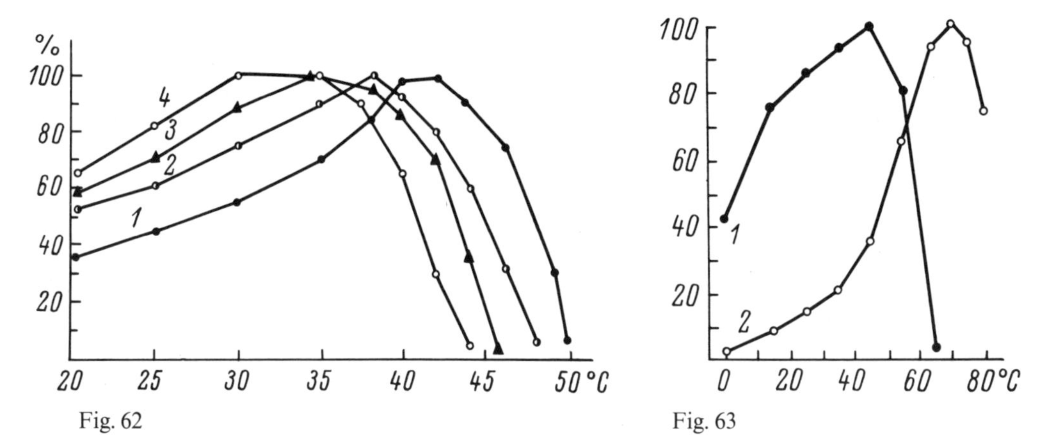

Fig. 62

Fig. 63

Fig. 62. Effect of temperature on relative activity of ATP-ase of 4 lizard species (Licht, 1967). *Abscissa:* assay temperature; *ordinate:* activity (% of maximal activity). *1: Disposaurus dorsalis* (preferred temperature 38.8° C), *2: Uma notata* (37.5° C), *3: Scleroporus undulatus* (36.3° C), *4: Gerrhonotus multicarinatus* (30.0° C)

Fig. 63. Effect of temperature on phenylalanyl-tRNA complex formation in an in vitro system derived from mesophile *Escherichia coli (1)* and thermophile *Thermus aquaticus (2)* (Zeikus and Brock, 1971). *Abscissa:* reaction temperature; *ordinate:* intensity of reaction (% of maximal intensity)

Lateral shifts of temperature-dependence curves of enzymes' activities are revealed also in genotypic adaptations of microorganisms. Fine examples of this regularity can be found in the works of Brock and his collaborators. Aldolase from the thermophile *Thermus aquaticus* has an optimum of its activity at 90° C and at 60° C the enzyme is practically inactive (Freeze and Brock, 1970). Figure 63 shows the temperature dependence plots of phenylalanyl-tRNA formation in an in vitro system derived from *Th. aquaticus* and *Escherichia coli* (Zeikus and Brock, 1971). Essentially similar laterally transposed plots have been reported also for the formation of isoleucyl-tRNA complexes. Likewise, the temperature-dependence plots of the activity of ribonucleotide reductase from the thermophile *Thermus X-1* and the mesophile *Lactobacillus leichmannii* are laterally shifted relative to each other (Fig. 64). An optimum for enzyme functioning in the case of the thermophile is more than 20° C higher than that for the enzyme derived from the mesophile. At a temperature optimum for the mesophilic enzyme's activity, the activity of its thermophilic counterpart decreases by 70% as compared to the level of activity at a temperature corresponding to its own optimum (Sando and Hogenkamp, 1973). The same result was obtained (Fig. 65), when comparison was made of the temperature dependence of a highly purified crystalline α-amylase activity for enzymes isolated from thermophilic and mesophilic bacteria (Ogasahara et al., 1970a, b). Stokes (1967) believes that, in terms of evolution, thermophily, mesophily and psychrophily of microorganisms are the results of an increasing heat sensitivity of the enzymes.

Consequently, in the process of evolution, adaptation to life under contrasting temperature conditions has very often been accompanied by evolution of enzymes

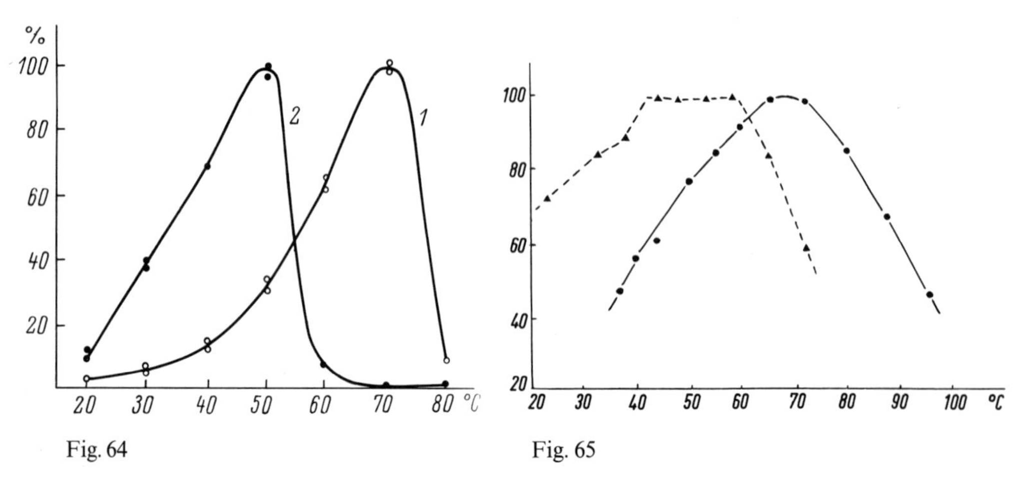

Fig. 64

Fig. 65

Fig. 64. Effect of temperature on maximal activity of ribonucleotide reductase of thermophile *Thermus* X-1 *(1)* and the mesophile *Lactobacillus leichmannii (2)* (Sando and Hogenkamp, 1973). *Abscissa:* reaction temperature; *ordinate:* enzyme activity (% of maximal activity)

Fig. 65. Effect of temperature on activity of thermophilic α-amylase and that from *B. subtilis* (Ogasahara et al., 1970a). — ● — *B. stearothermophilus* α-amylase; --- ▲ --- *B. subtilis* α-amylase. *Abscissa:* temperature; *ordinate:* relative activity (%)

in which dependence of their activity on temperature is found in correspondence with thermal ecology of a species. At the same time, acquisition of the ability to operate at an elevated temperature is, as a rule, associated with a loss of the ability to perform at low temperatures. The reverse takes place in adaptation to cooler environments. Such behavior of enzymes in evolution is in full accord with hypothesis stating the necessity of maintaining the enzymes in a semilabile state, i.e., in a state of sufficient, but not excessive conformational flexibility. As shown on p. 156, it follows from this hypothesis that a given enzyme can operate only within a limited portion of the temperature scale. This fact has been acknowledged by Hazel and Prosser (1974). They however, explain the restrictions imposed on the operating zone of the enzyme's performance by the fact that on both sides of that zone of temperature optimum, the affinity of the enzyme to substrate is drastically reduced. It is, however, impossible to explain in this way the limits of an optimum zone in those experiments where the temperature dependence of enzyme activity is assayed in vitro at substrate concentrations above the saturating ones. It might rather be assumed that diminiution of the enzyme-substrate affinity, observed outside the limits of an optimum zone, is itself due to the fact that, under these conditions an unfavorable shift of the conformational flexibility level of an enzyme occurs in this or that direction.

These assumptions somewhat simplify the real life procedure. Live nature disposes also of other devices for adaptation to the environmental temperature even at the level of enzymes. A reduction of enzymatic activity caused by a change in the temperature environment may be, to a certain extent, compensated through an increase of the amount of an enzyme or substrate. To this end, there may be a modification of the conditions for the enzyme functioning in situ, an alteration of the enzyme-substrate affinity, of the affinity to the co-enzyme and co-factors, as well as switching over to another metabolic route, etc. That is why we should not be surprised when facing negative results. For example, Licht (1967), who found a close correlation between the positions on the temperature scale of the temperature-dependence plots of ATP-ase activity of the enzymes isolated from four lizard species, and thermophily of these species (Fig. 62), was unable to detect a similar correlation for alkaline phosphatase. Such a reservation appears even more substantiated when we turn to analysis of the temperature dependence of complex physiological processes. Kelly and Brock (1969) studied two strains of *Leucothrix mucor*. The first strain thrived in a water pond, where temperature varied from 5 to 25° C depending on the season. Another strain lived at a constant temperature of 25° C throughout a year. These authors studied the growth-rate temperature dependence of the two strains and found that the temperature maxima for growth of both strains were the same—38° C. An optimum was also found to be within the same temperature range, 25–30° C. The temperature minima for these strains, however, were drastically different. That for the strain living under constant temperature conditions was 13° C, whereas for the organisms exposed to variable temperatures, the minimal temperature for growth was as low as 2° C. In this case, adaptation to a constant temperature has led to an one-sided narrowing of the temperature-dependence plot. This type of adaptation apparently occurs also in other cases of transition from eurythermy to stenothermy. The particular transformations of enzymes which underlie the narrowing or broadening of an optimum zone can hardly be defined at this time. In such cases, a certain role is probably played by a modification of the response of the enzyme–substrate affinity relationships to temperature, as follows from the works of Hochachka and Somero.

In any case, an important point is that a genotypic adaptation to the environmental temperature is quite frequently accompanied by a lateral shift of the entire curve of the temperature dependence of enzyme activity along the temperature scale. In addition, this is consistent with an assumption that for an enzyme to perform efficiently its protein macromolecule must be in a semistable state.

So far the discussion in this section has concerned genotypic adaptations. Extremely diverse evidence is available on modificational adaptations of the temperature optima (Bullock, 1955; Precht et al., 1955; Prosser and Braun, 1962; Precht, 1968). There are many investigators who have cultivated animals and plants as well as populations of microorganisms or cell cultures and exposed them to contrasting temperature environments. The temperature dependence of various processes was then estimated at organismal, cellular or molecular levels. Experiments with populations of microorganisms are often unsuitable for distinguishing whether a modificational adaptation or a genotypic one, the latter achieved through selection, had actually taken place. Sometimes, an assessment of the activity of this or that process has been performed in various study objects adapted to different temperatures, an assay being carried out at intermediate temperatures only. There are cases in which this approach revealed greater activity in the objects kept in the cold as opposed to those cultivated at higher temperatures. Thus, Freed (1965) kept goldfish (a favorite study object for experiments on temperature acclimation) at 5, 15, 25, and 30° C and the cytochrome oxidase activity was then assayed at 15° C in muscle homogenates. The lower the acclimation temperature, the higher the enzyme activity was. Similar data concerning a compensatory enhancement of activity of various functions and enzymes have been reported for other study objects (Segal and Dehnel, 1956; Carlisle and Cloudsley-Thompson, 1968; and others). Such a result may be a consequence of an increase in the amount of an enzyme without modification of the enzyme response to temperature.

Adaptation to cold frequently shifts metabolism toward anabolism that leads to an increase in the cell size, to polyploidization, to an increase in the mitochondria population, to an increase of the content of enzymes (see: Orgel, 1964; Černý et al., 1965; Mutchmor, 1967; Das, 1967; Jørgenson, 1968; Hazel and Prosser, 1974). Otherwise, in adaptation of the mesophilic bacterium *B. subtilis* to an elevated temperature (63° C), the protein content per cell increased three-fold as compared with bacteria grown at 37° C. Both the DNA and RNA content per cell, however, remained the same (Dowben and Weidenmüller, 1968)]. During adaptation of this kind, a compensatory intensification of metabolism in the organisms adapted to cold may be attained without accompanying changes in the positions of optima on the temperature scale. The data reported by Ushakov and Kusakina (1960) might illustrate the point. These workers kept medicinal leeches at 2–4° C, 18–20° C, and 31–33° C for 30 days. The cholinesterase activity was then evaluated in homogenates of the muscles, the assays being made at 3, 19, 32, 37, and 43° C. The enzyme activity at all of the assay temperatures proved to be higher, the lower the acclimation temperature had been. At the same time, the temperature optimum for all the groups of leeches was 37° C.

Referring to his studies on ATP-ase activity in insects acclimated to heat and cold, Mutchmor (1967) presents schematized graphs of the Arrhenius plots of adaptation of species and individuals. In the case of genotypic adaptations, the more warm-loving a species is the higher the temperature of an optimum for ATP-ase activity and the lower the activation energy of a given process. No shift of an optimum of the Arrhenius plot has been observed in individual adaptation, though the slope of the curve, i.e. the activation energy, was changed.

Three groups of desert shrubs *(Atriplex polycarpa)* were kept for 7–8 days at day-time temperatures of 16, 32, and 43° C, and at night-time temperatures of 5, 21, and 32° C, respectively. Following this exposure the respiration intensity of the leaves was measured at temperatures ranging from 5 to 40° C. The lower the adaptation temperature, the higher was the respiration intensity (Chatterton et al., 1970). Since respiration intensity increased with elevation of temperature up to 40° C, the position of an optimum has not been determined. It also remains unknown whether a shift of an optimum actually occurred.

The cases of the modificational adaptations to temperature, in which no positional shifts of optima and maxima could be seen, are fairly well consistent with the data on the conservatism of the level of cellular and protein thermostability that was discussed in Sections 1.2.1.3, 1.2.1.4, and 3.2.1.2. A number of observations can be found in the literature indicating that a temperature acclimation may also be accompanied by lateral transpositions of the temperature-dependence curves of enzyme activity. Thus, Mooney and West (1964) grew plants belonging to five species at three experimental stations located at altitudes of 1402, 2408 and 3094 m above sea level. In two of the five species the temperature optimum for photosynthesis was a function of the climatic conditions for growth, a maximal difference between the positions of optima being as big as 5° C. These two species (*Artemisia tridentata* and *Chamalbatiara millefolium*) differed from the other plants in that an area of their distribution was wider. The temperature-dependence plot of the Hill reaction in the chloroplasts isolated from *Deschampsia caespitosa* grown at 10° C is shifted several degrees Centigrade toward lower temperatures, unlike the curves for the plants grown at 20° C (Tieszen and Helgager, 1969). In some grasses grown at 15° C, the temperature optimum for photosynthesis is shifted on the average by 2.8° C toward low temperatures as compared with the plants cultivated at 25° C (Charles-Edwards et al., 1971). The temperature optimum for respiration of isolated ide muscles is 1° C higher for the organisms adapted to 22° C, as compared with that for respiration of the muscles from the fish acclimated to 15° C (Berkholz, 1966).

At present it is difficult to estimate the degree of controversy between the data of shifts of temperature optima for metabolic processes observed during acclimation of organisms, and those findings which indicate a strong conservatism in the level of thermostability of proteins and cells, that has been found to exist in animals and plants during fluctuations of the environmental temperature within the limits of the tolerant zone.

Frequently, the same functions in different tissues exhibit dissimilar behavior during acclimation of poikilothermal animals to different temperatures. Take, for instance, the goldfish adapted to 10° C and that kept at 30° C. In the former, the oxygen uptake by gill homogenates proved to be higher at all temperatures over the range of 10–30° C. The respiration intensity of brain and liver tissue homogenates, on the other hand, was practically the same in both the "warm" and "cold" fish (Ekberg, 1958). Respiration intensity of the muscle tissues of the crab *Pachygrapens crassipes* after a 4–6 week acclimation to 8.5, 16, and 23.5° C was found to be compensatorily enhanced in the animals adapted to a lower temperature. Conversely, the respiration intensity of the brain tissue showed no difference related to the acclimation temperature (Roberts, 1957). A large number of similar observations have been gathered by Precht and his colleagues (Precht, 1968).

A interesting and important illustration pertinent to the concept developed here can be found in a work of Chatfield et al. (1953). The body temperature of the silver seagull, which spends most of the year in northern seas, is 38.8° C. The temperature of the metatarsal part of its leg may be 6.1° C, and that of the tibial part, 33.3° C. Nerves in the tibial part fail to transmit excitation at 11.7° C, and those in the metatarsal part only at 2.8° C. What is important is that this difference in the temperature minimum for conductivity is matched to a lower temperature maximum for conductivity in the metatarsal portion of a peripheral nerve. There is no such difference in the hen, and in the seagull this difference fades away with prolonged exposure of the bird to warmth. These observations may be supplemented with those reported by Miller (1965), who has shown that peripheral nerves (tail, tibial) of some Alaskan mammals are still effective in mediating nervous stimuli during overcooling to −3.8–5.1° C. As a rule, mammalian nerves lose their ability to conduct stimuli at +5–15° C.

Summary. The evidence presented in the three preceding sections demonstrates that, in the process of adaptation of species and individuals to new temperature environments, various properties of enzymes may undergo modification. Among these properties are: activation energy, the enzyme-substrate affinity, and temperature dependence of the enzyme activity.

The accumulated findings concerning the first characteristic are extremely diverse. The majority of relevant observations, though, testify that a decrease in activation energy occurs in adaptation to a cooler environment, i.e., there are indications that a catalyst is more efficient at lowered temperatures. There are, however, some data revealing that an opposite regularity can also be seen. In addition, there are instances when this parameter undergoes no changes whatsoever. In each of these cases, the behavior of activation energy is likely to bear an adaptive significance. It is difficult, however, to formulate a general statement concerning the role given to activation energy or Q_{10} in the adaptation of organisms to the environmental temperature. Moreover, there are no data which might serve to relate changes in the activation energy of enzymes to the changes in their conformation or their conformational flexibility.

As for the changes of the enzyme-substrate affinities and their contribution to the genotypic and modificational temperature adaptations, it may be concluded, based on the works of Hochachka and Somero, as well as of other numerous workers, that alterations of the affinity are indeed important in the case of poikilothermal animals. In a number of instances the enzyme was shown to exhibit its maximal affinity to the substrate at temperatures very close to a temperature minimum for life. With elevation of the temperature in the environment, a decrease in the enzyme–substrate affinity counteracts the acceleration of a reaction, that acceleration being evoked by an enhancement of the kinetic energy of participating molecules. That is a mechanism of instantaneous action and it provides for a higher or lower degree of independence of the reaction rate upon the temperature in the zone of a temperature optimum. This mechanism operates only when a substrate concentration in the cell is below the saturating one. Adaptation to a new temperature environment is accompanied by a modification of the temperature dependence of the enzyme-substrate affinity. The temperature at which a maximal affinity value is reported is shifted to the right or left along the temperature scale, depending on the direction of the environmental temperature shift. In some organisms, this is achieved by virtue of the ability of an enzyme to exist in two forms, warm and cold depending on the environmental temperature. Difficulties arise when attempts are made to find out how widely nature employs this mechanism of adaptation. It is unknown if this mechanism operates in plants.

It would be important, in terms of the hypothesis developed in this monograph, to elucidate how the conformational flexibility and thermostability of enzymes are modified during transitions of enzymes from their warm form into the cold one. It would be interesting to know how a decrease in the enzyme-substrate affinity, provoked by a deviation of the temperature from the zone corresponding to a maximal affinity, and a shift of this zone relative to the temperature scale that occurs during temperature adaptations, are related to changes of the level of conformational stability of enzymes.

Frequently, in order to explain adaptation of organisms to variable ambient temperature, a reference is made to evidence concerning changes in isozyme composition. Pertinent findings are controversial and often negative. Somero (1975b) believes that more often than not the eurythermic, poikilothermic organisms show almost no difference if compared with the stenothermic ones in terms of greater polymorphism of their iso- and alloenzymes, although this difference might have been expected in case of substitution of certain enzymic forms for other forms as a prominent factor in temperature adaptations.

Investigations undertaken dealing with a modification of the temperature dependence of enzyme activity during adaptation to new temperature conditions are especially numerous. The observations accumulated leave no doubt as to the occurrence of lateral shifts of curves depicting this regularity manifested in numerous cases of genotypic temperature adaptations. In the case of species and intraspecific grouping which differ in their thermophilies, the temperature-dependence plots of the activity of homologous enzymes vary in the positions of minima, optima and maxima in reference to the temperature scale. Such mechanism of temperature adaptation of enzymes is fully consistent with the concept here developed; that is, if one assumes that active functioning of an enzyme requires that it be in a semilabile state (i.e., in a state of sufficient, but not an excessive conformational flexibility), then an adaptation to elevated temperature should be achieved through a reduction of the conformational flexibility of the enzyme. Such a modification of the enzyme would result in suppression of the enzyme activity at temperatures found in the region between the former optimum and minimum. In fact, many thermophilic enzymes are inactive already at temperatures of 40–60° C—they "feel cold". Just the opposite must be expected to occur during adaptation to cold: an intensification of the activity at low temperatures can be achieved at the expense of reducing the activity in the zone of temperatures between the former optimum and maximum.

Consequently, one of the points of application of the selection, which is responsible for maintaining a correspondence between the level of conformational flexibility and the environmental temperature conditions, might be visualized in distortion in the new environment of a favorable dependence of enzyme functioning upon temperature. There are findings indicating that a lateral shift of the temperature-dependence plot of enzyme activity can occur also in the case of an individual temperature adaptation.

The three facets discussed above certainly do not cover all the feasible mechanisms operating at the molecular level during adaptation of enzymes to temperature. Nature demonstrates an inexhaustible creativity in the process of its adaptation to the ambient world, and many of the contrivances it uses remain as yet entirely unknown. Adaptations to unfavorable environment and, in particular, to fluctuations of temperature, can be attained not only through a modification of the enzymes' properties, but also at the expense of an increase in the amount of enzymes and by adjusting appropriately the operating conditions in the cell. There are publications which indicate that in the course of temperature adaptations, a switch over to new metabolic pathways may occur (Hochachka and

Hayes, 1962; Irlina, 1963b; Christophersen, 1967; Prosser and Braun, 1962; Hochachka, 1968a; Hazel and Prosser, 1974; and others). This problem, however, is outside the scope of the present book.

5.4 Lifetime of Proteins in the Cell

The discussion in this section will focus on another possible site of application of natural selection that may be important in maintaining a correlation between the level of conformational flexibility of proteins and the temperature conditions of life. This point of application is not related to the performance of biochemical functions by proteins, but it is directly relevant to the mechanisms, which control the protein content in cells.

The enzyme functioning is known to be regulated in different ways. One of the most effective means for exerting regulatory functions is certainly a control of the enzyme content in the cell. Atkinson (1965) showed that maintenance of the cell's homeostasis is best achieved in the case of simultaneous regulation of both the catalytic activity of the enzyme and the synthesis of the enzyme. At the same time, an effective regulation of the enzyme content in the cell, by either acceleration or inhibition of its synthesis, would have been unrealistic had the enzyme stayed in the cell indefinitely. Under such circumstances, an acceleration or reduction of the enzyme synthesis in a non-dividing cell might result only in a more or less significant accumulation of the enzyme in the cell, whereas a complete inhibition of its synthesis would only serve to maintain the enzyme content at a constant level. It would be practically impossible to reduce the enzyme content in this situation. Obviously, such an imperfect mode of control could have no chance to be fixed in the process of evolution.

In fact, the investigations carried out on tissue cells (especially detailed studies have been reported in rat liver cells) demonstrated an extremely short life span to be characteristic of many enzymes. The lifetime values varied from scores of days to an hour (Table 20). Hence it follows that the concentration of an enzyme in the cell is a function of an equilibrium between its synthesis and degradation (Schimke and Doyle, 1970). In this case, the control of the enzyme-synthesis rate can be effective not only in maintaining or enhancing the enzyme content in a non-dividing cell, but also in reducing this content down to zero.

In dividing cells, for instance in a population of growing bacteria, each division halves the amount of every protein in the cell. Therefore, in spite of the ongoing protein degradation during reproduction of the cells, the amount of proteins in them is determined by the ratio of the protein synthesis versus division frequency. This equilibrium per se is already effective in regulating both accumulation and reduction of the protein content in the cell. It might be mentioned that there was a period of time when the process of protein degradation was not acknowledged as occurring in bacterial cells. Later, Mandelstam (1958) showed that protein degradation in growing *Escherichia coli* cells was negligible. A cessation of growth, however, reveals that protein turnover does exist with the rates of synthesis and degradation being approximately 4–5% per hour. That author provided evidence to confirm that a genuine intracellular protein degradation

Table 20. Rate of degradation of some proteins

Protein	Object	Halflife	Reference
Mitochondrial water-insoluble proteins	Rat brain	31.4 days	Beattie et al. (1967)
Mitochondrial water-soluble proteins	Rat brain	17.9 days	Beattie et al. (1967)
Mitochondrial cytochrome *c*	Rat liver	8.6 days	Beattie et al. (1967)
Structural proteins of mitochondria	Rat liver	8.8 8.8 days	Beattie et al. (1967)
Total protein	Rat liver	1.8–3.8 days	Swick (1958)
	Rat liver	3–4 days	Steinberg et al. (1956)
Arginase	Rat liver	4–5 days	Schimke (1964)
Microsomal proteins	Rat liver	3.5 days	Shuster and Jick (1966)
Total protein	Rat liver	4–5 days	Schimke (1966a)
Lactate dehydrogenase	Rat liver	3.8 days	Kuehl and Sumsion (1970)
Aldolase	Rat liver	2.8 days	Kuehl and Sumsion (1970)
Cytochrome-*c*-reductase NADP-H_2	Mouse liver	2.8 days	Jick and Shuster (1966)
Catalase	Rat liver	1 day	Price et al. (1962)
Cytochrome b_5	Rat liver	117–123 hrs	Omura et al. (1967)
Cytochrome-*c*-reductase NADP-H_2	Rat liver	75–83 hrs	Omura et al. (1967)
Tryptophan pyrrolase	Rat liver	2–2.5 hrs	Schimke et al. (1965a)
Tryptophan oxidase	Rat liver	2.3 hrs	Feigelson et al. (1959)
Tyrosine-glutamate-transaminase	Rat liver	2 hrs	Schimke (1967)
Tyrosine-α-ketoglutarate-transaminase	Rat liver	1.5 hrs	Kenney (1967)
Glutamate-alanine-transaminase	Rat liver	84 min	Schimke (1967)
δ-aminolevulinic acid synthetase	Rat liver	67–72 min	Tschudy et al. (1965)

occurred, but not lysis of the cells. It has already been proved that also occurred in bacteria there is a protein fraction (less than 10% of the total protein content) which decays at a fast rate in both growing and non-dividing starving cells. The half-lives of these proteins are around 60 min. With termination of growth provoked by starvation, the amount of decaying protein increases almost by one order of magnitude (Nath and Koch, 1971). A review of the data on protein turnover in microorganisms was made by Pine (1972) and Goldberg and Dice (1974).

Transfer of the slime mold *P. polycephalum* from a complete nutrient medium onto a starvation salts medium also results in enhancement of the protein degradation (Sauer et al., 1970). Thus, apparently, an intracellular protein breakdown is an intrinsic property of all types of cells. This phenomenon, unfortunately, has been poorly studied in poikilothermal animals and plants. The extent of the shortage of relevant information in plants can be seen by the fact that Filner et al. (1969), in their review paper dealing with enzyme induction in higher plants, in the section concerning "Control of enzyme degradation", discuss evidence that was almost exclusively obtained in animal cell enzymes. Some information on degradation of plant proteins can be found in papers by Glasziou et al. (1966), Trewavas (1972a, 1972b), and Boudet et al. (1975).

Since the protein content in the cell is a result of a steady state, i.e., it is determined by the balance of its synthesis and degradation, it is clear that the control of the enzyme content in the cell may be achieved through the effects of metabolites not only on the rate of the enzyme synthesis, but also on its degradation. And in fact, recent years have witnessed the appearance of an ever-increasing number of investigations providing supporting evidence. Among these, a mention should be made primarily of the extremely important contributions made by Schimke and his colleagues (Schimke, 1964; 1966a, 1966b, 1967, 1969, 1973; Schimke et al., 1965a, 1965b, 1968). This group of biochemists has established the following facts, among others. The tryptophan pyrrolase activity in rat liver can be elevated by hydrocortisone or by injections of tryptophan. The mechanisms of action of these agents are, however, drastically different. The stimulating action of the hormone is due to the acceleration of tryptophan pyrrolase synthesis. It can be blocked by actinomycin D. An enhancement of this enzyme activity by the substrate, tryptophan, is not precluded by actinomycin administration, since tryptophan elevates the enzyme content by suppressing degradation of the enzyme, rather than by stimulation of its synthesis (see also Greengard et al., 1963). A difference in the modes of regulation of tryptophan pyrrolase by hydrocortisone and tryptophan is shown also by an observation that tryptophan alone increases the enzyme content in the cell five-fold, and the hormone alone, eight-fold, whereas simultaneous administration of both compounds results in elevation of the enzyme content 25–50 fold. The mode of action of tryptophan testifies to the fact that a metabolite, by modifying the rate of enzyme degradation, can efficiently influence the enzyme function. Additionally, Schimke proved the importance of the changes in degradation rates of arginase for the adaptation of this enzyme to shifts in dietary patterns. Thus, a shift from a protein-rich to a low-protein diet results in a slow down of the rate of arginase synthesis, while degradation of the enzyme increases. Thereby, the arginase content in the cell is reduced.

Bojarski and Hiatt (1960) noticed an activation of thymidylate kinase in the rat liver and kidney following injection of thymidylate or thymidine. The authors believe the activation of the enzyme is, at least in part, related to the stabilizing action of the substrates on thymidylate kinase.

In some cases, metabolites affect both synthesis and degradation of proteins simultaneously. The work of Drysdale and Munro (1966) concerning the effect of iron on ferritin content in rat liver cells is of particular interest in this context. Ferritin molecules represent the molecular depot of iron. They bind excess iron and release it in case of shortage. An administration of ferric ammonium citrate elevates the ferritin content in liver cells. Iron was shown to stimulate ferritin synthesis, inhibiting, at the same time, degradation of this metalloprotein.

Phenobarbital prolongs the half life of a ribosomal protein in rat liver cells from 3.5 to 7 days. Concurrently, protein synthesis is stimulated thus leading to accumulation of this ribosomal protein (Schuster and Jick, 1966). In mice liver, phenobarbital completely inhibits degradation of the microsomal NADP-H_2-cytochrome-c-reductase. The half life of this enzyme is normally 2.8 days. Besides, phenobarbital stimulates synthesis of this enzyme that significantly increases its content in the cell (Jick and Shuster, 1966). α-tyrosine-α-ketoglutarate transaminase is characterized by a rapid turnover in rat liver, having a half-life time around 3 h (Kenney, 1962). However, no notable effects of inhibitors of RNA and protein synthesis were observed, such as actinomycin D, puromycin, cycloheximide on this enzyme content (Kenney, 1967). Analyzing this paradoxical phenomenon in experiments on ^{14}C-leucine incorporation, the author found that the inhibitors used in his previous studies inhibited not only the synthesis of α-tyrosine-α-ketoglutarate transaminase, but coincidently also its degradation. Similar findings have been obtained by Sussman (1965, cited in Kenney, 1967) in *Dictyostelium discoideum:* cycloheximide inhibited both synthesis and degradation of uridine-phosphate-polysaccharide galactosyl transferase.

The investigations concerning the effects of naturally occurring metabolites as well as of artificial agents on the rate of protein degradation have been gaining in scope only in recent years. The progress reported in relevant studies justifies the view expressed by Schimke (1966a, 1969; Schimke and Doyle, 1970), that the control of enzyme functioning, exerted through changes in the rate of degradation, is no less important than control of the enzyme synthesis. This holds true primarily for the enzymes having short half-lives (Schimke, 1967).

Apart from the "routine" maintenance of protein content at a certain level under balanced conditions of synthesis and degradation, proteolysis that operates on a wider scale may be a tool for disassembling the cellular structures when need arises. The latter appears during cell differentiation, mitotic divisions, and also when the cell dies. The mechanisms that degrade proteins obviously function also as "quality control operators", associated with "the protein-synthesizing machinery". Thus, the normal repressor of the *lac* gene in *Escherichia coli* is stable during growth. A mutation has been found in which this repressor becomes defective—it fails to form tetramers—in contrast to the wild-type repressor. In the mutant the repressor decays at a fast rate with a half life of around 20 min (Platt et al., 1970). Similar findings have been reported for another *E. coli* mutant, in which the defective β-galactosidase is rapidly degraded by intracellular endopeptidases, whereas the normal enzyme is stable (Goldschmidt, 1970). The same fate befalls anomalous hypoxanthine-guanine phosphoribosyltransferase in four mutant clones of mouse L cells grown in tissue culture: the defective enzymes are degraded 3 to 17.5 times as fast as those in the wild type cells (Capecchi et al., 1974).

The above observations indicate that intracellular protein breakdown is a vitally important process in the economics of the cell. The degradative rate for each protein is regulated according to the role given to the protein in cellular metabolism. Deviation of this parameter from the norm inevitably leads to unfavorable consequences, and that must be controlled by natural selection.

What factors control the rate of protein turnover? In order to obtain insight into this problem, one should turn to the information available concerning the mechanism of protein degradation. So far the situation has been rather obscure, though certain facts emerge that can be regarded as substantiated. Among the latter, it should be noted that an agreement exists that a steady state of an enzyme content is accounted for by the specific relationships established in a normally functioning cell between the rates of protein synthesis and degradation. It has been shown that degradation of a protein cannot be, as a rule, ascribed to excretion of that protein from the cell (Price et al., 1962; Schimke, 1964). Contrary to Hogness et al. (1955), a reduction of protein content in tissue cannot be related to autolysis occurring during natural death of cells. Thus, in liver cells the half lives of many enzymes amount to hours or several days, while the liver cells themselves live considerably longer—160–400 days (Buchanon, 1961). Consequently, it should be considered as proven that the rate of protein turnover in the cell is determined by the rate of its degradation inside the cells. This is indubitably an enzyme-dependent process realized through the action of intracellular proteinases—cathepsins (for reviews see Schimke, 1966a, 1966b; Bohley, 1968; Schimke and Doyle, 1970; Goldberg and Dice, 1974). In one and the same cell the period of half-life for one group of proteins does not exceed 1–2 h, while for other proteins it may be extended to scores of days (Table 20). This fact leads us to suggest that the rate of degradation of a given protein is determined by some properties of the protein as an object of proteolytic attacks. An alternative assumption, namely that each protein is destroyed by its specific proteinase, is unrealistic. Such wastefulness by the cell would have been suppressed by natural selection.

The role played by the lysosomes in the process of protein degradation in a normally functioning cell is unclear. Some writers tend to acknowledge the participation of these only in cases of extensive proteolysis (Schimke and Doyle, 1970), when entire organelles or parts of the cytoplasm are being lyzed. In this context, it is important to know whether degradation of proteins occurs in nuclei, since lysosomes have not been found in nuclei. There is no unanimous agreement on this point. At any rate, degradation of certain histone fractions has been detected in the Chinese hamster cells grown in tissue culture, and, for instance, the half-life of f1 histone fraction was found to be 74 h (Gurley and Hardin, 1969). In addition, certain proteinases have been found in liver and thymus nuclei which are effective in protein breakdown at neutral pH (Brostrom and Jeffay, 1972). Dice and Schimke (1973) found relatively fast degradation of acidic proteins and slow degradation of histones occurs in rat liver cell nuclei.

Protein degradation and its regulation are undoubtedly very complex processes. There are a number of findings indicating that enzyme degradation depends on normal functioning of protein synthesis and energy metabolism in the cell. Several investigators reported a slow down of degradation of various enzymes brought about by inhibitors of RNA and protein syntheses—Actinomycin D, puromycin, cycloheximide, o-fluorophenylalanine, chloramphenicol (Steinberg and Vaughan, 1956; Mandelstam, 1958; Schimke, 1966b, 1967; Grossman and Mavrides, 1966, 1967; Kenney, 1967; Nath and Koch, 1971; Pine, 1972). An inhibition of protein degradation has been reported to occur during anaerobiosis and during the action of inhibitors of energy metabolism such as dinitrophenol, NaCN, sodium azide (Steinberg and Vaughan, 1956; Mandelstam, 1958; Goldberg, 1972a).

Naturally, protein degradation is a temperature-dependent process. In a starving culture of thermophilic bacteria (unidentified strain), the rate of protein breakdown was at its peak at 45–55° C, being markedly lower at 35 and 75° C. In *Escherichia coli*, a maximal rate of degradation was observed at 35° C, going lower at 45° C, whereas at 55, 65, and 75° C protein breakdown was practically absent. The rates of degradation at the temperatures optimum for the thermophiles and mesophiles were, however, identical (Epstein and Grossowicz, 1969). By the way, these data contradict the dynamic concept of thermophily advocated by Allen (see p. 117), which states that protein breakdown in the thermophiles at optimum temperatures for these organisms must go at a considerably higher rate than in the mesophiles at respective optimum temperatures.

It must be admitted that the mechanism of intracellular degradation of proteins remains so far largely unknown. A degradation rate must depend upon several variables: amount and activity of intracellular proteases, their distribution within the cell, the probability of their encounter with the proteins to be degraded, and in addition, on the so-far unknown co-factors. At the same time, however, the rate of degradation of a protein should, to a considerable degree, be a function of the resistance of the protein itself to the action of intracellular proteinases (Schimke and Doyle, 1970; Somero and Doyle, 1973). There are some facts supporting this view. Bond (1971) studied the resistance of five enzymes isolated from rat liver to the action of four proteases in in vitro assays. These enzymes varied appreciably in their half lives: lactate dehydrogenase—3.5–16 days, arginase—4–5 days, catalase—1.1–2.2 days, serine dehydrogenase—3–20 h, and tyrosine aminotransferase—1.5–11.7 h. With respect to their resistance to tryptic digestion and α-chymotrypsin attack, these enzymes could be arranged in the same order as that for progressively diminished halflives. Bond concluded that susceptibility to proteolysis is an important property of a protein that determines its life span. Dice and Schimke (1973) found a correlation between relative rates of degradation of different nuclear acidic proteins and histones in rat liver in vivo and the resistance of these in vitro to pronase and tryptic digestion.

Goldberg (1972a, 1972b) showed that various treatments promoting catabolism of *E. coli* protein in vivo, enhance susceptibility of the proteins to digestion by trypsin, chymotrypsin, subtilisin and pronase in vitro. Among such modifying influences are: incorporation of amino acid analogues into proteins; prematurely terminated synthesis of polypeptide chains resulting from puromycin inhibition of translation; and errors in protein synthesis produced by mutations. Analyzing this evidence, Goldberg also deduces that variations in half lives of normal proteins are determined by differences in the tertiary structure of the protein, which, in their turn, affect sensitivity of a particular protein to proteases. At this point the important data obtained earlier by Schimke and his collaborators should be mentioned. As already said, they showed that administration of tryptophan to rats reduces the rate of degradation of tryptophan pyrrolase in liver cells. In addition, they found that tryptophan in vitro increases the resistance of tryptophan pyrrolase to trypsin, chymotrypsin and bacterial proteinases. Moreover, tryptophan stabilizes the enzyme to the denaturing action of heat, alcohol and urea (Schimke et al., 1965a; Schimke, 1966a). Likewise, Drysdale and Munro (1966) found ferritin breakdown to be suppressed by administration of iron to rats. This was accompanied also by an enhancement of the resistance of ferritin to tryptic digestion.

Interesting findings have been reported by Segal et al. (1974). These investigators injected ^{3}H-leucine and ^{14}C leucine into rats and the experimental design was such that it allowed for the proteins with longer half lives to be labeled by tritium, while short-lived proteins were labeled with ^{14}C by the moment the animals were sacrificed. Soluble proteins were then isolated from liver and subjected to proteolytic digestion with lysosomal proteinases. It was found that the proteins labeled with tritium, i.e., long-lived proteins, were digested at a lower rate than were those displaying shorter halflives. In their review paper dealing with intracellular protein degradation, Goldberg and Dice (1974) conclude that differences in the rates of degradation of various proteins correlate well with the differences in the resistance of the proteins to intracellular proteolytic systems, although not all aspects of the intracellular proteolysis may be accounted for by this phenomenon.

When comparing the half lives of proteins in the living cell with the resistance of the proteins to proteolytic digestion in vitro, one should take into account the fact that the resistance of a protein to digestion, like its thermostability, may differ by a wide margin if assayed in situ and in solution. The accessibility of ribosomal proteins to a proteolytic attack has been shown to be essentially different in the ribosomes and when in a free state. Free proteins isolated from the ribosomes were much more sensitive, and upon isolation various proteins enhanced their sensitivity in different degrees (Ostner and Hultin, 1968). Therefore, more emphasis should be placed on those data which reveal a parallelism between the protein half lives in vivo and their resistance to proteinases assayed in vitro rather than on negative results.

All these data taken together unambiguously indicate that the most important parameter determining the rate of protein turnover in the cell is the resistance of a protein to proteases. What factors are responsible for this resistance? First of all, it is necessary to find out which protein molecules become prey to proteinases—native proteins are far more resistant to hydrolytic digestion than denatured ones. According to Linderstrøm-Lang (Linderstrøm-Lang et al., 1938; Linderstrøm-Lang, 1950; Linderstrøm-Lang and Schellman, 1959), proteolytic enzymes attack only denatured proteins, so that an equilibrium between the native and denatured protein is shifted to the right. Sometimes a protease can catalyze transition of a native protein into a denatured state (Korsgaard, 1949). According to this point of view, the activity of intracellular proteinases is largely limited to the destruction of "corpses" of protein molecules.

There is no firmly grounded evidence available at present that might justify the acceptance of the hypothesis advanced by Linderstrøm-Lang. This author himself referred to some examples that were not consistent with the scheme he had proposed. Thus he believed that chymotrypsin, for instance, acts on β-lactoglobulin that had not been denatured (Linderstrøm-Lang and Jacobsen, 1941). Collagenase isolated from tadpole tissues splits a collagen molecule in two: a smaller fragment one third of the original molecule's size, and a larger one—the other two thirds of the molecule. Such fragments still fully retain the conformation of the original tropocollagen molecule (Kang et al., 1966; Sakai and Gross, 1967). Collagenase derived from a tissue culture of human skin is certainly capable of attacking native collagen molecules (Eisen et al., 1968). As a result of tryptic digestion, light chains of rat immunoglobulins yield halves of the immunoglobulins retaining all the antigenic properties characteristic of intact light chains. Consequently, they maintain their native spatial structure (Vengerova et al., 1972). This reveals that a tryptic attack does not lead to denaturation of protein molecules in this particular case. There is no doubt that denatured protein molecules are split by proteases at a considerably faster rate than are native proteins.

If one assumes that at a normal temperature for an organism's life a spontaneous thermal denaturation of proteins occurs, then the half lives of these proteins must be correlated with the rate of denaturation (Goldberg, 1972a). [It should be remembered that the assumption that the proteins in the cell are being continuously and spontaneously denatured and renativated served as a stimulus for starting our research in cytoecology (p. 131)]. There are two investigations known to be aimed at verifying this probability. Segal et al. (1969) estimated the thermal half-life of alanine aminotransferase from the rat liver and muscles. They measured the rate of thermal denaturation of the enzyme at different temperatures and the activation energy of this process. The data obtained were used for calculating the thermal half-life of the enzyme at rat body temperature (37.5° C), that turned out to be 400 days for the enzyme isolated from both tissues. At the same time, the half-life of the enzyme in the muscles was 20 days and in the liver—3–2.5 days, i.e., 1–2 orders of magnitude shorter. That is why the authors reject an explanation of the difference in the rate of the enzyme degradation in muscles and liver as being due to a difference in the rates of their thermal denaturation.

Another publication (Kuehl and Sumsion, 1970) offers a comparison of thermal half-lives of the rat liver enzymes at 37° C, which exhibit similar half-lives in the cells: lactate dehydrogenase—3.8 days and aldolase—2.8 days. In spite of this similarity in turnover times, these enzymes displayed markedly different thermal half-lives at body temperatures of the animal. In this work too, the half-life values have been calculated by extrapolating the Arrhenius plots of the temperature dependence of the denaturation rate measured in vitro. Thermal half-lives were found to be 12000 days for lactate dehydrogenase and 130 days for aldolase.

A somewhat different view is held by Bohley (1968), who believes that native protein molecules bind with their hydrophobic sites to the surface of lysosomal membranes, where they are denatured. Accordingly, half-lives of various proteins are dependent on the size of the hydrophobic regions on the surface of protein molecules and on the rate of denaturation.

The findings described above indicate that activity of intracellular proteases is not limited to "sanitation" functions. It may be concluded that the main objects of proteolytic attacks are native protein molecules in the cell. The rate of degradation of a particular protein must depend on the activity of the proteolytic system and on the resistance of the protein to the hydrolytic action of intracellular proteases. This resistance is a function of accessibility of the peptide bonds being hydrolyzed.

The intimate relationships between the proteolytic resistance of proteins and the conformational lability of their molecules were discussed (see Sect. 4.3.2). A decrease in the conformational lability invariably lowers the rate of protein digestion. An enhancement of the conformational mobility accelerates an enzymatic hydrolysis of the protein. Consequently, a change of conformational flexibility of protein macromolecules must affect the rate of protein breakdown and the latter, in its turn, will affect the control of the protein content in the cell. Interfering with this regulation may alter the normal functioning of the cell. It follows then, that the conformational flexibility of protein molecules, being related to the mechanism of regulation of the protein content in the cell, must be under the control of natural selection. Since conformational flexibility is a temperature-dependent variable, it must be brought into correspondence with the environmental temperature when the latter is changed. All of these considerations must hold true in the possible case, of proteolysis of proteins in the cell being preceded by denaturation of the protein, because the rate of the protein denaturation also depends on the conformational rigidity of its molecules. In this regard, special note should be taken of the fact mentioned earlier, namely, that the rate of proteolytic digestion of proteins in thermophilic bacteria at a temperature optimum for their life, is roughly equal to the rate of degradation of the proteins in mesophilic bacteria at their respective optimum, which is found several scores of degrees lower than that for the thermophiles (Epstein and Grossowicz, 1969).

It may be deduced from the above that in nature a correlation between the environmental temperature during an active period of an organism's life and the resistance of its proteins to proteolytic digestion should exist. Relevant evidence, unfortunately, is strikingly scarce, but such as is available (p. 167–168) testifies in favor of the existence of this correspondence.

Summary. The experimental findings discussed in Sections 5.3 and 5.4 reveal at least two channels along which natural selection (and perhaps also other mechanisms responsible for the purposiveness of nature) can affect proteins by maintaining a correspondence between the conformational flexibility of their macromolecules and the temperature conditions of a species' life. One of these channels is assigned to the temperature dependence of the functions which a given protein performs in the cellular metabolism. This channel serves also for careful preservation of the correspondence between the environmental temperature and the position of the temperature optimum for the protein functioning. The other channel accommodates the correlation between ambient temperature and the rate of protein breakdown in the cell. The regulation of the protein content in the cell depends on this parameter. Maintenance of a correspondence of both parameters to the temperature of environment requires the existence of a correlation between the latter and the conformational flexibility of proteins. Thus, the conformational flexibility is to be found under the vigilant control of natural selection.

There is absolutely no reason to believe that the temperature dependence of various functions and longevity of proteins in the cell are the only factors that account for the existence of a correspondence between the level of conformational flexibility of proteins and the temperature of environment. The functions realized by proteins and their regulation alike are extremely complex and multifaceted phenomena and they remain as yet largely unknown. Further investigations will show the additional points of application which can be used by natural selection for exerting control on the degree of conformational lability of proteins and for correlating it with ambient temperature.

It is highly likely that an effective regulation of enzyme activity, by changing the affinity to substrates, co-factors and effectors under given temperature conditions, also necessitates that a certain level of conformational flexibility of protein macromolecules be attained.

Chapter 6
Plausible Mechanisms of Regulation of the Level of Conformational Flexibility of Proteins

The vast amount of experimental evidence discussed in the first three chapters indicates that adaptation of organisms to changes in environmental temperature is usually accompanied by appropriate changes in the primary thermoresistance of cells and in the heat stability of proteins. Furthermore, arguments were offered in favor of the view that appropriate shifts in the level of conformational flexibility of protein macromolecules underly these changes and that the main adaptive significance is seen in the maintenance of a correlation between the temperature conditions of an organism's life and the level of conformational flexibility of its macromolecules. A correlation between the variations in environmental temperature and the modifications of the cellular and protein thermostability is a secondary phenomenon; this correlation per se may have no adaptive significance.

In this connection, an important biochemical problem arises, namely, what mechanisms are responsible for the realization of adaptive changes in the level of conformational flexibility of protein macromolecules during modificational and genotypic adaptations to temperature. Theoretically, one may suspect the following mechanisms to be the most likely candidates: changes in the primary structure of protein molecules that might occur as a result of a selection of appropriate mutations; a modification of the meanings of the codewords, elicited by a shift of the temperature beyond the zone of optimum; a substitution of one isozyme for another; binding of the ligands to proteins and release of the ligands; changes produced in the intracellular milieu. Other mechanisms may certainly exist that do not come to mind at this moment. In some cases, individual modifications of the rigidity of certain proteins probably occur; in other cases, one can imagine entire shifts of conformational lability, encompassing the entire protein moiety in the cell. Later the possibilities likely to occur in these or those forms of temperature adaptations of organisms will be discussed. A note of caution should be given, however, that the data needed to arrive at reliable conclusions are certainly far from being sufficient and a great deal of work is yet to be done in order to achieve a better understanding of this problem.

6.1 Modificational Changes of the Conformational Flexibility of Protein Macromolecules

This section will consider primarily the plausible causes of stabilization of proteins in heat hardening of plants. The information concerning mechanisms responsible for other forms of modificational changes of protein lability in plants and animals is so far strikingly poor and a detailed discussion of these mechanisms is hardly justified at the moment.

6.1.1 Heat Hardening of Plant Cells

6.1.1.1 Principle Features of the Mechanism Responsible for Heat Hardening

A search for the biochemical mechanism underlying the protein stabilization observed in heat hardening should be conducted to include the relevant data accumulated in cytophysiology. As shown in studies on heat hardening the stabilization extends apparently over the entire protein moiety in the cell. Of utmost importance is the fact that hardening can be obtained with exposures as short as several seconds (p. 40). This finding refutes any attempt to explain the effect of hardening by replacement of the preexisting proteins by those synthesized de novo with novel properties. Consequently, one is left with an alternative to seek evidence for a mechanism that enhances the stability of the pre-existing proteins. Judging from the fact that an elevated thermostability of proteins isolated from heat-hardened plant cells can be detected after dialysis and purification (p. 103), the assumption that stabilization is due to a change in the composition of the protoplasmic milieu (changes in pH, ionic strength, dielectric constant, etc.), must be discarded. There remains a guess that the biochemical basis for the state of heat hardiness is the appearance in the cell of a certain ligand, or ligands, which, having bound to a protein, reduces the conformational flexibility of that protein and thereby elevates the stability of the protein against heat as well as against some other denaturing agents. The speed with which the state of hardiness appears may be explained by the pre-existence of a ligand present in the cell in an inactive state, which is released in response to a hardening treatment.

Such ideas can be easily accommodated within the framework of the evidence accumulated in cytophysiological studies directed toward elucidation of whether hardening of cellular proteins is a "personal affair" of every individual protein, or whether there exists a unifying mechanism that is brought into action by hardening and which is responsible for a simultaneous enhancement of the rigidity of all the protein molecules, or of the major portion of these. The meaning of relevant experiments is as follows. Given one and the same length of hardening procedure, in order to obtain a maximal effect in different plants which differ in their thermophilies, one has to carry out hardening at a temperature that must be raised to correspond with a greater initial thermostability of cells (see Sect. 1.2.2.1c). This fact may be explained in two ways. An unknown, unique mechanism of hardening may be suspected to exist, which is set in motion at a certain dose of heat treatment, and this dose is expected to be higher, the more thermostable the cell is. The triggering of the hardening mechanism leads to an increase in the stability

Table 21. Thermoresistance of heat-stable and heat-labile functions in *Tradescantia fluminensis* leaf cells and the heat treatments required for producing heat hardening of these cells. (From Barabalchuk, 1969)

Cellular functions	Minimal temperature of a 5-min heating effective in complete suppression of a function (° C)	Temperature of a 3 hr heat hardening effective in elevation of thermoresistance equal to $^1/_2$ of the maximal hardening effect (° C)
Protoplasmic streaming	43.7	33
Phototaxis of chloroplasts	44.2	34
Selective permeability:		
capacity for plasmolysis	59.4	35
leakage of antocyan	60.9	34
Respiration	65.0	33

of many, if not all, cellular proteins. An alternative explanation is to assume that hardening is a reaction to a supraoptimal temperature characteristic of every individual protein, and that for this response to appear in a more thermostable protein, a more intensive heating is to be used. The latter supposition is not completely unsubstantiated, since there are indications in the literature on the ability of proteins to acquire stable modifications in response to the action of denaturing agents and, among others, to heat (Tsyperovich, 1954, 1956; Tsyperovich and Loseva, 1956; Privalov and Monaselidze, 1963; O'Brien et al., 1973; Brand and Andersson, 1976). The data obtained by O'Brien et al. may be explained otherwise; namely, that the protein preparations these workers have dealt with contained fractions exhibiting different thermostability.

The experiments carried out by Barabalchuk (1969) in the authors laboratory gave grounds to assess which explanation is more valid. In the cell there are functions which are known to be rather thermolabile (protoplasmic streaming, chloroplast phototaxis, photosynthesis) and other functions exhibiting higher thermostability (respiration, maintenance of selective permeability). Thus, for instance, a complete cessation of protoplasmic streaming and inhibition of chloroplast phototaxis in the *Tradescantia* cells heated for 5 min, occur at 44° C, and complete inhibition of photosynthesis at 48° C, while oxygen uptake is suppressed by half only at 65° C and a loss of the selective permeability occurs at 60° C. These functions are obviously realized by protein systems that differ appreciably in their thermostability. The next step in this research was to find out whether an elevation of heat resistance of these functions occurs at the same or at different hardening temperatures. If every protein responds to a hardening heat exposure by independently elevating its resistance, then it might be expected that for elevation of the heat stability of the more thermostable functions (such as respiration, selective permeability) higher hardening temperatures would be required as compared with the thermolabile functions (protoplasmic streaming, chloroplast phototaxis). As may be recognized from Table 21, the heat stability of both the thermolabile and thermostable functions starts to increase at comparable hardening temperatures. The findings obtained by Barabalchuk have been corroborated by those of Shcherbakova et al. (1973). These authors showed that a maximal increase of the heat stability of the functions and protein systems, which appreciably differ in their original thermostability, occurs during heat hardening at essentially similar doses of hardening heat treatments.

These observations better agree with the first version that presumes the existence of a single mechanism of hardening that can simultaneously increase the resistance of various cellular proteins, the thermolabile and thermostable alike. In view of the data obtained, the second version—visualizing an independent response of every cellular protein to heating—seems hardly plausible.

As pointed out above, the stabilization of proteins in heat hardening is most likely to be the result of the appearance in the cell of a stabilizer, an antidenaturing agent. The substances which might act as likely candidates for realization of these functions and the particular substances which might be responsible for eliciting the hardening response will now be studied.

6.1.1.2 Action of Anti-Denaturants on Proteins and Cells

A decrease of conformational flexibility of protein molecules and, as a consequence, an enhancement of their resistance to the denaturants and proteolytic enzymes can be produced by numerous substances of widely different nature. Among them, one should discern those substances which specifically stabilize only certain proteins and, on the other hand, those compounds which are characteristic of a wide spectrum of stabilizing actions, affecting various proteins in a similar fashion. Substrates, co-enzymes and allosteric effectors belong to the first group. They stabilize only that enzyme with which they interact specifically. It is impossible to explain the effect of heat hardening as due to interactions of this kind, because, as we have seen, hardening embraces simultaneously the whole bulk of the cellular proteins. This type of stabilization will remain outside the scope of this discussion.

The second group of stabilizers comprises both organic and inorganic substances, electrolytes and non-electrolytes. Former investigators made no distinction between the antidenaturing and anti-coagulative actions. For our purposes, only true anti-denaturants are of interest, and among the latter only those likely to be present in a normally functioning cell. A search for such agents should be based on the knowledge of whether the substances effective in stabilizing the proteins in vitro can also elevate the resistance of living cells to heat. If a given stabilizer is capable of protecting proteins from the denaturing action of a given agent, and, in addition, can increase the resistance of a living cell to this agent, then these properties taken together may serve as proof for the assumption that a given agent damages the cell by denaturing its proteins.

Among the first anti-denaturing agents to be discovered were the sugars. It was found that, if taken in high concentrations, sugars considerably elevate the temperature of denaturation of various proteins and also that temperature which leads to death of microorganisms (Beilinsson, 1929; Zilber, 1936; Kunitz and McDonald, 1946; Berger et al., 1946; Putnam, 1953; and others). Anti-denaturing properties of various sugars differ. In the experiments of Santarius (1973) with isolated spinach chloroplast membranes, raffinose was the most effective sugar to protect membranes from heat and cold denaturation, sucrose proved to be less potent, whereas glucose was found to be the last in the sequence. Santarius believes that accumulation of sugars in plant cells in wintertime is responsible for the simultaneous enhancement of the resistance of plant tissues to both cold and heat. Sugars exert their protective influence also during thermal injury of plant cells. As shown by Feldman (1962), the heat stability of *Tradescantia* leaf epidermal cells increases after a preliminary treatment of leaves with hypo- or isotonic sucrose, glucose or lactose solutions. Similar protective action of sucrose was found in four other plant species belonging to different families. Gorban (1968) reported an enhancement of the heat stability of wheat coleoptile cells upon immersion of these into 1% sucrose solution. Zavadskaya (1964) made an attempt to elucidate whether sugars are the natural stabilizers of proteins during heat hardening. In *Dactylis glomerata* and *Campanula persicifolia* leaves an increment of the sucrose content in the former plant after 10 sec hardening at 50° C was observed, while in the second one both sucrose and fructose content increased. However, following 18-h heat hardenings at 37.5° C of *D. glomerata*, *C. persicifolia*, *Hepatica nobilis* and *Leucanthemum vulgaris* leaves, the sugar content proved to be appreciably lower than in the control leaves, in spite of the fact than an elevation of thermostability, produced by the second version of hardening (18 h at 37.5° C), was the same as in the first one. This finding casts a shadow of doubt on the concept that the appearance of sugars in the cell is responsible for the stabilization of proteins during heat hardening.

Glycerol is also a stabilizer of both proteins (Beilinsson, 1929) and cells. Arronet (1964) found that in glycerol the resistance of the frog palate ciliary epithelium to heat and high hydrostatic pressure increased. Glycerol models of these cells also displayed higher resistance to both agents if the models were kept in glycerol during exposure. To my knowledge, there is no evidence concerning stabilization of plant cells by glycerol or any indications that glycerol accumulates during hardening of plant cells. Furthermore, data of this kind are so far unknown with respect to salts of fatty acids, which are thought to act as anti-denaturing agents during treatment of protein solutions with urea, guanidine and heat (Boyer, 1945; Rice et al., 1945; Boyer et al., 1946a, 1946b).

There are observations showing that certain amino acids protect various proteins from the denaturing action of heat, ultrasound and urea (Gordon, 1953; Tsyperovich and Loseva, 1960; Dietrich, 1962; Candlish and Tristram, 1964). Studies, in which amino acid stabilize those enzymes for which these amino acids serve as substrates, are not included in the report cited (see, for instance Chuang et al., 1967). The author (Alexandrov and Tabidze, 1972) studied the effects of some DL-amino acids on the heat stability of *Tradescantia* leaf epidermal cells, as well as on the heat resistance of the *R. temporaria* palate ciliary epithelium and of the glycerol models of ciliary epithelium. It was found that glycine in concentrations of 0.03 and 0.06 M, as well as 0.1–0.05 M alanine enhance thermostability of all three study objects mentioned. Serine in concentrations of 0.025–0.12 M had almost no effect, and above 1.2 M even lowered the resistance of the cells to heat. Tryptophan reduced thermostability of the cells in all the concentrations studied. This experiment indicates that an increase in the content of certain amino acids in the cell can enhance stability of the protoplasmic proteins. Protective effects of some amino acids were observed by Rudenok and Konev (1973) during heat treatment of yeasts. The contribution of a change in amino acid concentration to the development of heat hardening effect still remains to be seen.

There are numerous observation on the stabilizing effect of calcium and, to some extent, of magnesium ions on proteins. Certain proteins can be stabilized by manganese ions (Shapiro and Ginsburg, 1968). It was shown in the laboratory of Okunuki (Hagihara et al., 1956a, 1956b; Okunuki, 1961), that Ca^{2+} protects the *B. subtilis* α-amylase and taka-α-amylase from *Aspergillus aryzae* against denaturation by heat, urea and acetic acid as well as against digestion of the enzymes by a protease. The reports on the stabilizing action of calcium ions were confirmed later in a number of studies on amylase preparations isolated from other organisms (Vallee et al., 1959; Hatfaludi et al., 1966; Imanishi, 1966; Karpukhina and Sosfenov, 1967; Loginova and Karpukhina, 1968; Pfueller and Eliott, 1969; Isono, 1970; Sosfenov, 1972), as well as on proteases from various sources: on a protease from *Streptomyces griseus* (Nomoto et al., 1960), on a neutral protease from the mesophilic and thermophilic strains of *B. stearothermophilus* (Ohta, 1967; Sidler and Zuber, 1972), on thermolysin from *B. thermoproteolyticus* (Feder et al., 1971), on trypsin (Sipos and Merkel, 1968), and on *Micrococcus* nuclease in the presence of serum albumin and nucleotides (Sulkowski and Laskowski, 1968). Serum albumin is stabilized by calcium and magnesium ions against thermal denaturation and tryptic digestion (Gorini and Andrian, 1956). Hsiu et al. (1964) believe that Ca^{2+} makes α-amylase and some other enzymes rigid, by forming bonds inside the molecules, which resemble, in terms of function, the disulfide bridges. Conversely, according to Hippel and Schleich (1969) Ca^{2+} labilizes collagen and ribonuclease. On the other hand, Spichtin and Verzör (1969) regard Ca^{2+} as a stabilizer of collagen macromolecules.

The stabilizing action of calcium ions on proteins is confirmed by the numerous findings concerning the protective effects of Ca^{2+} observed during thermal treatments of microorganisms, animal and plant cells (Port, 1927; Scheibmair, 1938; Schlieper and Kowalski, 1956a, 1956b; Amaha and Ordal, 1957; Bandas and Bobovich, 1961; Grigoryan, 1964; Alderton et al., 1964; Ljunger, 1970, 1973; and others). In greater detail the effects of ions on the resistance of plant cells to various injurious agents have been studied by Barabalchuk (1970b). As for the effect of Ca^{2+} on the *Tradescantia* leaf epidermal cells, the following facts have been established. Thermostability of the cells increased by 1.8° C, if tested in 0.03–0.01 M $CaCl_2$, i.e. an increase in heat stability comparable with that observed in heat hardening. In 0.03 M solutions of $CaCl_2$, this effect was accompanied by a slowdown of the protoplasmic streaming and suppression of phototaxis of chloroplasts. The effects of the $CaCl_2$ stabilization and those of heat hardening are not additive either in the case $CaCl_2$ is given prior to heat hardening or when the leaves are placed in $CaCl_2$ after the hardening. $CaCl_2$ enhances the resistance of the cells not only to heat, but also to high hydrostatic pressure and acetic acid. However, no increase in the resistance to alcohol and ammonium was noted. The similarity in the effects of heat hardening and $CaCl_2$ on cells is a central feature of these experiments. Thermostability of *Tradescantia* cells could also be increased by $MnCl_2$, and, to a lesser extent, by $MgCl_2$. KCl only slightly stabilized the cells at concentrations of 0.06 M, whereas more concentrated solutions reduced the heat stability of the cells. In the literature can be found non-systematic observations concerning an increase in the resistance to heat produced by various ions; the available evidence, however, proves that calcium ions display the most pronounced and universal capability for stabilizing the proteins and cells.

6.1.1.3 Suppositions About the Biochemical Mechanism Responsible for Heat Hardening of Cells

The speed with which the heat-hardening state can be obtained rules out the possibility of the effect of hardening being due to replacement of cellular proteins by more stable proteins synthesized de novo. The suggestion that the effect of heat hardening is related to a selection of isozymes, expressed by Pandey (1972) is also hardly plausible. Another mechanism for explaining a fast heat hardening achieved in a matter of seconds or minutes, has been proposed by Levitt (1969). He believes that elevated temperatures modify the protein conformation in such a way that some hydrogen bonds break down, while hydrophobic interactions are established between those groups that had previously been separated. An increase in the contribution of hydrophobic interactions to the maintenance of the macromolecule's conformation should result in higher thermostability of this macromolecule, since the force of hydrophobic interactions increases as the temperature is elevated. It will be shown later (p. 228) that an attempt made by Levitt to explain thermophily as based on a higher contribution of the hydrophobic interactions is at variance with the data available at present. It is moreover difficult to imagine that such extensive conformational perturbations may be compatible with maintenance of the functional activity of proteins.

It seems that the fast and universal stabilization of cellular proteins in heat hardening is more realistically related to a release of unknown ligands which enhance the rigidity of the protein macromolecules. Suitable candidates for these roles were discussed in the preceding section. Among these, calcium ions appear to be a plausible agent because they stabilize many proteins of widely different structures as well as various cells of extremely diverse nature. The protective effect of Ca^{2+} on the cells exhibits a number of features resembling those characteristic of the effect of heat hardening. Still another argument in favor of this candidate is the ability of certain cells to retain calcium in a bound state, and this has been proven beyond doubt. Moreover, such cells can release Ca^{2+} in a free form in a fraction of a second at an appropriate moment, and later bind it again just as fast. These are the cells which are endowed with the ability to contract. Ca^{2+} is required for activation of the interactions between myosin, actin and ATP. In resting skeletal muscle fibers, some calcium is bound by the terminal elements of the sarcoplasmic reticulum. An excitation wave, running along a nerve, transmits an impulse via the myoneural synapse to the muscle fiber. The impulse then goes along the fiber, enters it and, inside the fiber, reaches the terminal branches of the sarcoplasmic reticulum. At that moment, in response to a stimulus, calcium is immediately released from its bound state into a free form, and passing through the reticulum membrane, the calcium ions reach myofibrils where they interact with the proteins of the contractile system. A contraction of the fiber ensues. The

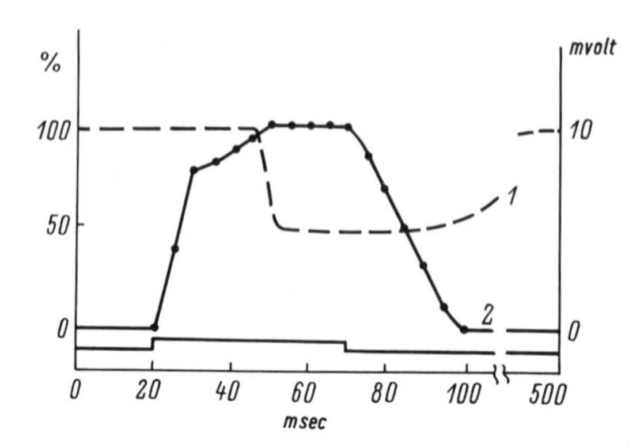

Fig. 66. Time course and magnitude of photon emission by equorine and contraction of the infusorian *Spirostomum ambiguum* in reference to initiation of electrical stimulation (Ettienne, 1970). *Abscissa:* time; *ordinate left:* length of infusorian body (% of initial value), *ordinate right:* intensity of light emission. *1:* changes in length of the infusorian body, *2:* intensity of light emission. Impulse lasted from 20th to 70th msec

contraction act being concluded, calcium is released from the contractile system and once again is taken up by the sarcoplasmic reticulum (Bianchi, 1968; Ebashi et al., 1969; Bendall, 1969).

Changes in the concentrations of calcium ions in the cytoplasm also regulate the contractile functions in non-muscle cells. A good illustration is a contraction of the body of the infusorian *Spirostomum ambiguum.* Either spontaneously or in response to an electric stimulus, the infusorian body contracts down to half its original size. This contraction is produced by the action of a network of fibrils located in the cortical protoplasmic layer. With this network are connected the vesicles accumulating Ca^{2+}. An electric impulse releases Ca^{2+} into the cytoplasm and provokes contraction of the fibrils. This has been shown by Ettienne (1970), who injected the protein equorine that fluoresces in the presence of calcium ions. Recording contractions of the cells and changes in the fluorescence intensity as a function of time, Ettienne established that fluorescence begins at the moment a maximal concentration of Ca^{2+} is observed (Fig. 66).

A reversible release of calcium from a bound state also accompanies some other types of cellular activity. It is proven that bioluminescence in various organisms is provoked by the appearance of calcium ions, which act as a trigger (Johnson and Shimomura, 1968; Morin and Hastings, 1971). The same initiating role is played by Ca^{2+} in the release of mediators by the nerve endings during a transmission of a stimulus to the next link (Radouco-Thomas, 1971).

It is appropriate to remember, in this context, the concept which Heilbrunn (1956) has been advocating over a period of many years concerning participation of calcium in the processes of injury and stimulation of cells. Heilbrunn believed that the cortical layers of the cells contain calcium bound by a protein. During the action of various physical and chemical agents on the cells, calcium is released in a free form, thus being able to initiate various events. Depending upon the circumstances, these events may either be of destructive nature or may result in various forms of cellular activity. The concept of Heilbrunn is outdated in many respects, but the idea of the significant role played by the transition of calcium from a bound state into a free form and vice versa in the cell life has proved to be valid. Corroborating evidence is apparently being accumulated concerning the idea of Heilbrunn on the retention of calcium in a bound state by a protein. MacLennan and his co-workers (MacLennan and Wong, 1971; MacLennan et al., 1973) isolated a specific protein from the rabbit sarcoplasmic reticulum membranes. This protein, called by the authors "sequestrin" contains 37% of dicarbonic amino acids and is capable of binding large amounts of Ca^{2+} (up to 970 nm per mg of the protein). This protein is obviously the site of storage of the bound calcium that is released from the protein at the moment of muscle stimulation.

These findings make more realistic the assumption that in plant cells too calcium exists in a bound form, probably associated with a protein that releases calcium in response to the action of supraoptimal temperatures and certain other agents.

It goes without saying that at this time we are unable to decide what biochemical processes are responsible for the stabilization of cellular proteins in heat hardening. The present discussion of the plausible mechanisms of hardening should be regarded only as an approach to understanding the strategy that should presumably be taken in studies relevant to this problem. It should be emphasized that this is a mechanism of major biological significance.

The experimental observations referred to earlier (p. 158) suggest that in heat hardening a mechanism is set in motion which regulates the level of conformational flexibility of cellular proteins, not only in response to changes in temperature. If the response had been limited only to an adaptive reaction to overheating, this mechanism could hardly be retained in the Arctic plants, and, in particular, in the algae from Franz-Joseph Land. In favor of the view that the mechanism of reactive changes of protein rigidity also controls this rigidity in the case of non-temperature shifts, which affect conformation of the protein molecules, one can say that a rapid reactive increase in the primary thermostability of the cells, similar to heat hardening, can be elicited by other actions: wound injury and high hydrostatic pressure.

6.1.2 Other Types of Modificational Changes of Conformational Flexibility of Protein Molecules

Along with rapid reactive changes in the rigidity of protein molecules of the heat-hardening type, in more slowly developing changes occur in animal and plant cells which can be assessed by changes of the primary thermostability of cells. These changes may be related to growth and developmental processes, to changes of active and dormant stages, to seasonal and diurnal cycles, or to reactions to variations in environmental factors (temperature, water content, salinity, etc.). When seeking an explanation of the slow changes of conformational flexibility of proteins, one should take into account not only the possibility of an increase or reduction of the concentrations of stabilizing agents in the cell, but also the probability of changes in the isozyme spectra of proteins in the process of protein synthesis. Isozymes are known to differ in their thermostability, while isozyme patterns are rather labile. A pattern may be changed during ontogenesis of animals and plants (Yakovleva, 1968; O'Sullivan and Wedding, 1972), with temperature fluctuations in the environment (De Yong et al., 1968; Somero and Hochachka, 1971; Hochachka and Clayton-Hochachka, 1973; Lagerspetz et al., 1973), and under other conditions. A transition of a given enzyme from one form into another, without substitution for another enzyme, has been reported during temperature acclimations (Fox and Burnett, 1962; Jacobson, 1968; Somero, 1969a; Hebb et al., 1972).

When modificational changes of the enzyme stability are explained as due to changes in isozyme patterns, it should be remembered that most frequently isozymes are not interchangeable because they often differ not only in the resistance of their macromolecules, but also in a number of other parameters, such as localization within the cell, relation to the co-enzymes, allosteric effectors, etc. It is difficult to visualize isozyme pattern modifications being responsible for changes in the stability involving the whole of protein moiety in the cell. In such cases, it is more reasonable to look for the cause of changes in variations of ligand concentrations which affect stability of protein molecules. Descriptions of cases can be cited, in which binding of a single ligand molecule suffices for an alteration of the conformational stability of a protein macromolecule to occur. If the primary thermostability of cells is increased following an appropriate treatment and the proteins isolated thereafter show no increase in their thermostability (see Sect. 3.2), it would only be natural to suspect that the cause of changes in the heat stability of cells lies in the ligands, which are lost on isolation of these proteins from cells. If, even after purification procedures, increased thermostability of a protein persists, a conclusion concerning non-involvement of ligands must be drawn with caution. Binding of a ligand to a protein may be rather firm: in that case, this association may withstand isolation procedures and subsequent purification steps.

The most striking examples of stabilization of cells, aside from desiccation of mosses, lichens and other organisms, are seen in the formation of bacterial spores. Thermostability of these may be scores of degrees higher than the heat resistance of vegetative forms. The stabilization of the spore proteins cannot be ascribed only to a transition of water from a free state into a bound one. Thus, for instance, Stewart and Halvorson (1954) showed that alanine racemase isolated from spores is appreciably more resistant to heating in a buffered solution than is the enzyme derived from vegetative forms of the same bacterium. It has been the aim of many investigations to find out which factors can be responsible for the high thermostability of spores (for a review see Loginova et al., 1966). Several workers believe that this high thermostability is related to an accumulation of dipicolinic acid as well as to the ratio of its concentration to that of divalent cations, primarily Ca^{2+} (Church and Halvorson, 1959; Black et al., 1960; Walker et al., 1961; Levinson et al., 1961; Mishiro and Ochi, 1966; Ochi, 1967). Kalakutsky et al. (1969) demonstrated a direct correlation between the heat stability of spores of various actinomycete species belonging to the *Actinobida* genus and the concentration of dipicolinic acid, Ca^{2+} and, to a lesser extent, Mg^{2+}. Another relevant finding: a mutant of *Bacillus cereus* has been obtained, the spores of which lacked dipicolinic acid. The spores of this mutant proved to be significantly more sensitive to heat than were those of the wild type. At 80° C they lost vitality in 1–2 min, whereas normal spores were not in the least affected by a 30-min exposure to this temperature. An addition of dipicolinic acid to a medium during sporulation of the mutant resulted in formation of spores resistant to 80° C heating (Wise et al., 1967). In the spores of some bacteria *(Clostridium rostrum)* no relationship could be detected between the dipicolinic acid content and thermostability of the spores (Byrne et al., 1960).

Another example is an elevation of the heat resistance of insect tissues occurring when the insects enter the state of a diapause. It has been shown in a number of studies that this is accompanied by an extensive accumulation of polyols, mostly glycerol, in the tissues (for reviews see Astaurov et al., 1962; Ilyinskaya, 1966). This serves primarily to increase the frost resistance of an organism and, at the same time, thermostability of the tissues is also elevated.

Winter dormancy of plants is very frequently accompanied not only by an increase in cellular resistance to cold, but also by an elevation of their primary thermoresistance (see Sect. 1.2.2.3), which fact indicates an increase in the stability of protoplasmic proteins. The biological significance of this phenomenon may be that an alteration of the higher order structures of protein macromolecules can be produced not only by elevated temperatures, but on exposure to low temperatures as well. Besides, in the state of a relative metabolic rest, it is apparently more thermodynamically profitable to be composed of more rigid proteins which degrade at a lower rate and which display reduced biochemical activity. These considerations are valid also in other forms of rest of plants, animals and microorganisms. Cold hardening of plants may be accompanied by an accumulation of various metabolites, which, along with protecting against cold, can also stabilize the proteins against thermal denaturation. Among these are sugars and their derivatives, amino acids, protective proteins, etc. In order to elucidate what particular metabolites are responsible for stabilization of proteins, it would be interesting to evaluate the difference, if any, between the composition of cells in cold-hardened plants, which exhibit an accompanying enhancement of heat resistance, and those plant cells whose heat stability has not been modified by cold hardening.

Little can be said about the biochemical bases of modification of protein stability during acceleration and retardation of growth processes in plants. As for the seasonal variations in the stability of animal cell proteins, the initial stages apparently involve endocrine shifts. No information, however, has been reported so far concerning intracellular mechanisms of this process.

The available evidence makes us believe that the modificational regulation of the level of conformational flexibility is hardly realized through a single mechanism. In various cases known, the stabilization of the protein occurs against a background of essentially dissimilar metabolic shifts, such as an accumulation of polyols in the diapause of insects, an appearance of dipicolinic acid in the sporulation of bacteria, an elevation of sugar concentrations during cold hardening of a number of plants, a reduction in the soluble sugar concentration during prolonged (several hours) heat hardening, etc. The diversity of mechanisms responsible for modificational stabilization of proteins is manifested also in those experiments in which various types of hardening procedures have been combined.

Shukhtina (1964) showed that after a one-time hardening of orchard grass and sedge leaves, produced under optimum conditions, a subsequent heat hardening never leads to further elevation of the primary thermostability of the cells. In winter, during a period of cold hardening, cellular heat stability in these plants is considerably higher than in summer. If, however, leaves of winter orchard grass are subjected to heat hardening, the level of their thermostability shows an increase identical to that observed with the heat hardening of the summer plants. The same results have been obtained by Shukhtina and Shcherbakova (unpublished data) in winter wheat. These findings create an impression that an unknown mechanism, and a specific one at that, operates in the cells during cold hardening, but not in hardening by heat. At the same time, the mechanism responsible for cold hardening does not preclude the mechanism of heat hardening from reacting to the action of an adequate heating. In those cells whose the heat stability has been elevated by dehydration, the heat hardening is again effective in yielding an additional increase in their thermoresistance (Zavadskaya and Shukhtina, 1971).

Summary. While performing their biochemical functions, the cellular enzymes interact with substrates, co-factors and allosteric effectors. Numerous investigations have shown that this is accompanied, as a rule, by changes in the conformational flexibility of proteins either toward their labilization or stabilization. Such interactions are usually strikingly specific. The experimental material discussed in this book proves that apart from separate changes of conformational flexibility of individual proteins, there may also occur in the cell total shifts of flexibility encompassing the whole cellular domain. A decrease in conformational flexibility, produced by heat hardening, high hydrostatic pressure, centrifugation and wound injury of plant cells, may develop within seconds or minutes. A slower stabilization which takes hours or days to develop, occurs during temperature adjustments, in cold hardening, water deficiency in plants, transition into a dormant state of bacteria (spores) and animals (diapauses in insects), in retardation of plant growth, and during seasonal cycles in animals. An increase in the conformational flexibility of protein macromolecules, as estimated by a diminished primary thermostability of cells and thermostability of proteins, is achieved within hours or days: it develops following elimination of reactive reductions of flexibility, after termination of rest stages, and at appropriate phases of seasonal cycles. Our experiments with heavy water (p. 159) show that an elevation of the protein conformational flexibility occurs also in adaptation of cells to a stabilizing agent.

Changes of the level of conformational flexibility of proteins and related changes in the protein stability as well as in the primary thermostability of cells, may be realized through variations of concentrations of the substances and ligands, which are capable of stabilizing protein macromolecules in the cell. In the case of slowly developing modifications, a change of isozyme pattern of proteins as well as post-translational chemical modifications of proteins may be suspected as responsible for altering the rigidity of protein macromolecules.

6.2 Genotypic Changes of the Conformational Flexibility of Protein Molecules

6.2.1 Some Information Concerning the Genetics of Thermostability

The evidence presented in Chapters 2 and 3 proves that a difference in the heat stability of the cells and proteins in species which differ in their thermophilies is determined genetically. Frequently, constant genotypic differences in thermostability can also be revealed between intraspecific groupings (subspecies, races, populations, clones). A large number of temperature-sensitive mutants has been obtained and studied in animals (Fatt and Dougherty, 1963; Suzuki et al., 1967; Fenstein et al., 1971), plants (Langridge, 1963a, 1965), fungi (Horowitz, 1956; Horowitz and Fling, 1956; Fincham, 1959, 1960; Horowitz et al., 1961; Hartwell and McLaughlin, 1968a, 1968b), microorganisms (Maas and Davis, 1952; Aharonowitz and Ron, 1972; Ingraham and Neuhard, 1972; and many others), viruses and phages (Krylov and Zapadnaya, 1965; Jockusch, 1966, 1968). The above listed are only a few of a large number of investigations on temperature-sensitive mutations, which are widely used in treating various aspects of molecular biology and genetics. Especially numerous are the studies performed on microorganisms. More often than not they deal with the thermolabile mutants (an increased sensitivity to high or low temperature); however, thermostable mutants have also been obtained, in particular, in psychrophilic (Malcolm, 1969a; Tai and Jackson, 1969) and mesophilic bacteria (DeCicco and Noon, 1973), and in tobacco mosaic virus (Aach, 1960). A single amino acid substitution may be sufficient to modify heat stability of a protein macromolecule (Henning and Yanofsky, 1962; Jockusch, 1966, 1968) and to alter its spatial structure (Greer, 1971a, 1971b). In this context very impressive findings have been made in human carbonic anhydrase. Two variants of this enzyme have been detected: (1) CAId Michigan—a considerably more thermostable enzyme than its normal counterpart, the only difference in them being a substitution of lysine in the position 100 for threonine, (2) CAIf London shows markedly lower thermostability than the normal enzyme. This mutant protein also differed from the normal enzyme in only one substitution of lysine in the position 102 for glutamic acid (Tashian and Osborne, 1975).

It is highly probable that most amino acid substitutions modify, to a greater or lesser extent, the rigidity of a protein molecule by either increasing or decreasing it. Igarashi (1966) obtained temperature-sensitive mutants of *Paramecium aurelia*. He showed that the loci responsible for such mutations are scattered over the entire genome of the infusorian.

In a genetically heterogeneous population of organisms and cells, variations of thermostability are determined by both modificational and genotypic differences. Altukhov and Ratkin (1968) showed that 50–60% of the variability of heat resistance of muscles in a heterogeneous population of the silkworm is determined by hereditary differences between individual organisms. The rest of the variability depends on modificational differences and on the inaccuracy of the methods employed. A selection for thermostability of tissue cells has been performed by Amosova (1967) on the meat fly. It was easier to select for low resistance than for high resistance. Ushakov and Chernokozheva (1963) exposed frog spermatozoa to heat treatment, which eliminated more thermolabile cells. Fertilization of the egg cells by the surviving spermatozoa gave rise to a population of tadpoles that contained less organisms characterized by lowered thermostability of their muscle tissue. A modification of thermostability of bacteria can be obtained through selection of the spores (Davis and Williams, 1948).

In order to gain some insight into the mechanism of heredity of cellular thermostability, one should turn to the interesting studies carried out in *Protozoa*. Amoebic clones that differ in their thermostability were selected. By transplanting nuclei from one clone into another, Sopina and Yudin (Sopina, 1968a, 1968b; Yudin and Sopina, 1970) proved that in amoebae, the level of thermostability is determined by the nucleus. No effects of the cytoplasm could be detected in their experiments.

Preer (1957) studied two lines of *Paramecium aurelia* which differed in their thermostability. Cross-breeding experiments revealed that the difference in their heat resistance was determined by one gene, and the dominant allele proved to be that which had been responsible for the elevated thermoresistance.

An important finding has been reported by Osipov (1966a, 1966b, 1966c) in his studies on *Paramecium caudatum*. This species is notable for major variations of thermostability observed in various clones, these differences being genetically fixed. Homozygous clones can be obtained by prematurely dissociating conjugating pairs of this *Protozoan*. The subclones obtained exhibited considerable differences in their heat stability in spite of the fact that they possessed identical genomes. These subclones differed only in that they belonged to different karyonides, i.e., their macronuclei have been formed from different anlage—derivatives of micronuclei. If, however, infusoria lines were obtained belonging to the same karyonide, then no differences in the heat stability of the organisms could be detected. Osipov concluded that "the hereditary mechanism of thermostability of a *Paramecium* clone is related to the pattern of differentiation of the macronucleus during its development and is epigenetic in nature" (1966c, p. 129). It is highly likely that epigenetic variability, as well as modificational and genotypic variability, is very important for adaptation of organisms to the environmental temperature. This aspect of the problem, however, has not been investigated at all.

In Section 1.2.1.1, a description was given of the cases where enhancement of cellular thermostability had been observed, if *Protozoa* were exposed to an elevated cultivation temperature. This elevation persisted throughout many generations, and also after a decrease in the cultivation temperature. It is not an easy task to decide whether these long-lasting modifications are in fact the result of modificational changes or whether they are the consequence of a selection of genomes.

Tai and Jackson (1969) reported an interesting phenomenon. Two mutants capable of reproducing at 30° C were obtained from the obligate psychrophile *Micrococcus cryophilus* with the aid of UV irradiation. The parent strain could not survive 25° C, but if DNA isolated from the mutants was introduced into the parent strain cells, transformed cells appeared capable of reproduction at 30° C. Approximately 1000 transformed cells appeared per 2×10^7 cells. No effect was observed if DNA from the wild-type cells, calf thymus DNA or DNA preparations treated with DNAse were supplied to the cells. What is important is that an increased thermostability of the transformed cells disappeared if the cells were transferred to a minimal medium at 20° C where they grew quite well.

Dissimilar results were reported earlier by Sie et al. (1961a, 1961b). They isolated a factor that they believed to be a nucleoproteid from the thermophile *B. stearothermophilus*. When this factor is added to a nutrient medium, it enables the mesophile *B. sphaericus* to grow at 55° C. One addition of this factor was already sufficient to induce thermophily in the latter organism which was maintained during an infinite number of transfers. A yeast autolysate and liver extract were also effective in endowing the mesophile *B. megatherium*, *B. subtilis* and *B. sphaericus* with the ability to grow at 55° C. For this ability to be preserved, however, these factors had to be continually present in the nutrient media. McDonald and Matney (1963) added a DNA preparation, isolated from a facultative thermophilic strain cultivated at 55° C, to a culture medium supporting growth of the mesophile *B. subtilis* strain (the upper temperature limit for growth is 51° C). This resulted in the appearance of transformed cells (at a rate $1:10^4$), which were able to reproduce at 55° C.

Lindsay and Creaser (1975) added *B. caldolyticus* DNA (optimum growth at 72° C) to a culture of *B. subtilis* (optimum growth at 37° C) that resulted in appearance of transformed cells, capable of growing at 70° C. The transformed cells contained proteins that exhibited an elevated thermostability. L-histidine dehydrogenase from the parent strain of *B. subtilis* was completely inactivated at 70° C, an optimum for activity being around 40° C, whereas at 20° C the enzyme's activity constituted approximately 40% of the maximal activity. This enzyme isolated from the transformed cells had an optimum around 90° C, and at 35° C its activity did not exceed 5% of the maximal activity. These authors believe that the genes, incorporated into the recipient cells, modified the mechanism of translation in the cells in such a way that a synthesis of the proteins with higher thermoresistance was initiated.

An analysis of temperature-sensitive mutations made in some of the investigations leaves no doubt as to the possibility of modifying by mutation the response to temperature of only a single protein. Therefore, only one function can be made exceedingly thermolabile in temperature-sensitive mutants. Moewus (1940), for instance, has shown that the temperature limits for life and copulation of *Chlamydomonas* are inherited independently of each other. A thermolabile mutant of *Salmonella typhimurium* (strain 4a) normally grows at 25° C. At 42° C a component required for late stages of division fails to perform, whereas DNA synthesis and accumulation of net weight retain their normal resistance to temperature (Ahmed and Rowbury, 1971). Loomis (1969) isolated three classes of thermolabile mutants of the slime mold *Dictyostelium discoideum*. There are strains in one class that grow at 22° C, but fail to grow at 27° C; their myxamoebae, however, can aggregate and produce multicellular sorocarp at either temperature. Strains of another class of mutations display a reverse reaction to temperature: they grow equally well at both temperatures, but their myxamoebae form a multicellular sorocarp only at 22° C. And, finally, mutants belonging to the third class fail to reproduce and form sorocarps at 27° C. All these mutations differ from each other.

Four classes of temperature-sensitive mutants have been obtained in the yeast. Transferring these mutants from 23 to 36° C results in alteration of diverse cellular functions (Hartwell and McLaughlin, 1968 b).

The results gained in studies carried out with temperature-sensitive mutants of higher and lower organisms indicate that during evolution a correlation between the level of conformational flexibility and the environmental temperature can be maintained during temperature fluctuations, by selection of such mutations which affect flexibility of individual proteins. This conclusion is supported by the observations discussed on pages 116 and 126, which show that when comparing thermostability of proteins within intraspecific groupings, characterized by different thermostability, one finds that there are proteins displaying adequate differences in their thermostability and also proteins which do not exhibit such differences. If interspecific comparisons are made, a larger number of different proteins show a correspondence between their thermostability and thermophily of a species.

To understand the mechanisms in evolution supporting a correlation between temperature environment and rigidity of protein molecules, it is imperative to find out whether there are mutations that affect flexibility of the protein in a pleiotropic fashion, i.e., mutations which simultaneously affect most, if not all, proteins in an organism. This facet of the problem, however, has not even been approached. Marré (1962), in his discussion of the origins of the differences between protein thermostability in the thermophilic and mesophilic algae, expresses doubts that the difference could have appeared as a result of modification of all the genes at once or over a period of time. Marré believes that thermoresistance of organisms is determined by the appearance of a key factor which stabilizes all the proteins and structures in the protoplasm. In this respect the work of Olsen and Metcalf (1968) is of interest. Irradiating the mesophilic bacterium *Pseudomonas aeruginosa* with UV light, they obtained a large number of psychrophilic mutants in which not only the temperature maximum was lowered, but the cells also acquired an ability for reproduction at 0° C. In addition, the authors succeeded in performing a transduction of psychrophily by using the phage Px4 and the psychrophile *Pseudomonas fluorescens* as a donor. The psychrophilic mutants and the transductants resembled the wild-type psychrophiles of *Pseudomonas* genus in terms of their response to temperature. The authors inferred that psychrophily is maintained by a limited number of genes. The same conclusion follows from the work of Lindsay and Creaser (1975) cited above.

The only known means for transfer of hereditary information (excluding the non-substantial information transfer through education and teaching) is the transfer of the information on the primary structure of proteins. The biological evolution amounts to modification of this information. An introduction of changes in the amino acid composition or in the order of amino acid sequences may, in various ways, result in shifts of the level of the conformational flexibility of a protein macromolecule. A priori one can envisage the following possibilities: (1) an amino acid replacement affects the level of the conformational flexibility of a macromolecule by modifying it in accord with the requirements imposed by a temperature shift in the environment; (2) an amino acid substitution per se either does not affect the conformational lability of a macromolecule or modifies it into an inappropriate direction. Such a substitution, however, modifies the affinity of the protein to a stabilizing or labilizing ligand or to neighboring macromolecules; consequently, the level of conformational flexibility is found to shift in the required direction; (3) a mutation also results in changes of the concentration of those substances in the cell which affect the level of conformational flexibility of protein molecules.

In the case of mutations of the first type, a change produced in the conformational flexibility can be detected in isolated and purified protein preparations by evaluating the changes appearing in the heat stability of the protein. If the second type of mutations is responsible for changes in the rigidity of a protein macromolecule, the situation may be different. Provided the binding of a ligand to a protein is strong enough, a shift in the protein thermostability can also be observed in a highly purified protein preparation. With more loose association, the chances for detecting this shift may be already lost in the process of isolation of the protein from the cell or later on at various stages of purification. Finally, in the third case, a difference in thermostability can be detected during heat treatments of the cells and may be absent in in vitro assays of the proteins, when the latter are found in an artificial medium similar to that for a protein isolated from non-mutant cells. The experimental observations discussed in Section 3.2 and those which follow, indicate that actually all of these possibilities are realized. The first two types of mutation affect the temperature response of individual proteins. If mutations of the third type occur, they can affect most, if not all, proteins in the cell.

Within the scope of the genetics of thermostability one certainly finds the problem of long-lasting modifications, which were discussed above in Section 1.2.1.1 when dealing with thermostability of *Protozoa*. A phenomenon resembling long-lasting modifications has been observed by Amosova (1971), who exposed meat flies to a short-term heat treatment and was able to observe an increased thermostability of muscles in the larvae of next generation.

6.2.2 Mechanisms of Genotypic Changes of the Conformational Flexibility of Protein Molecules

Let us now turn to the evidence concerning the relationships between genotypic differences in thermostability and the primary structure of proteins. The first attempts to discover a correlation between thermophily of a species and thermostability of its proteins and the amino acid composition of these proteins were made on collagen by Gustavson (1953, 1955, 1956) and, independently, by Takahashi and Tanaka (1953). In a number of vertebrates, these authors found a close correlation between thermostability of collagen and the percentage of the imino acids—proline and oxyproline. The latter strengthens the collagen structure by forming H-bonds between the chains of the molecule. This correlation was later confirmed by Fischman and Levy (1964) in fish, and by Rigby (1967a) in collagens from 13 species of warm- and cold-blooded animals (Fig. 67)[6]. In his other publi-

[6] Working with the chick embryo tendon tissue culture, Rosenbloom et al. (1973) added an inhibitor—proline hydroxylase—to growing cells. They demonstrated a direct proportionality between the percentage of hydroxylation of the imino acids and the melting temperature of tropocollagen.

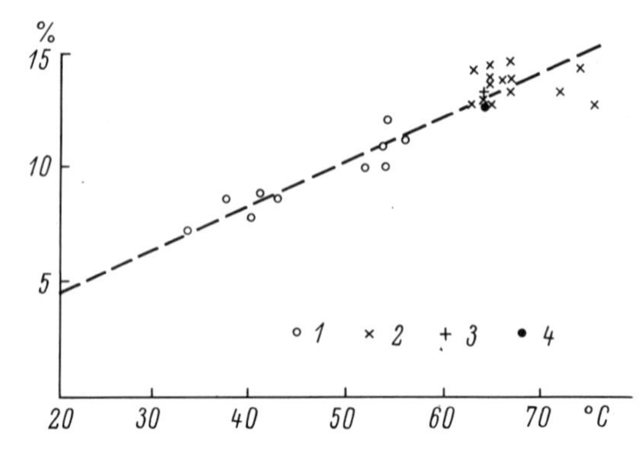

Fig. 67. Dependence of temperature of collagen denaturation (T_S) on its hydroxyproline content. (Rigby and Spikes, 1960). *Abscissa:* T_S; *ordinate:* hydroxyproline content (%). *1:* fish skin, *2:* human skin, *3:* rat tail tendon, *4:* bovine skin

cation, Rigby (1967b) reported that along with a positive correlation between thermostability of collagen in 11 animal species and the total content of proline and oxyproline in this protein, there exists a negative correlation between thermostability and the content of serine. Amounts of other amino acids correlated neither with T_S (the temperature of a half-contraction of collagen fibers), nor with T_D (the temperature of half-denaturation of tropocollagen). Further investigations performed on collagen preparations isolated from three species of the cold-loving Antarctic fish destroyed this initially perfect concept. It was found that both T_S and T_D are 10° C lower in the Antarctic fish than in the cod; however, with respect to the total amount of proline and oxyproline, the collagen from the Antarctic fish shows no difference when compared with the cod collagen (Rigby, 1968a). In addition, no difference could be found in the serine content. Rigby believed the sequence of the amino acids was the most significant. No correlation was found between the total amount of proline and oxyproline and the heat stability of the collagen preparations derived from eight worm species (Rigby, 1968b).

From the studies reported by Rigby it becomes clear that not all genotypic variations of thermostability of collagens can be reduced to differences in the imino acid and serine content. The findings obtained by Rigby and Mason (1967) in collagen from *Gastropoda* (p. 111) and by Andreyeva (1971) on collagens isolated from cod varieties (p. 114) show that the differences in the heat stability of the collagen fibers, being ecologically fully justified, may be determined not by the collagen primary structure itself, but by differences in the stabilization of collagen by unknown ligands. Rigby and Mason and Andreyeva alike reported differences in T_S, but not in T_D. This, however, does not mean that the difference in T_S is not genetically determined. Equally, these data do not exclude that in certain groups of animals, the level of thermostability of collagen is controlled by the imino acid and serine content. Based on collagen studies, it should be concluded that nature employs various means for adjusting the level of the conformational flexibility of protein macromolecules to the temperature conditions of a species' life.

Aside from collagen, various other proteins have been studied with the aim of gaining further insight into the mechanisms responsible for changes of the level of thermostability of proteins in evolution. In particular very detailed comparisons have been made on chemical and physical properties of the proteins isolated from thermophilic and mesophilic microorganisms; less information is available concerning psychrophilic proteins and those isolated from higher organisms. Differences in the primary structure of the proteins were the first to be sought. Initial studies along these lines concerned the effects of protein purification on protein heat resistance. Findings which reported isolation of proteins from cells and the effects of isolation on protein thermostability were referred to in Section 3.2. Purification of the already isolated proteins will now be discussed.

The facultative thermophiles *B. coagulans* and *B. stearothermophilus* were cultivated at 35 and 55° C by Campbell (1954, 1955). α-amylase from the bacteria grown at a higher temperature was found to be considerably more resistant to heat than was the enzyme from the cells cultivated at 35° C. Multistep purification procedures affected neither thermostability of the enzyme, nor this difference. Similar data have been obtained by Karpukhina and Sosfenov (1967). Again, the degree of purification did not affect thermostability of the *Neurospora* tyrosinase (Horowitz et al., 1961). Aldolase from *B. stearothermophilus* displayed identical thermostability at various stages of purification (Thompson and Thompson, 1962). ATPase and pyrophosphatase from this organism retained an elevated thermostability equally well when associated with the membrane and in a dissociated state. In contrast, phenylalanyl-tRNA-synthetase became thermolabile when separated from the membrane (Hachimori et al., 1970). That, however, was not the case with leucyl-tRNA-synthetase from *B. stearothermophilus*. In purified and unpurified preparations alike, this enzyme exhibited equally high thermostability (Vanhumbeeck and Lurquin, 1969). According to Vinogradova (1965b), myosin from *Rana ridibunda* is 3° C more thermostable than that from *R. temporaria*. It was shown, however, that this difference grows less with increasing purification of the preparation (Tartakovski et al., 1969, 1973). Later, additional evidence will be presented indicating that high thermostability is preserved in some proteins isolated from the thermophiles even when purification procedures yield a completely homogeneous preparation, as revealed by ultra-centrifugation and polyacrylamide gel electrophoresis.

The data discussed above show, however, that as with collagens, the level of the enzyme stability in different organisms may be determined not only by the primary structure, but also through complexing of the enzymes with other substances that may get lost in the course of purification. It should be remembered that the modification of protein thermostability, brought about by purification procedures, testifies to the fact that the level of protein thermostability depends on complexing of the protein with unknown compounds; the failure of the purification procedures to affect the protein thermostability may be related either to the fact that the rigidity of a protein molecule is predetermined by its primary structure, or to a firm binding of a stabilizing ligand to the protein.

Major efforts have been directed to deciphering the physico-chemical bases of thermophily. This bold experiment set up by nature seems to serve to facilitate our understanding of the mechanisms responsible for variations in the level of conformational flexibility of protein macromolecules during evolution. The dynamic hypothesis of thermophily, which explains the ability to live at high temperatures as being due only to rapid renaturation of the proteins being denatured (p. 118), is in its general form, unacceptable. As has been shown, high thermostability has been unequivocally shown for most proteins of thermophiles (see Sect. 3.2.2.3). Amelunxen and Lins (1968) heated the lysates obtained from *B. stearothermophilus* and the mesophile *B. cereus* and found that a number of thermophilic enzymes display an appreciably higher thermostability. By heating a mixture of lysates of the two bacteria, the authors proved that the difference in resistance is not associated with the presence of protective substances in the thermophilic lysate. Two enzymes, however, pyruvate kinase and glutamate oxalate transaminase, showed identical resistance to heat in both bacteria[7]. The authors admit that the dynamic hypothesis holds true in the case of these enzymes. At the same time, it may be that during preparation of the homogenates, certain factors which had stabilized these proteins in situ, were lost.

What actually endows proteins of the thermophiles with an ability to withstand such a high temperature? In order to acquire a better understanding of this problem, investigations on homologous proteins isolated from thermophilic and mesophilic organisms have been carried out along several lines: by studying the amino acid compositions, the various physico-chemical properties and the response of these proteins to the agents which destroy different linkages, supporting the structure of the protein macromolecules. To date, unfortunately, no substantial evidence is available about comparison of the primary structure and spatial organization of protein macromolecules with different thermostability.

Koffler and his colleagues (Koffler et al., 1957; Mallet and Koffler, 1957; Koffler, 1961) were among the first to investigate the factors responsible for protein thermostability in thermophiles, concentrating their efforts on flagellin—a protein from bacterial flagella. Having studied the response of this protein to various denaturing agents, they ascribed an elevated thermoresistance of the thermophilic flagellin to higher efficiency of hydrogen bonds and the hydrophobic interactions.

Campbell and his associates (Manning and Campbell, 1961; Manning et al., 1961; Campbell and Manning, 1961) studied α-amylase isolated from *B. stearothermophilus*. They crystallized the enzyme and discovered that it was electrophoretically homogeneous. An optimum for the enzyme's activity was within 55–70° C. Ca^{2+} was essential for the enzyme's activity. The resistance of this α-amylase was appreciably higher than was that of the mesophilic enzyme. The thermophilic α-amylase had a molecular weight of 15600, i.e., almost three times as low as α-amylases from the mesophiles. These authors also found significant differences between the amino acid compositions of the enzymes from the thermophiles and mesophiles. The thermophilic enzyme contains 4.5 times as much proline and is rich in glutamic acid, which according to the writers, precludes formation of the α-helix; no tryptophan was found, and S-S bondings were observed. Heat, urea and guanidine failed to affect the left-rotation of the thermo-

[7] It is interesting to note that Nowell et al. (1969), when comparing thermostability of glycolytic enzymes isolated from two thermophilic and one mesophilic *Clostridia* species, detected that ten enzymes from the thermophiles exhibited higher thermostability compared with the respective enzymes from mesophilic organisms, and only pyruvate kinase showed the same thermostability in either case.

philic enzyme to any great extent. Considering all these data, Campbell and his colleagues concluded that the thermophilic α-amylase represents, in contrast to regular proteins, a semi-randomly or randomly folded coil with S-S bonds scattered here and there. In this context, the exceptional resistance of the thermophilic α-amylase to heat and other denaturants could be accounted for by the fact that its catalytic activity is determined by the primary structure of the macromolecule, rather than by the higher-order structures which are supported by fragile bonds.

According to Ohta and his co-workers (Ohta et al., 1966; Ohta, 1967), just the opposite structural features can be detected in a thermostable protease produced by *B. thermoproteolyticus*. Having recrystallized the enzyme, the authors studied it with various physico-chemical techniques, including absorption and fluorescence spectra, an assessment of the degree of ionization of the tyrosine residues by a spectrophotometric titration, the polarization properties, and the resistance to heat, urea, alcohol and detergents. The results obtained were interpreted to indicate that the thermophilic protease is a compact coil with a low helical content, without S-S bonds. Its high stability is due primarily to the hydrophobic interactions as well as to hydrogen linkages. Calcium ions contribute significantly to the stability of the thermophilic protease. Removal of Ca^{2+} with the aid of EDTA inactivates the enzyme and accelerates its thermal inactivation. Bigelow (1967) compared the amino acid composition of phycocyanin from five mesophilic and one thermophilic algae and found that hydrophobicity of the protein from the thermophilic alga is roughly 10% as high.

Although the structural patterns of the thermophilic enzymes, α-amylase and protease, were essentially different, these investigations nevertheless left hope that the specific features of thermophilic proteins can be revealed and, thereby, the basis of their exceptional thermostability can be understood. The idea that in thermophilic proteins the hydrophobic interactions are especially abundant was also very appealing, since the strength of such interactions does not grow less as the temperature is elevated to between 50 and 70° C, but, on the contrary, increases. It might be expected that in thermophilic proteins, the content of hydrophobic amino acid residues (alanine, leucine, isoleucine, valine, methionine, phenylalanine, tyrosine, tryptophan and cysteine) is higher than that in mesophilic proteins.

Further attempts to solve the problem of thermophily, however, brought only disappointment. A report from Pfueller and Elliott (1969) on the results of their studies of the α-amylase isolated from the same strain as that used by Campbell and his collaborators strangely enough had nothing in common with those reported previously for this enzyme. Thus, it was found that the molecular weight of the *B. stearothermophilus* α-amylase, in fact, does not differ from that typical of amylases and is 56000 rather than 15600. No excess proline and glutamic acid could be found, there were no S-S bonds, and tryptophan could not be detected. According to these authors, there were no reasons to regard the conformation of the thermophilic α-amylase as a semi-random one. The stabilization of the enzyme essentially depended on Ca^{2+}, and removal of calcium increased sensitivity of the enzyme to heat as well as to the digestion with papain, trypsin, chymotrypsin and pronase. These data were found in good agreement with those of Ogasahara et al. (1970a, 1970b) also obtained on the *B. stearothermophilus* α-amylase.

Table 22. Degree of hydrophobicity of glyceraldehyde-3-phosphate dehydrogenase, α-amylase and pronase from different organisms. (From Singleton et al., 1969)

Enzyme	Object	Degree of hydrophobicity estimated by three different indicators		
		I	II	III
Glyceraldehyde-3-phosphate dehydrogenase	A thermophilic bacterium	1060	0.33	0.96
	Rabbit muscles	1060	0.33	0.81
α-amylase	Crawfish muscles	1100	0.34	0.89
	A thermophilic bacterium	1210	0.41	0.84
	A mesophilic bacterium	1070	0.34	1.28
Protease	A thermophilic bacterium	1077	0.35	1.28

Note I: An indicator based on the data reported by Tanford concerning the changes in free energy during transfer of the side chains of amino acid residues from an organic environment to water (Bigelow, 1967); II: an indicator of the frequency distribution of apolar regions in a polypeptide chain (Waugh, 1954); III: an indicator of the ratio of the volume occupied by polar amino acid residues to that taken up by apolar residues (Fisher, 1964)

Contrasting results have been communicated by Karpukhina and Sosfenov (1967), Loginova and Karpukhina (1968) and Sosfenov (1972). They studied α-amylase preparations isolated from three bacteria featuring different optima for reproduction: *Bacillus circulans* 186 (65° C), *B. subtilis* 110 (60° C) and *B. subtilis*, Japanese strain (56° C). The molecular weights of these α-amylases were, respectively: 14200, 31000, and 48000. According to the reported findings, the thermophilic enzymes were enriched in glutamic and aspartic acids and were able to bind more calcium ions, which stabilize an α-amylase molecule. Thus, one molecule of the *B. circulans 186* α-amylase can bind 90 calcium ions. Removal of Ca^{2+} reduced the stability of α-amylase.

The attempt made by Koffler and his associates to relate thermophily of bacteria to hydrophobicity of the protein molecules has failed. Matsubara (1967) determined the amino acid composition of the *B. stearothermophilus* protease, earlier studied by Ohta, and found no reason to account for its thermophily to an increased hydrophobicity. He believes Ca^{2+} is an important stabilizing factor. Ljunger (1970, 1973) too visualizes the cause of thermophily in a high content of calcium ions. Singleton et al. (1969) studied highly purified recrystallized preparations of glyceraldehyde-3-phosphate dehydrogenase from *B. stearothermophilus*. This protein displayed an enhanced thermostability as opposed to analogous proteins from mesophiles. Nonetheless, neither the molecular weight nor the amino acid composition showed any appreciable differences when compared with those for analogous proteins isolated from the lobster and the pig muscles. The calculations reported by Singleton and his colleagues on the degree of hydrophobicity of glyceraldehyde-3-phosphate dehydrogenase, amylase and protease isolated from different organisms are worthy of note. As seen from Table 22, the thermophilic proteins do not show an increased hydrophobicity when compared with analogous proteins from the mesophiles. The authors leave the problem of thermostability of thermophilic proteins unsolved. Further investigations carried out on the *B.stearothermophilus* glyceraldehyde-3-phosphate dehydrogenase failed to uncover the mystery of thermophily (Amelunxen et al., 1970). TPN-dependent isocitrate dehydrogenase from this thermophile also shows no difference with respect to the total amount of apolar amino acids when compared to analogous enzymes from the mesophiles (Howard and Becker, 1970).

Likewise, only modest results crowned research conducted by other investigators. Japanese scientists (Hachimori et al., 1970; Sugimoto and Nosoh, 1971) made detailed studies of the physico-chemical properties and the composition of highly purified preparations of the *B.stearothermophilus* ATP-ase, fructose-1,6,-diphosphate aldolase and glucose-6-phosphate isomerase, and compared their findings with the data reported for analogous proteins of the mesophiles. Having completed their comparative studies, they had to conclude that protein thermostability in the case of thermophiles is not related to extraordinary structure or peculiarities of the amino acid composition of these proteins. Campbell and his collaborators arrived at the same conclusion concerning ribosomal proteins of *B.stearothermophilus* (Ansley et al., 1969; Bassel and Campbell, 1969). A mechanism responsible for the heat stability of ribosomes, according to these authors, remains yet to be found; they believe, however, that it is the interaction of proteins and nucleic acids that determines the thermophilic properties of thermophilic ribosomes.

Interesting findings bearing on the effects of macromolecular interactions influencing thermostability have been gained in hybridization studies involving ribosomal subunits isolated from mesophiles and thermophiles (Altenburg and Saunders, 1971). In terms of their thermostability, the ribosomal subunits may be arranged in the following sequence: the *B.stearothermophilus* 50S > the *B.stearothermophilus* 30S > the *E.coli* 50S > the *E.coli* 30S. The combination of the 50S subunits from the *B.stearothermophilus* and the 30S subunits from *E.coli* results in the formation of thermostable ribosomes. Reverse hybridization obtained with the *B.stearothermophilus* 30S subunits and the *E.coli* 50S subunits gave rise to ribosomes featuring low thermostability. Consequently, the 50S subunits from the thermophile exert a stabilizing effect on the 30S subunits of the mesophile.

The data cited above were used to compare the proteins from thermophilic bacteria with those from mesophiles. It is seen that an attempt to link thermophily to an increased hydrophobicity of proteins has failed. An important study comparing protein hydrophobicity was made by Goldsack (1970). Using the Bigelow index, he calculated the degree of hydrophobicity of 13 proteins: actin, α-amylase, cytochrome c, insulin, lactate dehydrogenase M, malate dehydrogenase, myoglobin, myosin, phycocyanin, triose-phosphate dehydrogenase, glutamate dehydrogenase, hemoglobin and tropomyosin B. Each protein was taken from organisms markedly different in the temperature conditions of their life. This study comprised proteins isolated from birds, mammals, reptiles, amphibia, fishes, plants and bacteria. A total of 135 proteins were analyzed for their hydrophobicity. The hydrophobicity index for different proteins varied from 809 to 1298. Indices for a given protein isolated from different organisms varied only within

10%, and these variations showed no correlation whatsoever with the temperature conditions of the organisms' life. For example, the hydrophobicity index of the chicken hemoglobin was 1186, of the Chinook salmon—1211; of the chicken lactate dehydrogenase—1095; that of the frog lactate dehydrogenase—1143, etc. From the data presented by Goldsack it follows unequivocally that differences in protein thermostability are created in evolution, not through changes in the content of hydrophobic amino acids in protein macromolecules.

Bull and Breese (1973) determined the melting temperature for 14 proteins with known primary structure and compared it with the hydrophobicity index and the mean volume of an amino acid residue. They found that the larger the mean volume, the higher was the heat stability of a protein and concluded that "hydrophobic interactions not only do not stabilize a protein against temperature, but actually tend to destabilize a protein in respect to temperature" (p. 684).

Converse relationship between thermostability and hydrophobicity was established in comparative analyses of the amino acid composition of enolases from four sources: rabbit muscles, yeasts, the thermophilic bacterium *Thermus* X-1 and the extreme thermophile *Thermus aquaticus* YT-1. The temperature producing a 50% inactivation of highly purified enolase preparations during 5-min exposures were, respectively: 40, 45, 74, and 90° C. A calculation of the hydrophobicity, according to Bigelow, revealed a negative correlation between thermostability and hydrophobicity of the enzyme, and a positive correlation between the number of amino acid residues effective in the establishment of hydrogen linkages with the side chains in a molecule (Barnes and Stellwagen, 1973). These data are consistent with the findings reported by Privalov and Khechinashvili (1974) concerning the forces that stabilize globular structure of the protein. A thermodynamic analysis of thermal denaturation of five proteins demonstrated that the central role in the stabilization of the structure of these proteins is given to hydrogen bonds. Barnes and his colleagues found (1971) fructose-1,6-diphosphate aldolase from thermophilic *Clostridia* species to be 30° C more thermoresistant than analogous enzymes from a mesophilic species. Sensitivity of the enzyme to urea and guanidine chloride treatment, however, proved to be identical in both cases. These findings made these authors believe, in contrast to what had been known about enolase studies, that high thermostability of fructose-1,6,-diphosphate aldolase is not determined by hydrogen bonds.

The data obtained by Goldsack are in fair agreement with the evidence concerning the evolution of the primary structure of some proteins. Margoliash and Schejter (1966) analyzed the evolution of the polypeptide composition of cytochrome c, covering a time period of more than $1 \cdot 10^9$ years, and concluded that hydrophobic groupings are extremely conservative. This conservatism is due to the fact that the hydrophobic residues of amino acids, by concentrating in specific regions of polypeptide chains, play a decisive role in the establishment of a specific tertiary structure which is responsible for the functioning of the protein. Lim and Ptitsyn (1970) analyzed data concerning the primary structure of a large number of myoglobins and hemoglobins and found an extremely high constancy of volumes of hydrophobic cores in the macromolecules of these proteins, in spite of the fact that a number of amino acid replacements were found to be present. These authors also believe the maintenance of a specific native structure of a macromolecule is achieved on account of the constancy of the volumes of the hydrophobic core.

Evidence is available concerning variations in the heat stability of different lysozymes, these variations being related to different cysteine residue content. Thermostability of the lysozymes may be arranged in the following order: hen egg white > duck egg white > human milk = human tears > goose egg white. Accordingly, the lysozymes from these sources contain 8, 8, 6, 6, and 4 cysteine residues respectively (Jollès et al., 1966). Tanaka and his associates (1973) found that ferredoxin from the thermophilic *Clostridium thermosaccharolyticum* differs from the less thermoresistant *C. tartarivorum* (see Fig. 43) ferredoxin only in the substitution of the glutamines in positions 31 and 44 for glutamic acid. Aach (1960) has reported that a protein, molecular weight 17500, isolated from a thermophilic mutant of tobacco mosaic virus, has one threonine residue more and one isoleucine or leucine residue less that analogous protein in the wild type virus.

Consequently, an analysis of the proteins from thermophilic and mesophilic organisms so far has provided no answers as to the mechanisms governing the level of conformational flexibility in the process of evolution. One should make use of another natural experiment—the creation of organisms, whose temperature optimum lies around zero. To my knowledge, however, there are few reports available concerning the molecular mechanisms underlying the exceptional thermolability of psychrophilic proteins. A mention should be made of the work performed by Szer (1970) who detected, in the psychrophile *Pseudomonas* sp. 412, a protein—factor P—that enables the ribosomes in this organism to synthesize polyphenylalanine on poly-U template at 0° C. This factor P can be isolated from the ribosomes. Devoid of this factor, the ribosomes fail to perform at low temperatures, retaining their activity at 25–37° C. If the factor P of the psychrophile is added to the mesophile *Escherichia coli* ribosomes or to the ribosomes isolated from a *Pseudomonas* mutant deprived of the factor P and unable to function in the cold, the latter ribosomes start to synthesize polyphenylalanine actively at 0° C. The factor P is not required for the psychrophilic ribosomes to work at 15–20° C. It evidently resides in the 30S subunits.

Summary. To recapitulate, there is every reason to believe that genetically determined changes in the stability of protein molecules occur in evolution. Genetic mechanisms affecting the protein stability are, obviously, diverse and they may be of both genetic and epigenetic nature. Amino acid substitutions themselves may affect the level of conformational mobility of macromolecules, as well as influence the post-translational interactions of protein molecules with stabilizing or labilizing ligands or with neighboring macromolecules. Finally, modifications of various parameters of intracellular milieu affecting the stability of conformational structure of protein molecules, may also be inherited. Mutations of this type may produce a pleiotropic effect.

The available experimental material does not allow at present any detailed conceptualization of the biochemical mechanisms controlling the level of rigidity of protein molecules. An attempt to explain an elevated thermostability as due to enhancement of the hydrophobic interactions gained no support. The search for specific amino acid substitutions which correlate with a given thermostability level has led to the establishment of certain regularities valid only for collagen (direct dependence of thermostability on the imino acid content), and, probably, for lysozyme (direct relationship with the number of cysteine residues). The dependence of the genetically determined resistance of various proteins to heat and other denaturing agents upon the calcium ion content has been shown. One gets the impression that, in phylogenesis, in order to achieve a correlation between the environmental temperature of a species' life and the level of conformational flexibility of a protein, nature follows diverse routes depending on the properties of a concrete protein, and in respect of a given protein, depending also on its origin from a given species. The latter view is fully illustrated in the case of collagen. There are some organisms in which the differences in the resistance of collagen are related to the primary structure of the protein itself, while in other organisms, to the ligands which are removed from the proteins during purification procedures (*Gastropoda*, the intra-specific differences in the cod species). Moreover, so far no answers have been provided concerning the biochemical bases of thermostability of proteins in thermophiles and for the exceptional heat sensitivity of some enzymes in psychrophiles. It appears that in order to obtain solutions to all of these problems, more sophisticated techniques are required. From analysis of amino acid compositions one should turn to comparisons of primary structure and the spatial arrangement in homologous proteins that differ in their thermostability. A simpler approach would be to begin with an analysis and comparison of thermostability of those proteins whose primary structure is known for a large number of species.

Chapter 7
Thermostability of Nucleic Acids and the Temperature Environment of Species' Life

The main reason for writing this monograph was the fact that protein macromolecules have to undergo conformational changes while performing their biological functions. These changes amount to the disruption and establishment of weak bonds, which are temperature-dependent processes. That is why the level of conformational flexibility of macromolecules is to be correlated with the temperature conditions of the protein functioning in the cell. The level of conformational flexibility affects not only the performance of a given protein, but also its longevity. The latter parameter is important for the control of the protein content in the cell. In fact, all of these regularities are also applicable to another class of macromolecules—nucleic acids. It has been established that conformational changes occur in DNA during replication and transcription, and in RNA during translation.

The backbone of nucleic acids is a chain made of alternating phosphate residues and pentoses. One heterocyclic base is linked to one pentose: in the deoxyribonucleic acid (DNA) these are: adenine (A), guanine (G), thymine (T) or cytosine (C); in the ribonucleic acid (RNA)—uracil (U) instead of thymine. Elements of the backbone are bound together by firm covalent bonds. As in proteins, this primary structure is not everything. All nucleic acids also possess, in addition to this, structures of a higher order which are supported by numerous although weak bonds. In DNA, such a structure is formed by two polynucleotide chains wound one around the other, so that the following rule for base pairing is observed: G in one chain is located in front of C in the opposite chain; three hydrogen bonds are established between these. In DNA, A is paired with T, and in RNA—with U. Such a pair is fixed by two hydrogen bonds. Consequently, the strength of the G-C pairing is higher than that of A-T or A-U. In addition to hydrogen bondings, the DNA helical structure is supported by electrostatic, hydrophobic interactions, as well as by the interactions between neighboring bases arranged in parallel planes. When exposed to high temperatures, the hydrogen bonds between the complementary polynucleotide chains comprising the DNA helix, break down. The chains separate, the helical structure disappears and the molecule transforms into a disordered coil, consisting of polynucleotide chains. This helix-coil transition is associated with modification of a number of physico-chemical properties.

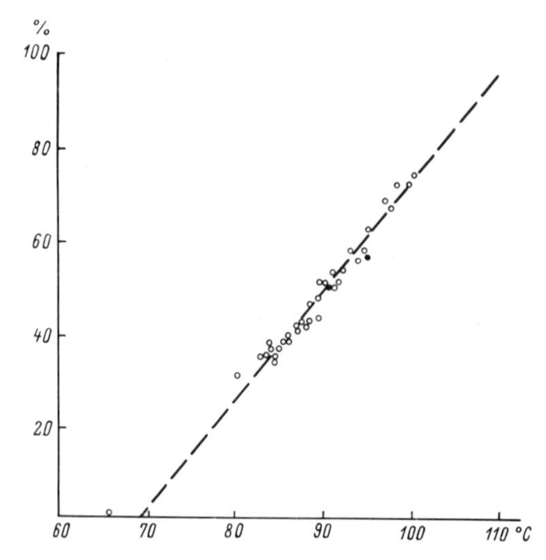

Fig. 68. Effect of G-C pair content in DNA from various sources in denaturation temperature (T_m) (Marmur and Doty, 1962). *Abscissa:* T_m; *ordinate:* percent of G-C pairs. Assayed in 0.15 M NaCl plus 0.015 M Na-citrate

Among these, of greatest practical importance is the so-called hyperchromic effect: as a result of destruction of the DNA or RNA helical structure, i.e. when nucleic acids are denatured, their optical density dramatically increases. By evaluating the changes in the light absorption at 260 nm (maximal absorption), one can quantitatively estimate the process of denaturation of nucleic acids. The resistance of nucleic acids to thermal denaturation is usually expressed by the temperature at which a 50% destruction of the helical structure is achieved. This temperature is used as a measure of the stability or rigidity of nucleic acid molecules and is designated T_m.

In studies on nucleic acid denaturation, investigators most frequently estimate the hyperchromic effect, though various other physico-chemical indicators may be used as well: optical activity, viscosity, sedimentation constant, etc.

The conformational stability of nucleic acids is determined by various factors. Most importantly, it depends on the primary structure of nucleic acids. The more G-C pairs there are, the higher is the temperature of denaturation of a given nucleic acid. Marmur and Doty (1962), having compared the G-C pair content and the T_m of 41 DNA samples, found a direct, linear dependence between these parameters that could be expressed by the following equation: $T_m = 69.3 + 0.41$ (G-C), where G-C is the relative content of this nucleotide pair on a percent basis (Fig. 68). The stability of nucleic acids is very sensitive to the ionic strength of a solution. The *Escherichia coli* DNA in 1.0 M KCl shows T_m that is 15° C higher than that in 0.05 M KCl (Marmur and Doty, 1962).

Polycations (spermine and spermidine) and divalent metal ions play a significant role in the stabilization of nucleic acids. For instance, T_m of melting RNA in the *E. coli* ribosomes is 64° C when assayed in a buffer containing Mg^{2+}. In the absence of Mg^{2+}, T_m goes down to 44° C (Friedman et al., 1967).

Most of their life nucleic acids exist as complexes with proteins, i.e. as nucleoproteins. This holds true for DNA, ribosomal RNA and messenger RNA; the latter is associated with proteins thereby forming informosomes or informofers (Spirin, 1966; Georgiev and Samarina, 1969), prior to interaction with ribosomes. Complexing with proteins exerts pronounced stabilizing effects on nucleic acids. Deproteinization considerably reduces the temperature of their denaturation. Concurrently, nucleic acids become more accessible to nuclease attacks. According to Gorovsky (1964), T_m of DNA in the sea urchin sperm chromatin is around 80° C, whereas that of the DNA dissociated from the protein is about 50° C.

According to current views, local fluctuational openings of the DNA double helix appear continually at physiological temperatures. An area of an opening, as a rule, spans over only 1–2 nucleotide pairs. The probability of a disconnection of the bonds inside a helix is about 10^{-4}, and it is appreciably higher for A-T pairs as opposed to G-C pairs. The frequency of disconnections and connections correlates with the G-C pair content and T_m. This type of fluctuation, called "breathing" of DNA, is important for realization of the processes requiring exposure of heterocyclic bases. Such processes are replication and transcription (Frank-Kamenetsky and Lazurkin, 1974). Consequently, the percentage of G-C pairs and the melting temperature (T_m) reflect the reaction capacity of DNA that is realized at a normal functioning temperature.

Following these brief and schematic remarks, the available evidence concerning the relationship between thermostability of nucleic acids and temperature for life of a species will be studied. Again, it is assumed that differences in denaturation temperatures, in particular in those temperatures which produce half-complete denaturation (T_m), reflect a difference in the conformational stability of nucleic acids at physiological temperatures, although they are considerably lower than the temperature effective in separating the complementary chains manifested in the hyperchromic effect. There is similarity between thermal denaturation of nucleic acids and proteins in that in both cases a disruption of weak bonds takes place, and in both cases this is a cooperative process in which preceding disruptions of bonds facilitate subsequent ones.

Marmur (1960) was apparently the first to be interested in this problem. He compared a curve of thermal denaturation of deproteinized DNA from the thermophile *B. stearothermophilus* and DNA from the mesophile *E. coli*. The results he obtained were opposite to those he expected. T_m of the thermophilic DNA was 87.5° C, that is somewhat lower than that of the mesophilic DNA at 90° C. This difference in heat stability of the DNAs corresponded to a difference in their primary structures, because the G-C content of the *B. stearothermophilus* DNA was 46.7%, whereas that of the *E. coli* DNA was 50%. No correlation could be found between the resistance of DNA to heat and an optimum temperature for growth in other thermophilic, mesophilic and psychrophilic bacteria (Mangiantini et al., 1965; Sugiura and Takanami, 1967; Pace and Campbell, 1967). Table 23 presents convincing data in favor of the absence of any correlation between thermophily of microorganisms and the G-C content of their DNAs.

Dissimilar results have been reported by Stenesh et al. (1968). They compared the nucleotide composition of DNA in three strains of *B. stearothermophilus* and in three mesophilic strains of the same genus (*B. pumilus*, *B. sp.* and *B. licheni-*

Table 23. Thermophily of microorganisms, T_m of ribosomes and percent of G+C in ribosomal RNA and DNA. (From Pace and Campbell, 1967)

Organism	Temperature maximum for growth (° C)	T_m of ribosomes (° C)	G+C content (%)	
			rRNA	DNA
V. marinus (15381)	18	69	52.1	40
7E-3	20	69	50.0	36.0
1–1	28	74	57.2	—
V. marinus (15382)	30	71	51.6	42
2–1	35	70	50.7	43.7
Desulfovibrio desulfuricans	40	71	52.1	56.3
D. vulgaris	40	73	56.5	60.2
E. coli (B)	45	72	55.6	52
E. coli (013)	45	72	54.1	52
Spirillum itersonii (SI–I)	45	73	57.0	55
B. megatherium (Paris)	45	75	53.1	38
B. subtilis (SW–25)	50	74	55.1	43
B. coagulans 43P)	60	74	56.4	47.4
Desulfotomaculum nigrificans (8351)	60	75	55.7	44.7
Thermophile T–107	73	78	58.5	41.0
Thermophile 194	73	78	56.6	44.6
B. stearothermophilus (1503R)	73	79	59.2	52
Thermophile B	73	79	58.7	44
B. stearothermophilus (10)	73	79	59.3	51.5

formis) and found a parallelism between the maximal temperature for growth and the G-C content. Full confirmation of this observation can be seen in Table 23, if comparison is made within closely related species: five species of *Bacillus* genus and two species belonging to *Vibrio* genus. An exceptionally high G-C content (68%) and high melting temperatures were found for DNA from the extreme thermophile *Thermus thermophilus* (Oshima and Imahori, 1974). Interesting data have been reported by Barenboym et al. (1971). They gave the plots of thermal stability of deproteinized DNA and of deoxyribonucleoprotein (DNP) isolated from spermatozoa of two sea-urchin species: the more warm-loving *Strongylocentrotus nudus*, and *S. intermedius* which lives in a cooler environment. The melting curves for DNA from both species coincide (Fig. 69, curves 1 and 2). The melting curves for DNP, however, reveal that when complexed with protein, the *S. nudus* DNA is stabilized to a higher degree than is the *S. intermedius* DNA, the difference being approximately 8° C (curves 3 and 4). According to the authors this ecologically justified difference in the stability of the DNAs from the two sea-urchin species is accounted for by histones. The amino acid compositions of the histones from these two species are different. Besides, the primary thermostability of the spermatozoa themselves is higher in *S. nudus* than in *S. intermedius*. The same holds true also for the egg cells (Andronikov, 1963) and muscle fibers (Ushakov, 1959a).

The stabilization of DNA by histones suppresses its template activity at the same time. Bonner and his collaborators (Huang and Bonner, 1962; Bonner and

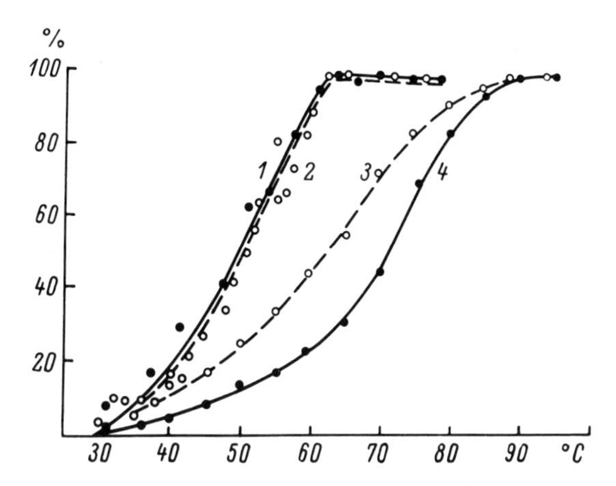

Fig. 69. Thermal denaturation of DNA *(1* and *2)* and deoxyribonucleoprotein complexes *(3* and *4)* from sea urchin *Strongylocentrotus intermedius* sperm *(2* and *3)* and from the more warm-loving species *S. nudus (1* and *4)* (Barenboym et al., 1969). *Abscissa:* temperature of heating; *ordinate:* extent of denaturation (% of maximal denaturation)

Huang, 1963; Huang et al., 1964) found that association of a purified chromosomal DNA from peas with histones results either in suppression or complete inhibition of the DNA capacity for directing in vitro DNA-dependent RNA synthesis. Four of the calf thymus histone fractions studied elevated thermostability of the DNA to various extents. The stabilizing efficiency of histone fractions could be arranged in the following order: Ib > IIb > III > IV. These fractions are found in the same sequence in their ability to suppress the template activity of DNA in vitro. Similar data have been obtained also with isolated calf thymus nuclei (Allfrey et al., 1963) and later were confirmed by other investigators. In addition, it has been shown that histones make calf thymus DNA less accessible to deoxyribonuclease (Padayatty et al., 1968).

Actinomycin D elevates the T_m of the phage T_2 DNA by crosslinking the chains of the double helix. At the same time, this inhibitor reduces the ability of the phage DNA to genetic transformation in *Bacillus subtilis* (Permogorov et al., 1965). Consequently, an increase in the stability of a native DNA molecule, i.e., making the separation of complementary chains more difficult, interferes with the realization of its genetic functions. An excessive reduction of stability of the native DNA structure should also result in adverse consequences. *Thus, the principle of semistability—not too rigid and not too labile—must be applicable also to nucleic acid molecules. Hence, it follows, that in evolution, the level of conformational stability of nucleic acids should be correlated with the environmental temperature of a species' life and natural selection must be responsible for the control of this correspondence.*

These ideas can be corroborated by a number of facts reported for ribosomal RNA (rRNA) as well. Mangiantini et al. (1965), studying the hyperchromic effect, compared the stability of the ribosomes and rRNA from *B. stearothermophilus*, grown at 64° C, and those from *E. coli*, cultivated at 37° C. As seen from Figure 70, rRNA in ribosomes is more heat-resistant in the thermophile than in the meso-

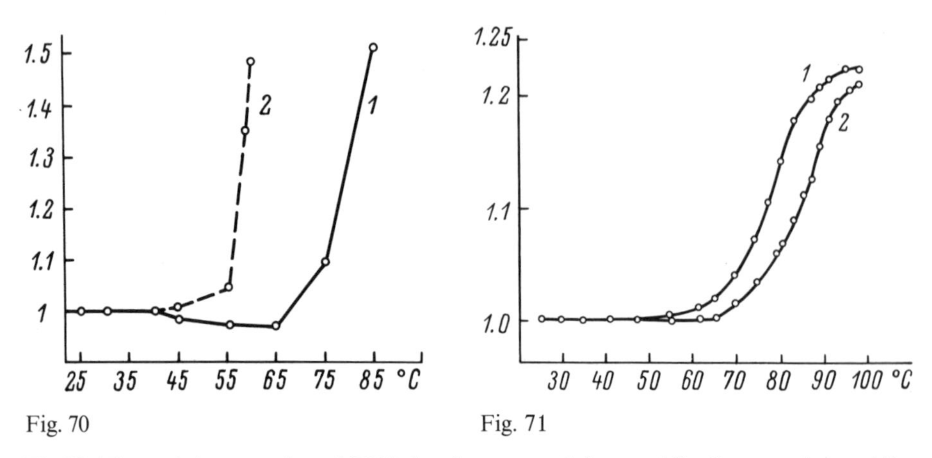

Fig. 70 Fig. 71

Fig. 70. Thermal denaturation of RNA in ribosomes of thermophile *B. stearothermophilus (1)* and mesophile *E. coli (2)* (Mangiantini et al., 1965). *Abscissa:* temperature of heating; *ordinate:* extent of denaturation, estimated by light absorption at 260 nm

Fig. 71. Thermal denaturation of purified transfer RNA from *Escherichia coli (1)* and thermophilic bacterium *Thermus aquaticus (2)* (Zeikus et al., 1970). *Abscissa:* temperature of heating; *ordinate:* relative absorption at 260 nm

phile by approximately 20° C; rRNA in the thermophilic ribosomes is more stable against urea too. After isolation of rRNA from the *B. stearothermophilus* ribosomes, the melting curve shifts toward lower temperatures and its shape changes. An analogous curve for the *E. coli* rRNA also shows modification of its shape; however its shift to the left is less pronounced. Thus, the difference between the heat stability of rRNAs from the thermophile and mesophile lessens by 10° C. When isolated, the thermophilic rRNA also displays higher resistance to urea. Thermophilic rRNA contains more G-C pairs than that from *E. coli.* In the former case the A+U/G+C is 0.78, whereas in the latter it is 1.04. The authors, however, could not detect any differences in the heat stability of soluble RNA in these two bacterial species. The soluble RNAs from both bacteria contain roughly identical amounts of G-C pairs. The A+U/G+C ratio for the thermophilic soluble RNA is 0.72 and that for the mesophilic soluble RNA is 0.66. Thermostability of ribosomes, according to the Italian authors, may be a limiting factor for life of microorganisms at elevated temperatures.

The facts referred to above have found support in subsequent investigations of a number of workers, in particular those of Friedman and his colleagues (Friedman and Weinstein, 1966; Friedman et al., 1967; Friedman, 1968, 1971). The T_m of the ribosomes from *B. stearothermophilus* grown at 65° C, as estimated by the hyperchromic effect, is 81° C, and that of the *E coli* ribosomes from cells cultivated at 37° C, is 71° C. In the absence of Mg^{2+}, the stability of ribosomes is strikingly lower, but the difference just mentioned becomes even more pronounced: the T_m of the thermophilic ribosomes is 64° C and that of the mesophilic ribosomes is 44° C. Moreover, it has been shown that higher thermostability is characteristic not only of ribosomes from that thermophile, but also of the 30S and 50S ribosomal subunits. T_m of rRNA isolated from the *B. stearothermophilus* ribosomes is

60° C and that of rRNA from the *E.coli* ribosomes is 50° C. The amino acid composition of the ribosomal proteins in both cases is essentially the same. These authors, like the Italian investigators, could not find any notable difference in the heat stability of soluble RNA from the thermophiles. They concluded that the secondary structure of soluble RNA is stabilized in vivo not by virtue of the peculiarities of its primary structure, but by the formation of complexes with polycations. Stenesh and Holazo (1967) demonstrated that when the ribosomal RNA, isolated from three mesophilic *Bacillus* species is compared with the ribosomal RNA from three thermophilic strains of the same genus, one can notice that the lower G+C/A+U ratio in the former (1.25 versus 1.49) is accompanied by a lower resistance to heat (65° C versus 70° C) and 8 M urea.

A comparison of the ribosomes and rRNA from *B.stearothermophilus* and *E.coli* can also be found in the works of Campbell and his colleagues (Saunders and Campbell, 1966; Ansley et al., 1969; Bassel and Campbell, 1969). In their studies, the heat stability of intact ribosomes and of ribosomal fragments was estimated by the hyperchromic effect. In addition, the amino acid composition of ribosomal proteins and electron microscopy of the ribosomes were compared. Their observations yielded no information about any dissimilarities between chemical characteristics of the ribosomes from the thermophile and mesophile. These authors agree that the factors responsible for thermophily of *B.stearothermophilus* ribosomes remain yet to be discovered, but they believe that thermal stability of thermophilic ribosomes depends on the interaction between rRNA and proteins.

At the same time, Pace and Campbell (1967) present tables summarizing data taken from the literature and obtained in their studies. They depict temperature maxima for growth of the organisms, T_m of ribosomes (evaluated by the hyperchromic effect) and the relative content of the G-C pairs in rRNA from 19 microorganisms (thermophilic, mesophilic and psychrophilic) (Table 23). A correlation coefficient has been calculated by the author based on the data from that table, and it was found that there is an intimate relationship between a temperature for life of a microorganism and thermostability of its RNA in the ribosomes: $r_1=0.92$. A less pronounced, though undoubtedly existing, correlation can be detected between thermophily of the organisms and the relative content of the G-C pairs: $r_2=0.52$. The difference between the correlation coefficients r_1 and r_2, estimated by the method of z-transformation, proved to be trustworthy with a 95% probability. The existence of this difference can be explained by the fact that an approximate correlation between the temperature maxima for life of bacteria and thermostability of their ribosomal RNA, the latter being a function of its primary structure, becomes appreciably closer when the rRNA are complexed with proteins, and, probably, with other ingredients of the ribosomes. In other words, not only the interaction of rRNA with proteins is responsible for a difference in the levels of thermostability of rRNA, but its primary structure is an important factor too.

The significance of both the association of RNA with proteins and of an increased content of the G-C pairs for thermophily can be inferred also from a comparison of the ribosomes and rRNA from *E.coli* with those from the extreme thermophile *Thermus aquaticus*, growing at 79° C (Zeikus et al., 1970). T_m of the

T. aquaticus ribosomes is 86° C, that of the *E. coli* ribosomes is 71° C. Isolation of rRNA from the ribosomes decreases their T_m and diminishes but does not eliminate the difference between the T_m values for the thermophile and mesophile. In the former, T_m is 64° C, in the latter it is 58° C. Moreover, ribosomal RNA from *T. aquaticus* had a higher G-C content.

Neither Mangiantini et al. (1965), nor Friedman and Weinstein (1966) could detect any difference in thermal stability of soluble RNA (tRNA) from *B. stearothermophilus* and *E. coli*. A difference, however, was found when the plots of thermal denaturation of the tRNA from *E. coli* and *T. aquaticus* (Zeikus et al., 1970) were compared. Figure 71 illustrates the considerably higher thermostability of transfer RNA from the thermophile: the difference in T_m is 6° C (86° C and 80° C, respectively). Removal of Mg^{2+} lowers T_m of both tRNAs, but the difference persists: 81° C and 70° C, for respective transfer RNAs. Consequently, an elevated thermostability of tRNA of the thermophile is determined by its primary structure. This is supported by higher G-C content in the *T. aquaticus* tRNA as opposed to that in the transfer RNAs from both *B. stearothermophilus* and *E. coli*.

It has been seen that an elevated thermostability of proteins is usually associated with higher stability of the macromolecules against proteolytic digestion. This regularity apparently holds true also for nucleic acids. According to Stenesh and Yang (1967) thermophilic ribosomes are more resistant than mesophilic ones not only to heat, but also to ribonuclease. Three species of mesophilic and three species of thermophilic bacteria, belonging to one *Bacillus* genus, have been studied. A five-hour exposure of the mesophilic ribosomes to ribonuclease at 37° C resulted in 80% degradation of the ribosomes, while the ribosomes from the thermophiles were only 55–70% degraded. Histones make RNA less sensitive to the destructive action of nucleases, similar to that observed with DNA (Padayatty et al., 1968).

Summary. Publications dealing with the relationships existing between the temperature ecology of species and the properties of nucleic acids are so far extremely few in number and are limited almost exclusively to the world of microorganisms. Despite this, there are data available that allow a realistic conclusion to be drawn that for this class of macromolecules, similar to proteins, there is a correspondence during evolution between the level of stability of higher-order structures and the temperature conditions of a species' life. This has been very convincingly shown for ribosomal RNA still present in the ribosomes of many bacteria that differ in their thermophily, and in some cases also for the rRNA isolated from ribosomes. Relevant data for DNA are more fragmentary, although in two sea-urchin species a correlation has been shown to exist between the stability of DNA within the chromatin and the respective thermophilies of these organisms. In several closely related species of bacteria, this correlation could also be demonstrated with isolated DNA. There is one case known in which an elevated thermostability of tRNA from an extreme thermophile has been proven. Illustrative examples were given to emphasize that a difference in the resistance of nucleic acids to heat accompanies respective differences in their resistance to urea and nuclease treatment. The mechanisms responsible for determining the level of stability of nucleic acids are more readily explainable than those operating in proteins. Among these are: the relative G-C pair content (the higher it is the stronger the binding of the chains in a double helix), complexing with proteins (in particular with histone), complexing with polycations and divalent metal ions (in particular with magnesium ions). All these three possibilities are probably used in evolution to maintain a correspondence between a temperature optimum regime for the cell's life and the degree of rigidity of its nucleic acids.

To this end, evolution might have made use of variations in the nucleotide composition of the coding sequences in DNA and messenger RNA, without being afraid of unfavorable genetic consequences, by operating within the limits of the degeneracy of the genetic code. Such possibilities are ample indeed, since one and the same amino acid is coded by several different triplets. In fact, alanine may be coded by GCA and GCC, valine-GTA and GTC, and leucine by TTC and CTC, etc. In the instances listed, each second version for coding the same amino acid provides the possibility to incorporate $^1/_3$, or even $^2/_3$, more nucleotides with triple hydrogen bonds, and this allows a stronger association to be established with the complementary partner. Bovine ribonuclease, for example, contains 124 amino acid residues. Using degeneracy of the code, one can compose a gene for ribonuclease in two extreme versions: with a maximal and minimal content of G-C pairs. The first case would be 64.0% of G-C, the second, 50.0%. By solving the Marmur and Doty equation (p. 234) with these values, the T_m in the first case is 95.5° C, in the second it is 89.3° C, namely, a difference of 6.2° C. It may be that one of the biological destinations of the code degeneracy is that an appreciable modification of the conformational rigidity of DNA can be achieved without altering the primary structure of proteins. It was noted above that Oshima and Imahori (1974) found in the extreme thermophile *Thermus thermophilus*, an exceptionally high G-C pair content although the total amino acid composition of proteins from that thermophile differed only slightly from that of the mesophile *E. coli* proteins. The authors write in this connection: "It is probable that the thermophile utilizes predominantly (G+C) rich codons among the degenerate candidate codons for each amino acid" (p. 182).

In order to verify this assumption, it is important to know whether stability of the messenger RNA varies with the temperature ecology of a species. Data of this kind appear to be unavailable. In the case of the sea urchin DNA, a specific difference between organisms which vary in their temperature ecologies in respect to thermostability of DNA is apparently entirely induced by the proteins, which form complexes with nucleic acids. As for the ribosomal RNA in bacteria, this difference is accounted for only partially by such complexation of the nucleic acid with appropriate proteins. The availability or absence of polycations considerably enhances or respectively lowers the resistance of nucleic acids; the significance of these for the ecological regulation of resistance levels is, however, unclear so far.

A new facet of the taxonomy of animals, plants and microorganisms is being developed progressively, based on studies of the primary structure of DNA, called gene systematics (Belozersky and Antonov, 1972; Antonov, 1974). The facts discussed in this section attract attention to the possiblity that, aside from phylogenetic relatedness of organisms, the primary structure of DNA may be affected by temperature conditions of a species' life. In order to detect the difference in the

DNA structure determined by adaptation to temperature, it is imperative to compare DNA structures in *closely related* species whose temperature ecologies vary greatly.

All of the evidence discussed in this section has concerned the genotypic interspecific differences in stability of nucleic acids. Investigations dealing with modificational changes of the nucleic acid stability brought about by exposure of individuals of one species to different temperature conditions obviously have never been carried out.

A number of writers state that T_m of RNA in the ribosomes limits the ability of microorganisms to live at high temperatures. Judging from the data included in Table 23, however, this might be the case only for thermophiles. In fact, a difference between a maximal temperature for growth of thermophiles and T_m of their ribosomes is limited to only 5–6° C. On the other hand, in microorganisms having a temperature maximum of 50° C and lower, this difference varies from 26° C to 50° C. Nonetheless, also in these latter organisms, a correlation is maintained between their habitat temperatures and the heat stability of RNA in the ribosomes. It is hard to believe that the significance of this correlation is in providing for the heat resistance of ribosomal RNA. It would be more logical to acknowledge, by analogy with proteins, that the correlation between T_m of RNA and the habitat temperature indirectly reflects a correlation between the level of conformational flexibility, and, concurrently, the metabolic activity of nucleic acids and the environmental temperature. In this regard, the observations, so far rather few in number, that concern the reduction of nucleic acid activity occurring with stabilization of their spatial structure are of major interest.

Chapter 8
Saturation of Fatty Acids and the Temperature Conditions of Life

In this chapter the discussion will be focused not on macromolecules, but on the molecules of a lower rank—on lipids. A justification for taking up this topic may be seen in the following considerations. In no other sphere of cellular activity can one witness such an efficient, rapid and perfect reaction to changes in the temperature as that which takes place in the lipid moiety of the cell. The necessity for adaptive rearrangements of the lipids with temperature fluctuations is dictated by the same general principle that underlies the adaptive behavior of the macromolecules. Similar to proteins and nucleic acids, lipids are to be found, while performing their physiological tasks, in a semilabile state, and are directly dependent on the temperature. Moreover, numerous traits in the behavior of the membrane proteins reflect the changes occurring in the state of the lipids.

The main field of action of lipids is the membrane. Here they are associated with proteins in a most intimate way, both in terms of structure and function. An important part of a lipid molecule is the hydrocarbon tail of a fatty acid residue and the following discussion will be concerned essentially with the response of the physical state of fatty acids to temperature. The cardinal factors responsible for the physical state of fatty acids are the degree of their saturation, i.e., the number of double bonds (C=C) in a hydrocarbon chain, the remoteness of a double bond from the acid group—carboxyl, cis- or trans-configuration of the double bond, the length of chain and extent of its branching. The higher the degree of unsaturation of a fatty acid, the higher is its molecular mobility, the lower the melting point and the larger the area a fatty acid molecule occupies on the surface of a monomolecular layer. These parameters shift in this direction when a double bond is in a cis-configuration as opposed to a trans-configuration; if the double bond is located farther from the carboxyl; and also with branching or shortening of the chain. The degree of mobility of various parts of a lipid molecule differs. Long chains of fatty acids may be in a liquid state at body temperature, whereas ionic end groups of lipid molecules form crystalline structures melting at about 200° C (Byrne and Chapman, 1964). The cells are able to vary the degree of unsaturation of the fatty acids in their lipids. The composition of the fatty acids in lipids can be modified by supplying the required fatty acids with food or by addition of a nutrient medium. Experiments with artificial mixtures of fatty acids have shown that at a certain ratio of saturated versus unsaturated fatty acids, the freezing point of a mixture may be considerably shifted (by 15° C) with as minor a change in the content of an unsaturated fatty acid as 5% (Lyons and Asmundson, 1965).

Here are several examples illustrating the effect of the physical state of fatty acid residues in membrane lipids on the performance of membrane-bound enzymes. Hazel (1972a, 1972b) compared various parameters of the activity of the muscle succinate dehydrogenase built into mitochondrial membranes, with those of the enzyme found in a water-soluble state after extraction of the lipids. Removal of the lipids reduced the specific activity of the enzyme three to four times, and lowered the activation energy and the Michaelis constant in respect to succinate. If phosphatides are added to the delipidated enzyme, a partial restoration of the original properties of succinate dehydrogenase occurs. The reactivation will be more complete if lipids containing large amounts of unsaturated fatty acids are added to the enzyme. Towers et al. (1972) studied the kinetics of protein synthesis by isolated liver ribosomes and by the ribosomes bound to the endoplasmic reticulum membranes. The Arrhenius plot of ^{14}C-leucine incorporated into the protein synthesized by isolated ribosomes showed a straight-line dependence from 8 to 37° C. The activation energy was 25.5 kcal/mole. The membrane—bound ribosomes exhibited markedly different kinetics of protein synthesis. In this case, the Arrhenius plot has a sharp break around 20–22° C. The activation energy in the region from 22 to 37° C was 16 kcal/mole, and between 20 and 8° C—43 kcal/mole. These authors detected by the electron paramagnetic resonance technique, a phase transition in the membrane lipids occurring in the interval between 22 and 23° C, i.e., at a temperature corresponding to a discontinuity in the Arrhenius plot of the protein synthesis by membrane—bound ribosomes. These observations were interpreted by the authors to indicate that the cause of a change in the kinetics of protein synthesis by the membrane—bound ribosomes at 22° C was a phase transition in the membrane lipids.

Raison and his colleagues (1971b) reported discontinuities in the Arrhenius plots of the temperature dependence of a maximal rate of succinate dehydrogenase, succinate oxidase and cytochrome-c-oxidase activities in rat liver and batata mitochondria. The discontinuity persisted even when the mitochondria were destroyed by hypotonic solutions, freezing and thawing or ultrasound. The break could not be observed, however, and the Arrhenius plots were represented by a straight line when the mitochondria had been treated with detergents to extract the lipids. The authors concluded that the conformational changes of these enzymes, occurring at temperatures corresponding to the breaks in the curves, were induced by temperature—dependent phase transitions in the membrane lipids. Raison and his associates communicated similar results in their later publication dealing with other enzyme systems (McMurchie et al., 1973).

Wilson and Fox (1971) used an *Escherichia coli* strain auxotrophic for unsaturated fatty acids. They studied the effect of temperature on β-galactoside and β-glucoside transport in bacteria fed on various unsaturated fatty acids. The Arrhenius plots of the β-galactoside and β-glucoside transport showed breaks which occurred at the same temperature for both sugars. The temperature corresponding to a discontinuity was grossly changed (from 7 to 30° C) depending on what

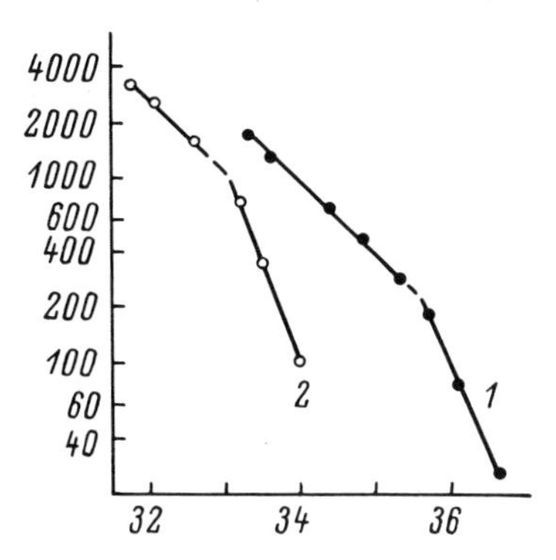

Fig. 72. Arrhenius plot of temperature dependence of ϱ-nitrophenyl-β-glucoside transport in cells of *Escherichia* strain auxotrophic for unsaturated fatty acids (Wilson and Fox, 1971). *Abscissa:* log of reciprocal absolute temperature, $\times 10^4$; *ordinate:* rate of β-glucoside transport (nm per hour per 2×10^9 cells, log scale). *1:* cells fed on linoleic acid (18_2), *2:* elaidic acid (18_1)

particular fatty acid had been fed to the auxotroph as an indispensible additive. Fatty acids with a lower melting point shifted the position of the break in the Arrhenius plot toward lower temperatures (Fig. 72). At temperatures below the discontinuity, the activation energy of the sugar transport sharply increased. This phenomenon is explained by the authors as due to a transition of fatty acid chains from a disordered, "liquified", state into a more structured hexagonal arrangement at temperatures below the break point; thereby restrictions were imposed on the performance of the protein carriers.

The yeast *Saccharomyces cerevisiae* grown under anaerobic conditions is unable to synthesize unsaturated fatty acids or sterol, but it can incorporate these in its membranes. Cobon and Haslam (1973) cultivated yeasts on media supplemented with different concentrations of ergosterol. Sterols lower the temperature of the transition of membrane lipids into a liquid—crystalline state. Accordingly, the higher the ergosterol content in the yeast mitochondria, the lower was the temperature at which a discontinuity occurred in the Arrhenius plots of the mitochondrial ATPase activity.

Esfahani et al. (1971) studied the membranes of an *E. coli* strain auxotrophic for unsaturated fatty acids, by supplying different unsaturated fatty acids to the cells. They found, using an X-ray diffraction technique, that the mean temperature of a phase transition of membrane lipids varied considerably as a function of the fatty acid added to the medium. For instance, if it is oleate (C_{18}), the temperature is 24° C (19–29° C), whereas in the case of myristoleate (C_{14}) it is 41° C (36–46° C). Below the melting point the apolar chains of lipids form a dense hexagonal packing. Conversely, above the transition point they are found in a less structured, liquid-crystalline state. The transition area is broad and spans over about 10° C; this results in overlapping of the transition zones for fatty acids with different melting points.

An average melting point of lipids in a membrane, in the case of exogeneously supplied different fatty acids, may differ from the melting point of the fatty acids added to a medium for the following reasons. When an auxotroph is given fatty acids with an excessive number of double bonds or with relatively short chains, the cell compensatorily increases the concentration of saturated fatty acids in the membrane. The lowering of the cultivation temperature will have an opposite effect (Esfahani et al., 1969). Moreover, the phase state of the fatty acids in the membrane may be affected by non-lipid components such as proteins. Esfahani and his colleagues studied the kinetics of proline uptake and the succinate dehydrogenase activity as a function of added unsaturated fatty acids supplied to an auxotrophic *E. coli* strain. Like Wilson and Fox, they also found discontinuities in the Arrhenius plots of these processes that could be traced to the addition of a particular unsaturated fatty acid. However, no correlation between the temperature at which a break occurred and the mean temperature of melting of the membrane lipids could be noted. The authors give an analysis of the possible reasons for this discrepancy, but irrespective of the solutions proposed, one can draw an unambiguous conclusion that the physical state of fatty acid residues in membrane lipids is of the utmost importance to the operation of the enzymes associated with the membranes, as well as to the pattern of the temperature dependence of these enzymes. This contention might be substantiated with much additional evidence (e.g. Cobon and Haslam, 1973), but that already referred to suffices.

The temperature for normal functioning of cells is evidently somewhat higher than, or in the region of, the melting point of the bulk of hydrocarbon chains of the fatty acids in the membrane lipids (Luzzati and Husson, 1962; Chapman, 1967; Engelman, 1970). This, in fact, is demonstrated by the data concerning modification of the heat capacity of the *Mycoplasma laidlawii* membranes by increasing the temperature. In the zone of the melting of fatty acids, the heat capacity increases, while hydrocarbon chains acquire new degrees of freedom (Melchior et al., 1970). The same inference has been made in studies concerning the effect of temperature on X-ray diffraction by *Escherichia coli* membranes (Esfahani et al., 1971), and in numerous other investigations. It follows from those studies that *the physical state of fatty acid residues in membranes is semistable.* Shifts toward liquefication or to higher orderedness and reduction of the molecular mobility must influence the activity of the membrane-bound enzymes as well as flexibility of the entire membrane. Any change in ambient temperature will inevitably affect the physical state of the fatty acid residues. That is why in plant and animal cells and in microorganisms, surprising mechanisms have evolved which adaptively modify the fatty acid composition of lipids when the environmental temperature is shifted for a sufficiently prolonged period of time above or below an optimum. A decrease in temperature brings about a compensatory increase in the relative content of unsaturated fatty acids in the cellular lipids. This prevents "solidification" of the fatty acid chains and excessive enhancement of the membrane rigidity. Conversely, the cell responds to an elevation of the temperature by reducing its unsaturated fatty acid composition. That serves to avoid the excessive "liquefication" of the hydrocarbon chains of fatty acids and to prevent abnormal labilization of the membranes. Not only do the membrane lipids react to temperature changes. by modification of the fatty acid composition, but the reserve fats are also modified in this way. It is probable that their physical state determines their accessibility to enzymes.

The literature concerning modificational changes in fatty acid composition provoked by temperature variations is both abundant and unequivocal. In the earlier papers, one can find reports about changes in the total content of unsaturated fatty acids estimated by the amount of iodine bound by these. Each double bond binds two iodine atoms. The higher the iodine number (grams of iodine bound by 100 g of fatty acids), the higher the degree of unsaturation. In the sixties of this century, more sophisticated techniques for analyzing the fatty acid composition of lipids were widely introduced, and among these were the gas-liquid chromatography and thin-layer chromatographic techniques. They made it possible to evaluate what particular fatty acids and in what lipids are responsible for changes in the degree of unsaturation in each particular case.

As early as in the second half of the last century it became evident from studies on the iodine number that the degree of unsaturation is higher in the oils derived from plants living farther to the North. Later this regularity was found to hold true for individuals of the same species living under contrasting temperature conditions. Ivanov (1926), for instance, estimated the iodine number of flax oil obtained from plants grown in 24 different geographical localities. A higher percentage of unsaturated fatty acids was found in oil produced from the plants grown either in areas located farther to the North, or at higher altitudes. That is an example of a modificational reaction of the fatty acid composition of plants to the habitat temperature. Similar data have since been repeatedly reported for plants. Alfalfa leaves, for instance, contain almost twice as much fatty acid in December after a cold hardening than the leaves in summer, and this is due primarily to an increase in the amount of fatty acids having two and three unsaturated bonds (Gerloff et al., 1966). Kuiper (1970) determined the lipid composition and the fatty acid composition of the lipid extracted from alfalfa leaves in the plants grown at 15, 20, and 30° C. In those grown in the cold, the monogalactosyl diglyceride fraction contained larger amounts of linolenic acid, which has three double bonds; phosphatidylcholine fraction from the "cold" plants contained more linoleic acid (two double bonds) than did that from the "warm" plants. An increase in the relative content of unsaturated fatty acids also occurs in yeast with a decrease of the cultivation temperature, as estimated by the iodine technique (Christophersen and Precht, 1950b), and by analyzing the fatty acid compositions with the aid of thin-layer and gas-liquid chromatography (Greshnykh et al., 1968; Brown and Rose, 1969; Kates and Paradis, 1973). The same results have been reported in fungi—*Aspergillus niger*, *Rhizopus nigricans* (Pearson and Raper, 1927), *Sterigmatocystis nigra* (Terroine et al., 1930) and in a number of thermo- and psychrophilic species of the *Mucorales* order (Sumner et al., 1969). The alga *Cyanidium caldarium* was cultivated at 20 and 55° C. The total amount of lipids in the algae kept at 55° C was twice as low compared with those maintained at 20° C. In addition, the ratio of unsaturated versus saturated fatty acids was three times as low in the former. Furthermore, the "warm" algae membrane could be altered by considerably more intensive heating than that of their "cold" counterparts (Kleinschmidt and McMahon, 1970a, 1970b).

Similar results have been obtained in studies on modificational changes of the degree of unsaturation of fatty acids in animals exposed to different temperatures. Fraenkel and Hopp (1940) cultivated the larvae of two meat fly species at different temperatures. As depicted in Figure 73, the degree of unsaturation of the fatty

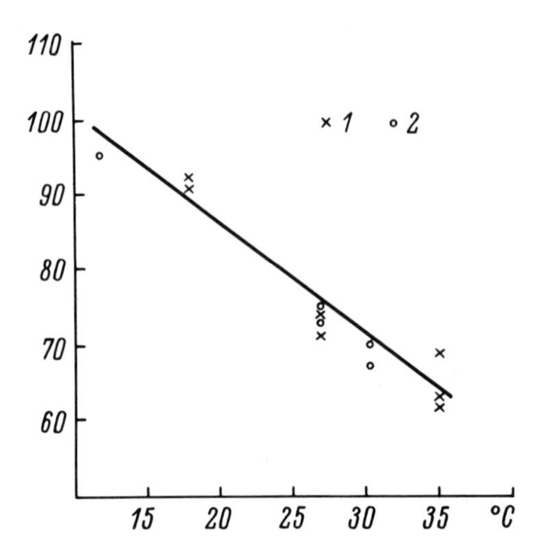

Fig. 73. Iodine number of phosphatides in *Calliphora erythrocephala* larvae *(1)* and those of *Phormia terra-nova (2)* as function of cultivation temperature (Fraenkel and Hopp, 1940). *Abscissa:* cultivation temperature; *ordinate:* iodine number

acids is inversely proportional to the cultivation temperature. These findings were later confirmed by Amosova (1963). She reported a higher degree of unsaturation of fatty acids in the muscle lipids obtained from the meat fly larvae kept at 13–14° C as compared to the larvae exposed to 33–34° C. Phospholipids in the livers of frogs kept at 7° C contained 41.8% of saturated and 58.2% of unsaturated fatty acids; in frogs kept at 25° C, the figures were 49.9% and 50.1% respectively (Baranska and Wlodawer, 1969).

The goldfish is a convenient study object. It is capable of acclimating over a wide range of temperatures. Hoar and Cottle (1952) adapted the fish to 5, 15, 20, and 35° C. The iodine number of the total lipid fraction became higher, as the acclimation temperature dropped. A series of studies employing chromatographic techniques has been carried out in the laboratory of Prosser (Johnston and Roots, 1964; Roots, 1968; Roots and Johnston, 1968). The effects of the acclimation temperature on brain lipids was especially pronounced. Lowering the acclimation temperature resulted in a decrease of the total amount of brain lipids, however the phospholipid content was not changed. This was accompanied by an increase in unsaturated fatty acids in the main phospholipid fractions. Brain plasmalogens (phospholipids containing a simple vinyl ether group) were found to be modified in the same fashion. Caldwell and Vernberg (1970) studied the changes occurring in 15 fatty acids of the mitochondrial lipids isolated from the gills of goldfish living at 10 and 30° C. The lipids of the "cold" fish showed a lowered content of saturated fatty acids (palmitic, stearic), the total content of fatty acids with one double bond decreased; the amount of fatty acids with four and five unsaturated bonds increased, however, and, on the whole, the lipids from goldfish acclimated to 10° C contained a higher percentage of unsaturated bonds as compared to the lipids from those kept at 30° C. Similar results were reported in that paper for another fish, *Ictalurus natalis.*

Table 24. Percentage of fatty acids with different number of double bonds in intestine mucose membranes of the goldfish acclimated to different temperatures. (From Kemp and Smith, 1970)

Acclimation temperature (° C)	Relative content of fatty acids (%) with different number of double bonds (0–6)						
	0	1	2	3	4	5	6
3	28.1	12.1	16.1	3.9	13.7	3.0	18.6
14	34.9	11.1	15.0	4.5	13.0	4.3	11.7
23	36.9	10.5	21.0	5.4	8.5	3.6	10.5
32	36.8	14.0	25.2	6.4	6.0	2.9	5.5

Kemp and Smith (1970) determined the content of 11 fatty acids in different lipid fractions of the mucose membranes isolated from the intestine of the goldfish acclimated to temperatures of from 2 to 32° C. Comparing the data they obtained with those reported by other investigators for goldfish, Kemp and Smith conclude that the modification of fatty acid composition during acclimation to different temperatures varies with different organs. Lipids respond less vigorously in the muscles than in the brain and intestine mucose membranes, to the temperature variation to which the organisms are exposed. An increase in the degree of unsaturation that has been detected in different membranes with a lowering of the acclimation temperature, is achieved at the expense of different fatty acids and different lipid fractions. Changes in the percentage of fatty acids in the intestine mucose membranes of the goldfish are illustrated in Table 24.

Above (p. 244) reference was made to the work of Hazel (1972b) in which the author described a reactivation of the succinate dehydrogenase isolated from the goldfish muscle mitochondria, which occurred when lipids were added to the delipidated enzyme. It is interesting that a higher reactivation capacity was recorded with mitochondria lipids isolated from fish acclimated to 5° C, compared with those from fish acclimated to 25° C. This difference is accounted for by a higher content of unsaturated fatty acids in the mitochondria from the "cold" fish.

Similar unambiguous results testifying to an increase in the relative content of fatty acids with unsaturated bonds, that occurs when the cultivation temperature is lowered, have been obtained in bacteria (Terroine et al., 1930; Gaughran, 1947b; Marr and Ingraham, 1962; Daron, 1970; Ray et al., 1971 and others). In the case of *Mycoplasma laidlawii* grown at 37, 22, and 15° C, an elevation of the content of unsaturated fatty acids in isolated membranes was noted in the cells cultivated at lower temperatures. In addition, electron paramagnetic resonance revealed that this was accompanied by an enhanced mobility of spin labels (Rottem et al., 1970). This work demonstrates most convincingly the augmentation of molecular mobility with the elevation of the content of double bonds in membrane lipids.

The ability of microorganisms to modify their fatty acid composition in response to fluctuations of the ambient temperature is an extremely effective mechanism of adaptation. This is illustrated by the observations reported for *Pseudomonas fluorescens* (Cullen et al., 1971). Bacteria were grown at temperatures from

Table 25. Viscosity of lipids extracted from *Escherichia coli*, grown at different temperatures. (From Sinensky, 1974)

Growth temperature (° C)	Assay temperature (° C)	Relaxation time (nsec)	Viscosity coefficient (poise)
15	15	2.8	1.8
30	30	2.7	1.9
37	37	2.6	1.8
43	43	2.7	2.0
43	15	13.8	15.0

5 to 33° C. Above 34° C they could not grow. The lipid composition did not change markedly, however the degree of unsaturation of fatty acids increased as the temperature of cultivation was lowered. Using the X-ray diffraction technique, the authors investigated the structure of phosphatidyl ethanolamine which constitutes approximately 75% of the total phospholipid content. It was found that the structure of this phospholipid is identical in the bacteria grown at 5 and 22° C, if the structure is compared at respective growth temperatures. Consequently, the state of fatty acids is adjusted by the cell to ambient temperature via modification of the degree of saturation.

Very impressive findings have been communicated by Sinensky (1974). Working with *Escherichia coli* grown at 15, 30, 37, and 43° C, he isolated a phospholipid fraction and a membrane fraction into which a spin label had been introduced. By estimating the relaxation time with the aid of electron paramagnetic resonance, Sinensky calculated the coefficient of viscosity of the fractions. The latter strongly depended on the temperature during an assay. If, however, measurements were performed on each fraction at a temperature corresponding to the respective cultivation temperature for the cells, then the viscosity in all cases proved to be constant (Table 25). The temperature of a phase transition of lipids also varied as a function of the cultivation temperature and has always been 16–14° C lower than the latter. Sinensky believes that the "homeostasis" of viscosity is maintained through an adaptive change in the length and saturation of fatty acids in the lipids, depending on the temperature for the life of the bacteria concerned.

With the spin label method and the electron paramagnetic resonance technique, Esser and Souza (1974) studied the state of lipids in spheroplast membranes isolated from the thermophilic strain of *Bacillus stearothermophilus* growing at a maximal rate at 72° C, and from a thermosensitive mutant unable to grow beyond 58° C. Fluctuations of the cultivation temperature resulted in changes of the fatty acid content, especially for the fatty acids with a melting point above 55° C. In this way the cells maintained the same state of their phospholipids in the membranes at different temperatures. That state corresponded to the lateral separation of phases. The thermosensitive mutant, in contrast to the thermostable strain, fails to adjust the composition of its fatty acids to that required for life at temperatures above 52° C. These investigations made the authors believe that the maximal and minimal temperature for bacterial growth is limited by the beginning and completion of a phase separation of the membrane lipids.

Changes of the fatty acid composition may occur within a rather short period of time. After transfer of *E. coli* from 37 to 10° C, a new pattern of the fatty acid composition, characteristic of the lower temperature, is established as early as in 24 h (Shaw and Ingraham, 1965).

In sum, the data discussed above testify to the capacity of cells for reactive modification of the degree of unsaturation of the fatty acids in their lipids and fats, when the ambient temperature is changed.

The problem of genotypic determination of fatty acid composition has been investigated much less. One of the earliest studies was carried out by Pigulevsky (1916). He analyzed the seed oils from 25 plant species of *Rosaceae*, *Ericaceae*, and Coniferae from different temperature zones and showed that the iodine number was higher in the species characteristic of more northern areas. Considering, however, an unusual phenotypic lability of fatty acids, a comparison of the degree of unsaturation between different species living at different temperatures does not warrant any conclusions to be drawn as to the genotypic determination of the observed dissimilarities. On the other hand, investigations involving comparison of different species or varieties, exposed for a sufficient length of time to the same temperature, are very few. The studies on two alfalfa varieties: the cold-resistant Vernal and cold-sensitive Caliverde, may be mentioned (Kuiper, 1970). Despite identical temperature conditions for growth, the lipid composition in the leaves was different. Thus, the monogalactosyl diglyceride fraction of linolenic acid (three unsaturated bonds) comprised 88% in the cold-resistant variety, whereas in Caliverde it was 57%. The author explains this difference as due to a higher cold resistance of the Vernal variety. Shen et al. (1970) compared the fatty acid composition in three mesophilic and three thermophilic species of bacteria. They concluded that the thermophiles contained less fatty acids with a low melting point than the mesophiles. Taking into consideration the fact that the mesophiles were grown at 37° C and the thermophiles at 55° C, one can hardly believe the difference observed in these studies to be unambiguously determined genetically. On the other hand, Kates and Baxner (1962) found that psychrophilic strains of the yeast *Candida* contain more unsaturated fatty acids than do the mesophilic strains, also when both strains are grown at the same temperature of 10° C.

Interesting though rather indirect data are available on the differences in membrane lipids associated with different heat and cold resistance of species. Above (p. 244), evidence was discussed indicating that a discontinuity in the Arrhenius plots of a number of reactions catalyzed by membrane-bound enzymes is induced by phase transitions of hydrocarbon chains of the fatty acids in membrane lipids. Lyons and Raison (1970a, 1970b) found that the Arrhenius plot of temperature dependence of the maximal rate of oxidation of succinate by mitochondria shows a break in case of the mitochondria from homeothermal animals (rat, marmot) and mitochondria from cold-sensitive plants (batata, cucumber, tomato). No discontinuity is observed in the case of mitochondria from poikilothermal animals (fish) and cold resistant plants (cauliflower, potato tubers, beetroots). The authors explain this difference as being due to the absence of a phase transition of lipids because of a high content of unsaturated fatty acids in the membranes of poikilothermal animals and of the cold resistant plants over the whole range of temperatures encompassed by an Arrhenius plot. In the marmot, during hibernation when the body temperature of the animal goes down to 4° C, a discontinuity in the Arrhenius plot of succinate oxidation by mitochondria disappears (Raison and Lyons, 1971). The reason for this may be either an increase in the relative amount of unsaturated fatty acids, or appearance of an unknown cryoprotective substance.

Consequently, the whole body of evidence testifies to the primary importance of maintaining a certain level of mobility of hydrocarbon fatty acid residues in lipids to preserve normal activity of the cell. The degree of mobility is a temperature-dependent parameter. At a certain temperature, a phase transition occurs from a highly mobile state to an ordered packing of chains. A temperature-zone optimum for functioning of the cells is above the transition temperature. Considering the fact that the cell regulates its fatty acid composition adaptively, one might conclude that the cell is hurt not only by an excessive orderedness, but also by an excessive labilization of the fatty acid changes. The latter must be found in a semilabile state, otherwise catastrophic situations may ensue.

According to some investigators, the lower and upper temperature limits for life may be determined by a modification of the state of fatty acids. Earlier data in favor of this point of view may be found in a monograph by Belehrádek (1935). Later, these views were shared by a number of workers. The fatty acid composition of the larvae of the diptera insect *Pseudosarcophaga affinis* was changed by supplying a mixture of fats with different degrees of saturation. The heat stability of the larvae was then tested by heating them to 45° C. It was found that the lower the iodine number, the higher was the heat resistance of the larvae (House et al., 1958). Similar experiments, however, aimed at disclosing the effect of fatty acid composition on cold resistance, have been carried out on albino rats (Huttunen and Johansson, 1963). One group of albino rats was fed on maize oil, the other one on coconut oil; the latter contains more saturated fatty acids and has a higher melting point. After three months of this dietary regime, the animals were cooled and the temperature at which the heart beat stopped was determined. An arrest of cardiac activity in the animals fed on the more saturated fat occurred at 12.9° C, whereas in those maintained on the less saturated oil—at 7.2° C.

In the study objects mentioned above and under the heating or cooling conditions employed, the limiting factor was evidently a change in the state of the fatty acids in lipids. In a number of other instances, however, one can prove that it is not a rule that lipids are responsible for determining limits for the cold and heat resistance of the organisms concerned and their cells. Fraenkel and Hopp (1940) could not find a difference in the iodine number of the lipids in two meat fly species, although the latter differed considerably in their thermoresistance. A series of experiments was performed in the laboratory of B. Ushakov (Ushakov and Glushankova, 1961; Glushankova, 1963a, 1963b; Amosova, 1963; Kusakina, 1963b; Sleptsova, 1963), which clearly show that the heat stability of muscle fibers is determined by thermostability of the proteins rather than by the degree of saturation of the fatty acids. At this point one example from this series of studies may be given. In the meat fly larvae grown at 33–34° C, the iodine number of the lipids in the muscle tissue was lower than that in the larvae kept at 13–14° C. In spite of this difference, the heat resistance of the muscles in the "cold" and "warm" larvae was absolutely the same (Amosova, 1963). Shaw and Ingraham (1965) advanced arguments against the importance of unsaturation of fatty acids to the control of lower temperature limits for life of *Escherichia coli*. Likewise, doubts have been expressed as to the reasons for growth limitation of the yeast at low temperatures (McMurrough and Rose, 1971).

Consequently, extreme temperature shifts may, in some cases, affect the physical state of the fatty acids that may be the *locus minoris resistentia*, whereas in other cases, the role of the Achilles' heel is played by other systems. This fact in no way diminishes the importance of the level of fatty acid lability to the functioning of cells in optimum temperature environments. Actually, had this not been the case, the surprising mechanisms that regulate the composition of fatty acids during temperature variations would not have evolved in cells.

By what means does the cell achieve an adaptive modification of the fatty acid composition in response to a deviation of the temperature from an optimum? Not much effort has been exerted solving this problem, which is, however, a very interesting one indeed, since it is associated with one of the most important mechanisms involved in cellular homeostasis. In the first place, clarification was necessary as to whether the temperature itself was a trigger that started a mechanism for adaptive modification of the degree of saturation of fatty acids. Brown and Rose (1969) grew the yeast *Candida utilis* in a chemostat culture. The proportion of different saturated and unsaturated fatty acids was varied with both changes of temperature and of oxygen concentration. At a given oxygen pressure, lowering the temperature increased the percentage of unsaturated fatty acids. Provided the temperature was kept constant, the same effect could be obtained with elevation of the oxygen pressure. Harris and James (1969) showed that in *Narcissus* bulbs deprived of chlorophyll, the lowering of temperature resulted in an enhancement of synthesis of fatty acids with unsaturated bonds. An increase in the oxygen concentration had a similar effect. In the chlorophyll-containing cells (*Chlorella*, castor beans and spinach leaves) lowering of temperature in the dark led to an elevation of the ratio of unsaturated fatty acids to the saturated stearic acid. This was not observed, however, when the illumination was adequate. The authors explain the negative result as due to the production of oxygen in the light and conclude that the ratio of unsaturated/saturated fatty acids is regulated by oxygen concentration in the water phase of the cell. The feasibility of mediation of temperature effects on fatty acid composition via oxygen concentration has been acknowledged also by other workers (Sumner et al., 1969; Kleinschmidt and McMahon, 1970a, 1970b). Further investigations will help to evaluate the validity of the assumptions in the cases cited above that the cells use oxygen concentration as a thermometer.

Changes in the ratio of unsaturated to saturated fatty acids in the membranes can be achieved by various means: (1) by varying the rate of synthesis and breakdown of the acids with a different number of unsaturated double bonds; (2) by varying the composition of the fatty acids incorporated into membranes; (3) by varying the activity of the desaturating enzymes that catalyze the formation of double bonds. It is a difficult task at this moment to evaluate the roles played by these processes in the control of fatty acid composition. I will refer only to some of the interesting investigations relevant to this topic. Fulco (1967, 1972) showed that a number of *Bacillus* species are able to desaturate palmitic acid, which has no double bonds, thus transforming it into hexadecenic acid with one unsaturated bond. This desaturation occurred when palmitic acid was added to a medium at 20° C. No desaturation could be observed at 30° C. The temperature of a preliminary exposure was of no importance. Fulco studied the effect of temperature on the amount of the desaturating enzyme and using an inhibitor of protein synthesis, he showed that the rate of synthesis of the desaturating enzyme strongly depends on temperature. This synthesis is induced by lowering the temperature. Moreover, at an elevated temperature around 34° C, the desaturating enzyme is irreversibly inactivated.

Cronan (1974) isolated an *Escherichia coli* clone, in which the cells carried the double *fab*A gene that codes for β-hydroxydecanoyl thioester dehydrase. This enzyme increases the amount of unsaturated fatty acyls in phospholipids. During continuous growth of the cells at 15, 37, and 42° C, the amount of unsaturated fatty acyls in the clone, bearing the double *fab*A gene, only slightly exceeded that in a normal clone exposed to the same temperature. With transfer of the cultures from 37° C or 42° C to 15° C in both the normal cells and those carrying the double gene, the content of unsaturated fatty acyls in phospholipids increased so that in the latter the amount of unsaturated fatty acyls doubled. With time this difference between the normal and clonal cells containing the double gene grew less. Cronan concluded that unsaturated fatty acids are not synthesized in the cell until they can be incorporated into phospholipids. Therefore, one must admit a mechanism exists in the cell that controls the amount of unsaturated fatty acids in phospholipids, apart from their biosynthesis.

Such a mechanism was detected by Sinensky (1971) in *E. coli*, when he studied phospholipid synthesis in this organism. He showed that a temperature-dependent control is exerted upon the reactions of transacylation during the synthesis of phosphatidic acid from α-glycerophosphate. If this reaction is run in vitro in the presence of the co-enzyme-A derivatives of palmitic (16_0) and oleic (18_1) acids, then, along with elevation of the reaction temperature, the acylation of palmitic acid increasingly predominates over that of oleic acid.

Summary. The level of mobility of hydrocarbon chains of fatty acids in lipids is important to the functioning of the enzyme systems in cellular membranes and to the membrane permeability. The fatty acid residues must be in a semilabile state in order that the enzymes might function normally. Lowering or elevating the temperature beyond the optimum zone leads, respectively, to either excessive solidification or liquefication of fatty acids. Such deviations from an appropriate level of lability adversely affect the enzyme activity. A modification of the phase state of fatty acids, by lowering of temperature in particular, yields a dramatic increase in the activation energy of various enzymatic processes that is expressed in a discontinuity of an Arrhenius plot. Animal and plant cells as well as microorganisms are capable of modifying the fatty acid composition of their lipids adaptively and of varying the proportions of lipid fractions in response to an elevation or lowering of the environmental temperature. At temperatures below the optimum, the relative content of unsaturated fatty acids with a lower melting point increases. Elevation of the temperature above the optimum induces cells to increase compensatorily the relative amount of those saturated fatty acids which melt at higher temperatures. It is not entirely clear which mechanism produces these adaptive reactions. However, it is obvious that cells use various means for temperature-dependent control of the fatty acid composition of lipids. In addition no solution can so far be provided to the problem of whether the agent initiating a compensatory modification of the fatty acid composition is a temperature shift itself, or whether the immediate trigger is a shift in oxygen concentration, which, in its turn, is evoked by temperature fluctuation. An excessive structuring or liquefication of fatty acids in some cases may be a factor in determining the resistance of organisms and cells to the action of cold or heat.

The aggregate state of lipids is affected, aside from higher or lower degrees of saturation of fatty acids, also by a substitution of a long-chain fatty acid for a short-chain one, and of a straight one for a branched fatty acid; as well as by incorporating higher or lower amounts of cholesterol in lipids. Dearth of information concerning the relation of these characteristics to temperature ecology of organisms precludes any detailed discussion.

Epilogue

When considering the structure of the most sophisticated machines made by man, one is overwhelmed by the intricate and most ingenious solutions found in tackling various technological challenges. Only during the several hundred years of the evolution of such machines has the human mind created and made use of a multitude of excellent ideas. All this, however, appears primitive and faint, when compared with the picture that is disclosed to the investigator trying to gain insight into a cell's life. And the more we understand about the construction and functioning of a cell, the more we are impressed by the extraordinary complexity of the system and the perfection of its control, combined with its astonishing miniaturization. It has taken hundreds of millions of years for the unhurried genius of nature to create this system. The end result, however, is an object of unfathomable complexity, that has enough secrets in stock to give away gradually and grudgingly under the torture of experiments set up by cytologists, geneticists, biochemists and biophysicists for hundreds of years to come.

Among the vast multitude of consummate and sophisticated mechanisms discovered in studies of the cell, one is most impressed by adaptations of the cell to various environmental factors, and, primarily, to the omnipresent and all-pervading factor—temperature. The most diverse mechanisms of temperature adaptations, operating at the supracellular levels of organization of the living, have been left outside the scope of the present book. Only adaptations at the cellular and molecular levels have been dealt with. Even these constraints imposed on the topics allowed us to consider only some facets of the problem. The striking variety of means utilized in temperature adaptations at the cellular and molecular levels is well illustrated in a review by Hazel and Prosser (1974). On the other hand, an approach employed in this book to analysis of temperature adaptations, both genotypic and modificational ones, disclosed regularities of general biological significance which extend far beyond the problem of adaptation of organisms to the environmental temperature. The wealth of evidence gained in studies on animal and plant cells, microorganisms, proteins, nucleic and fatty acids, which is discussed in this monograph, testifies to the existence of a rather general principle that may be designated the principle of semistability or semilability of main cellular constituents: proteins, nucleic acids and lipids. In all cases we are dealing with maintenance in a state of intermediate stability of various constructions supported by numerous weak bonds. When speaking of proteins and nucleic acids, this means the maintenance of the secondary, tertiary and quaternary structure of macromolecules in a state of neither too high nor too low conformational flexibility. Turning to lipids, this principle applies to the maintenance of fatty acid residues in a neither too solid, nor too liquid state. A state of certain semistability is apparently indispensible for the normal functioning of these components in their performance of the functions for which they are responsible, as well as for the purposes of regulation.

It has been shown in many proteins that changes in the conformation of their macromolecules accompany their activity. Again the same holds true for nucleic acids during reproduction, transcription and translation. The degree of liquefication of the fatty acid residues influences the activity of membrane-bound enzymes. The level of stability of all these three components depends on temperature. An increase in temperature above an optimum results in excessive augmentation of lability, a decrease leads to excessive stability. A significant deviation from the level of semistability of cellular components established in evolution must interfere with the cell's activity.

Cyto and molecular ecological investigations reveal modificational and genotypic mechanisms, which regulate the level of stability during fluctuations of the habitat temperature. Modificational adaptations of the level of protein stability underlie the temperature adjustment of *Protozoa* and algae as well as heat and cold hardenings. Changes in the degree of saturation of fatty acid residues in lipids should also be regarded as modificational adaptations. On the other hand, among genotypic adaptations one finds the maintenance of a correlation between the level of conformational flexibility of protein molecules and of nucleic acids in nucleoproteins, and the temperature of a species' life.

In the case of fatty acids, a conclusion has been drawn from direct measurements of the degree of saturation that determines the melting temperature of a fatty acid. In contrast, conclusions as to changes in the level of the conformational flexibility of proteins and nucleic acids are based on indirect evidence: by the melting temperatures of these macromolecules, i.e. the temperature of denaturation. In addition, protein research has widely employed a still more indirect approach to comparative evaluation of the level of their conformational flexibility: comparisons of the primary thermostability of cellular functions. The latter method yields information concerning the resistance of proteins which are associated with a chosen function, against the denaturing action of heat in situ, and this offers an advantage over the estimation of heat stability of proteins isolated from the cell. In some cases, one had to be content with a comparative assessment of the general cellular thermostability that was not always correlated with the primary thermostability.

It can readily be seen that a physical approach to the estimation of the conformational flexibility level of individual proteins in a living cell would be the most convincing one. Such a method, however, has not yet been made available. If, some day, such techniques are devised, one should bear in mind that it is only the functional mobility of macromolecules, i.e. the mobility associated with the performance of intrinsic biochemical functions, that is biologically significant. Only that mobility is expected to be controlled by natural selection. At the same time, functional mobility may not always coincide with the total mobility of macromolecules, that we can so far estimate only by their resistance to the denaturing action of heat, and that might be evaluated in a more direct way by the rate of exchange of slowly exchangeable hydrogen atoms for deuterium or tritium. In this regard, a method designed for the evaluation of protein lability based on the resistance of proteins to proteinases is more informative. By using the latter approach, one obtains information on the resistance of protein macromolecules which is important for assessment of their halflives in the cell. In other words, this method serves to estimate one of the variables of the functional lability of the protein.

Despite the great complexity of the problem, a large number of facts obtained in various indirect ways make one believe that the observed correspondence of thermostability of cells and proteins in animals, plants and microorganisms to the environmental temperature conditions of life reflects, with a high degree of probability, the necessity to maintain the functional conformational flexibility of proteins at a certain semilabile level. The evidence accumulated in studies on thermostability of nucleic acids in nucleoproteins indicates this conclusion is valid also for nucleic acids. Anyhow, this concept sheds light on a number of facts hardly explainable otherwise. It becomes clear why, in species which differ in thermophily, there is a difference in heat stability not only of thermolabile, but of extremely thermostable proteins as well, which play no part in the determination of the level of thermostability of the cell, or of the organism itself for that matter. The melting temperature of nucleic acids in nucleoproteins is also a case in point, although it is correlated with the thermophily of a species. In the course of evolution species have adapted both to warmer and cooler climates. Therefore, confronted with the fact of a correlation between thermostability of the proteins and the thermophily of a species, one should be able to provide answers as to why the more warm-loving species display higher thermostability of proteins as well as why the less warm-loving species exhibit lower heat resistance of their respective proteins. In acknowledging an adaptive significance of the level of thermostability per se, answers can only be given to the first question. If, however, it is assumed that the level of conformational flexibility, which is indirectly reflected in thermostability, is of adaptive significance, it becomes clear that an excessive resistance, rigidity in the cold-loving forms, is unacceptable, just as a too dense lubrication for a machine. Inappropriateness of low conformational stability for the warm-loving forms also becomes readily apparent. This kind of reasoning is obviously also applicable to the nucleic acids which have been less studied in this respect. It also applies fully to the modificational adaptations of the fatty acid composition of lipids.

In my childhood I was frequently told by my parents in a didactic way: "Too much is no good". I could not understand at that time why it was bad to be too beautiful, too kind or too strong. Now the wisdom of those words seems more obvious to me.

Poikilothermal organisms are frequently exposed to temperatures varying over a wide range during an active period of their lives. In related species that differ in their thermophilies, these variations overlap. On the other hand, in related species which differ only slightly in their respective temperature ecologies, one often observes a marked difference in thermostability (flexibility) of their proteins. It is surprising how nature, in the process of evolution, can feel the "average" intensity of a highly variable environmental factor and produce adaptation to exactly that level of intensity. This capability shows most explicitly in the creation of protective coloration and forms of animals. In those cases too, an average color and structure are found.

A correspondence of the level of conformational flexibility of macromolecules to the environmental temperature in phylogenesis may be maintained via selection only when a deviation from this correspondence may inflict damage on a species. In the search for possible points of application of natural selection, so far only two processes have been found which are obviously dependent on the level of conformational flexibility of macromolecules. These are first, the rate of proteolytic breakdown of proteins that is associated with quantitative regulation of the protein content in the cell, and second, a position of the curve of temperature dependence of a maximal rate of enzyme activity. It is highly likely that there is a certain relationship between variations in the conformational flexibility of an enzyme and its affinity to the substrate, co-enzyme and effector. As shown by Hochachka, Somero and a number of other investigators, a change of this affinity may play a major role in both modificational and genotypic temperature adaptations. It can be expected that further research will disclose other channels through which natural selection may exert its "quality-checking control" on the level of conformational flexibility of proteins.

The situation is less clear with nucleic acids. It has been found that an enhancement of rigidity of the DNA structure via formation of complexes with different histone fractions reduces the template activity of DNA so that suppression is deeper, the higher the level of rigidity. Most clearcut results have been gained in studies of the correlation between temperature conditions of a species' life and the melting temperature of RNA in ribosomal nucleoproteins. The function of ribosomal RNA is unknown, but the very existence of a correspondence of their melting temperatures to the environmental temperature makes one suspect that the functioning of rRNA is associated with changes in the conformation of these macromolecules.

As far as fatty acids are concerned, it has been convincingly shown that the melting temperature of their hydrocarbon chains correlates with the activity of membrane-bound enzymes. A deviation of their aggregate state from an optimum level provoked by a temperature shift, sets in action the mechanism that modifies the fatty acid composition of lipids in an appropriate direction.

In the foregoing the correspondence of the level of conformational flexibility of macromolecules to the temperature of environment was covered, because flexibility is a temperature—dependent parameter. The conformational flexibility of a macromolecule is also affected, however, by numerous other intracellular factors. Among these are: ionic strength, concentration of hydrogen and other ions which may affect the rigidity of macromolecular structures in a specific fashion, concentrations of apolar compounds, and hydrostatic pressure—a factor in the life of abyssal forms, etc. Consequently, regulation of the level of conformational flexibility of macromolecules, both genotypic and modificational, must be correlated, not only with the temperature, but also with other factors which affect the forces responsible for the conformation of a macromolecule. This warrants the contention that variations in the level of conformational rigidity of molecules may be related to changes of non-temperature factors as well. Actually, it has been found in studies on modificational changes of the resistance of cellular proteins that this resistance may be enhanced not only by temperature adjustments, but also by water deficiency, wound injury and, as discovered most recently, may be produced by the action of high hydrostatic pressure in moderate doses and by centrifugation (experiments of E. I. Denko, unpublished). On the other hand, a decrease in stability of protoplasmic proteins could be obtained not only in *Protozoa* and algae during their adjustment to low temperature, but also in the process of adaptation of plant cells to heavy water.

These facts make it appear likely that there is a mechanism(s) in the cell, which, depending on circumstances, is able to modify the level of conformational flexibility of cellular proteins in both directions, i.e. either lowering or elevating it. The view that the mechanisms regulating the level of flexibility have been designed during evolution not only for temperature changes, is confirmed by the discovery of such mechanisms also in organisms that live practically under constant temperature conditions, for instance in the algae taken near the shore of the Franz-Joseph Land, where the water temperature is essentially constant around $-1.8°$ C. These algae increase their primary thermoresistance in response to heating at 30–35° C for several minutes. The plants have never been exposed to such temperature stresses, and it is concluded that a thermal shock in this case plays the role of an unspecific stimulus, setting in motion a mechanism designed to react to other natural stimuli.

Emphasis should be placed on the peculiarities of the reactive mechanisms regulating the level of flexibility of protein macromolecules, in particular, the specificity of heat hardening. Diverse mechanisms regulating the activity of individual links in the metabolic network have been found to operate in the cell. These mechanisms are based on strictly specific individual associations between an effector and a concrete protein. One example is the allosteric suppression of enzyme activity by the end-product of that reaction sequence, which is initiated by the enzyme. If, at the beginning of the branching of a metabolic sequence, different isozymes are involved, then each of these reacts to the end product of its respective branch. This type of regulation may be compared with orders given by a responsible administrator to an appropriate executive officer by telephone or by a letter designed for "personal delivery". Such a "telephone" mode of regulation is extensively employed by the cell for diverse tasks under the control of both enzyme activity and protein synthesis. Besides, we find in the cell various regulating factors which simultaneously and in like fashion affect many metabolic links. Take, for instance, ATP, the concentration of which in the cell influences numerous energy-dependent processes, as well as the cyclic AMP that controls synthesis of a number of proteins. This type of regulation resembles the transmission of an order through a broadcasting system for information and coordination of many, if not all, citizens. Among such "broadcasting" regulations one finds an increase in the protein stability produced by hardening. In fact, heat hardening has been shown to elevate simultaneously the stability of numerous proteins associated with diverse cellular functions.

Based on the principal concept exposed in this book, it should be acknowledged that an overall enhancement of the stability of protoplasmic proteins under given conditions must result in a deceleration of the rate of cellular metabolism.

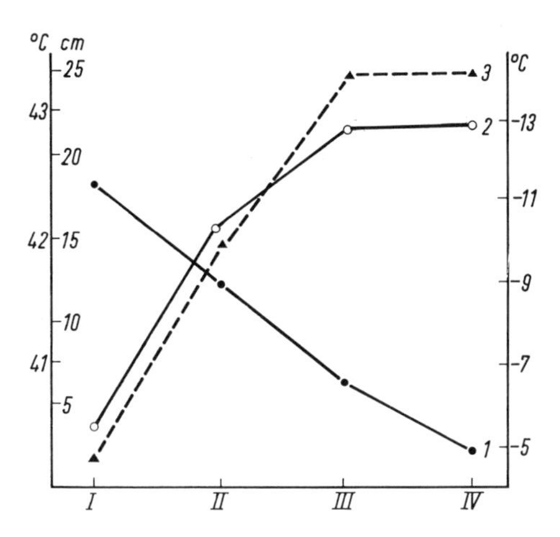

Fig. 74. Changes in resistance of *Gagea lutea* epidermal leaf cells to cold *(1)* and heat *(2)*, and changes in length of leaves *(3)* in the course of the plant's development (Alexandrov et al., 1964). *Abscissa:* phases of development (*I:* appearance of seedlings, *II:* flower bud formation, *III:* florescence, *IV:* fruit bearing); *left ordinate left:* maximal temperature at which protoplasmic streaming persists after 5-min heating (° C); *left ordinate right:* length of leaf (cm); *right ordinate:* minimal temperature (° C) at which protoplasmic streaming persists after 5-min cooling

An increase in their conformational lability within certain limits, on the contrary, would stimulate the metabolism. It is conceivable that these events might be involved in determining the higher thermolability of proteins in growing cells and tissues, and be responsible for a decrease in the lability with decline of growth. That might also be the case with the high stability of protoplasmic proteins during transition of the cells into a state of more or less complete rest (diapauses, cysts, spores, etc.). In many plants, an elevated cold resistance during the winter rest is associated with high thermostability. In the early spring plants, however, active growth occurs against the background of high resistance of their cells to cold, the resistance to heat being rather low (Alexandrov et al., 1964). As the weather becomes warmer and the growth slows down, the cold resistance also decreases, while the resistance to heat increases (Fig. 74).

Association of high sensitivity to the action of various injurious agents with the intensive metabolism of a living being has been repeatedly demonstrated in various fields of biology. This idea was developed in the classical works of Child (1915, 1928, 1941). Referring to a large number of examples taken from the animal and plant kingdoms, he showed that a gradient of sensitivity and metabolic intensity exists along the axis of bodies or organs of many organisms. A decrease in the sensitivity is paralleled with a reduction in the intensity of metabolism.

It is evident from the above that the biological significance of the regulation of the level of confarmational flexibility of macromolecules goes for beyond the limits of the problem of adaptation of organisms to the environmental temperature. On the other hand, this regulation is not the only one possible for adaptation to the environmental temperature at the cellular and molecular levels. A temperature adaptation may involve such processes as variations in the rate of synthesis of various compounds, a switch over to other metabolic routes, a modification of the affinity of enzymes to the metabolites with which they interact, changes in the activation energy of enzymatic reactions and, probably, other processes which at this time can hardly be linked to the shifts in conformational flexibility of macromolecules.

Compensatory changes in the stability of protein macromolecules may be used in temperature adaptations to the extent determined by the influence of temperature variations on the level of their stability. A temperature increase that reduces the stability of protein macromolecules may be counterbalanced by the strengthening of the bonds supporting the secondary, tertiary and quaternary structures. A decrease in temperature should be counteracted by a reverse process. At the same time, as follows from the data discussed on page 146 and illustrated by Figure 45, a maximum of protein stability corresponds to a certain temperature. Both lowering and elevation of this temperature below or above this point results in labilization of the protein. In the zone of a maximum, the curve of temperature dependence of protein stability is rather flattened. Within this zone, temperature shifts show almost no effect on the level of stability of a protein macromolecule. It is clear that adaptations to temperature fluctuations within this zone should not be accompanied by changes in the level of flexibility of protein macromolecules. Consequently, both genotypic and modificational adaptations of organisms to temperature variations over this range of temperatures should not be accompanied by shifts in the heat stability of proteins and primary thermostability of cells. Supporting evidence for the view that such a situation might indeed occur emerges from investigations, so far of a rather preliminary nature, carried out with plants of the Franz-Joseph Land. It has been found that the primary cellular thermoresistance of those plants proves to be no lower than that observed in related plants growing on the Taimyr Peninsula, where the climate is much warmer (experiments of Shukhtina and Alexandrov unpublished). Adaptation of the plants to the exceptionally severe climatic conditions on the Franz-Joseph Land has proceeded apparently not by way of further labilization of protoplasmic proteins, but must have utilized numerous other cellular, organismal and biocenotic possibilities. With a lowering of the temperature below the point of the protein's maximal stability (Fig. 45) a labilization of the protein molecule occurs. Hence it follows that adaptation to such a low temperature must be accompanied not by a reduction, but by an enhancement of the rigidity of the protein molecule. This is probably the reason for the elevation of thermostability of cells and proteins that is observed in some plants in winter. The reasoning given above is purely theoretical. In order to substantiate or repudiate these logical considerations, the temperature dependence of protein stability in a number of poikilothermal animals living in cold regions should be investigated, followed by an evaluation of the temperature corresponding to a maximum of protein stability.

In spite of these reservations, the presence of a correspondence of the level of conformational flexibility of proteins to the environmental temperature is to be acknowledged as an important factor in the temperature adaptations of organisms. So far little information has been gained as to what biochemical means are used to maintain this correlation. To discover this secret comparative investigations of the primary and spatial structure of homologous proteins of thermo-, meso- and psychrophilic organisms are needed. Nonetheless, indirect evidence demonstrates that nature has again displayed an extraordinary ingenuity. Modifications of amino acid compositions and sequences have apparently been used, as well as changes in ligand compositions and variations of numerous physico-chemical parameters of the intracellular milieu.

The literature abounds in evidence proving that a substitution of a single amino acid is frequently effective in modifying the stability of a protein macromolecule. This fact should be taken into consideration in arguments advanced by the advocates and opponents of the so-called "non-Darwinian" evolution (for a review see King and Jukes, 1969; Kirpichnikov, 1972). Those who share "non-Darwinian" views on evolution admit that in a population of organisms accumulations of those mutations may occur which have no selective value. At the same time, most mutations, i.e., amino acid substitutions, should, in one way or another, affect the level of conformational flexibility of protein macromolecules, which fact must have biological consequences. This might be taken as another argument against the concept of "non-Darwinian" evolution, a critique of which has been given by Kirpichnikov (1972).

The present book offers a concept of the primary importance of maintaining a correspondence between the level of conformational flexibility of protein macromolecules and nucleic acids, on the one hand, and the intensity of the factors affecting their flexibility, on the other. The same holds true also for the aggregate state of lipids. The most important factor among these is temperature. This concept and a number of corollaries are substantiated with various more or less indirect arguments. Much in this concept still needs further corroborative evidence. However, the author consoles himself with the thought that the stimulating power of an hypothesis is, in fact, reciprocally proportional to the degree of its substantiation. In any case, the experimental material discussed herein proves that time is ripe for a combination of both cellular and molecular approaches to attack various ecological problems. There are alle grounds at present to speak not only of cytecology, but of molecular ecology, which should be regarded as one of the branches of molecular biology. The facts presented justify still another important conclusion, in terms of strategy of research: it is not only in ecology that the molecular biologist must give a hand in solving various problems. A number of questions relating to operation and evolution of biological macromolecules cannot be answered without consideration of the ecology of the study objects, from which the macromolecular material is taken for research.

References

Aach, H. G.: Vergleich des Eiweißanteiles des Wildstammes vom Tabakmosaikvirus mit dem einer Temperaturmutante. Z. Vererbungslehre **91**, 312–315 (1960)

Abrahamson, Y., Maher, M.: The effects of temperature upon pancreatic amylase in selected reptiles and in amphibians. Can. J. Zool. **45**, 227–232 (1967)

Adensamer, E.: Über den Verlust der Leitfähigkeit des Nervus ischiadicus durch Erwärmung bei Lacertiliern. Z. vergl. Physiol. **21**, 642–645 (1934)

Adye, J., Koffler, H., Mallett, G. E.: The relative thermostability of flagella from thermophilic bacteria. Arch. Biochem. Biophys. **67**, 251–253 (1957)

Ageeva, O. G., Lutova, M. I.: The influence of heat hardening of peas *Pisum sativum* L. on photosynthesis and photochemical reactions. Bot. Zh. **56**, 1365–1373 (1971)

Aharonowitz, Y., Ron, E. L.: A temperature sensitive mutant in *Bacillus subtilis* with an altered elongation factor. G. Molec. Genet. **119**, 131–138 (1972)

Ahmed, N., Rowbury, R. J.: A temperature-sensitive cells division component in a mutant of *Salmonella typhimurium*. Gen. Microbiol. **67**, 107–115 (1971)

Aizenshtat, B. A.: Some data on the cotton leaf temperature. In: Trudy Tashkentskoi geophizicheskoi observatorii (Collection of works Tashkent Geophys. Observ. **7**, Tashkent, 1952, pp. 63–70

Alden, J., Hermann, R. K.: Aspects of the cold-hardiness mechanism in plants. Bot. Rev. **37**, 37–142 (1971)

Alderton, G., Thompson, P. A., Snell, N.: Heat adaptation and ion exchange in *Bacillus megatherium* spores. Science **143**, 141–143 (1964)

Alexandrov, S. N., Galkovskaya, K. F., Lozina-Lozinsky, L. K.: On thermostability of the isolated tissue and the organism of the lake frog found in the waters of the hot spring in Zhelesnovodsk. Tsitologia **2**, 442–447 (1960)

Alexandrov, V. Ya.: Denaturative alterations of proteins during physiological processes. Usp. sovr. biol. **24**, 45–60. Corrections of errors in the paper, ibid **24**, 3 (1947)

Alexandrov, V. Ya.: Denaturation and the reaction of cell to external influences. In: The Collection of Works (Sbornik) Soveshchanie po belku (Conf. on Proteins), Moscow and Leningrad: 1948, pp. 95–101

Alexandrov, V. Ya.: A method for vital staining of tissues and organs of mammals by basic dyes. Tr. Akad. Med. Nauk SSSR **3**, 10–15 (1949)

Alexandrov, V. Ya.: On the relationship between thermostability of protoplasm and temperature conditions of life. Dokl. Akad. Nauk SSSR **53**, 149–152 (1952)

Alexandrov, V. Ya.: A simple method of infiltration of plant tissues. Bot. Zh. **39**, 421–422 (1954)

Alexandrov, V. Ya.: Cytophysiological estimation of different methods of determination of plant cells vitality. Trudy Botan. Inst. Akad. Nauk SSSR, ser. 4, Exp. Bot. **10**, 309–355 (1955)

Alexandrov, V. Ya.: Cytophysiological analysis of thermoresistance of plant cells and some problems of cytoecology. Bot. Zh. **41**, 939–961 (1956)

Alexandrov, V. Ya.: Die Eiweiß-(Denaturations-)Theorie der Erregung. 2. Intern. Symp. über den Mechanismus der Erregung. Berlin: 1959, pp. 12–21

Alexandrov, V. Ya.: Application of the media improving a microscopic picture in the study of plant cells *in vivo*. Tsitologia **4**, 84–88 (1962)

Alexandrov, V. Ya.: Cytophysiological and cytoecological investigation of resistance of plant cells to the action of high and low temperatures. Trudy Botan. Inst. Akad. Nauk SSSR, ser. 4, Exp. Bot. **16**, 234–280 (1963)

Alexandrov, V. Ya.: A study of the changes in resistance of plant cells to the action of various agents in the light of cytoecological consideration. In: Kletka i temperatura sredy [Collection of works (Sbornik) The Cell and Environmental Temperature) Moscow, Leningrad: Izd. Nauka 1964a, pp. 98–104

Alexandrov, V. Ya.: The problem of autoregulation in cytology. II. The reparatory capacity of cells. Tsitologia **6**, 133–151 (1964b)

Alexandrov, V. Ya.: The problem of autoregulation in cytology. III. Reactive increase in cell resistance to the action of injurious agents. Tsitologia **7**, 447–466 (1965a)

Alexandrov, V. Ya.: Editor's preface to a Russian edition. In: Biebl, R.: Cytological Basis of Plant Ecology, Moscow: Izd. Mir 1965b, pp. 5–8

Alexandrov, V. Ya.: On the biological significance of the correspondence of the level of protein thermostability to temperature conditions of a species existence. Usp. sovr. biol. **60**, 28–44 (1965c)

Alexandrov, V. Ya.: Protein thermostability of a species and habitat temperature. In: Molecular Mechanisms of Temperature Adaptation (ed. C. L. Prosser). Washington: Am. Ass. Advan. Sci. 1967

Alexandrov, V. Ya.: Conformational flexibility of proteins, their resistance to proteinases and temperature conditions of life. Curr. Mod. Biol. **3**, 9–19 (1969)

Alexandrov, V. Ya., Alekseeva, N. N.: Thermostability of oxidative phosphorylation and its role in the development of thermal injury of plant cells. Tsitologia **14**, 591–597 (1972)

Alexandrov, V. Ya., Andreyeva, A. P.: Comparison of collagen stabilities of two frog species differing by their thermophily. Tsitologia **9**, 1288–1293 (1967)

Alexandrov, V. Ya., Andreyeva, A. P.: Comparison of collagen thermostability in three species of frog with different thermophilies. Tsitologia **16**, 235–237 (1974)

Alexandrov, V. Ya., Arronet, N. I.: Adenosine triphosphate causes the movement of cilia in ciliary epithelium killed by glycerine extraction (a "cellular model"). Dokl. Akad. Nauk SSSR **110**, 457–460 (1956)

Alexandrov, V. Ya., Arronet, N. I., Denko, E. I., Konstantinova, M. F.: The influence of heavy water (D_2O) on the resistance to several denaturating actions of plant and animal cells, cellular models and proteins. Tsitologia **6**, 666–679 (1964a)

Alexandrov, V. Ya., Arronet, N. I., Denko, E. I., Konstantinova, M. F.: Influence of D_2O on resistance of plant and animal cells, cell models and proteins to denaturing agents. Nature **205**, 286–287 (1965)

Alexandrov, V. Ya., Arronet, N. I., Denko, E. I., Konstantinova, M. F.: Effect of heavy water (D_2O) on resistance of plant and animal cells, cell models and proteins to certain denaturing factors. Fed. Proc. **25**, 128–134 (1966)

Alexandrov, V. Ya., Barabalchuk, K. A.: Repair of thermal injury of *Tradescantia* leaf cells after exposure to heat hardening. Tsitologia **14**, 1328–1334 (1972)

Alexandrov, V. Ya., Denko, E. I.: Adaptation of plant cells to heavy water. Stud. Biophys. **4**, 135–143 (1967)

Alexandrov, V. Ya., Denko, E. I.: On the factors responsible for the resistance of cells to the action of high hydrostatic pressure. Tsitologia **13**, 319–328 (1971)

Alexandrov, V. Ya., Denko, E. I., Kislyuk, I. M., Feldman, N. L., Shukhtina, H. G.: Seasonal changes in resistance of cells to the action of different agents in winter-green plants and spring ephemeroids. In: Tsitologicheskiye osnovy prisposobleniya rastenii k faktoram sredy [Collection of works (Sbornik) Cytological Aspects of Adaptation of Plants to the Environmental Factors], Moscow, Leningrad: Izd. Nauka 1964b, pp. 103–127

Alexandrov, V. Ya., Dzhanumov, D. A.: The effect of heat injury and heat-hardening on a photoinduced lasting after-luminescence of *Tradescantia fluminensis* Vell. leaves. Tsitologia **14**, 713–720 (1972)

Alexandrov, V. Ya., Feldman, N. L.: Investigation of the increase of cell resistance, as reaction to high temperature. Bot. Zh. **43**, 194–213 (1958)

Alexandrov, V. Ya., Lomagin, A. G., Feldman, N. L.: The responsive increase in thermostability of plant cells. Protoplasma **69**, 417–458 (1970)

Alexandrov, V. Ya., Lutova, M. I., Feldman, N. L.: Seasonal changes in the resistance of plant cells to the influence of different agents. Tsitologia **1**, 672–691 (1959)

Alexandrov, V. Ya., Shukhtina, H. G.: The state of the protoplast of plant cells in winter. A criticism of the theory of P. A. Henkel and E. Z. Oknina: The Separation of Protoplasm. In: Tsitologicheskiye osnovy prisposobleniya rastenii k faktoram sredy [Collection of works (Sbornik) Cytological Aspects of Adaptation of Plants to the Environmental Factors]. Moscow, Leningrad: Izd. Nauka 1964, pp. 137–154

Alexandrov, V. Ya., Tabidze, D. D.: The effects of some amino acids on thermoresistance of plant and animal cells and on cell models. Tsitologia **14**, 1513–1518 (1972)

Alexandrov, V. Ya., Vitvitsky, V. N.: The serum albumin thermoresistance from two different heat-loving frog species and some problems of cytoecology. Tsitologia **12**, 596–601 (1970)

Alexandrov, V. Ya., Vitvitsky, V. N., Levdikova, G. A.: Effect of heavy water (D_2O) on the resistance of procollagen to the thermal denaturation and to the action of collagenase. Dokl. Akad. Nauk SSSR **220**, 1445–1449 (1975)

Alexandrov, V. Ya., Yazkulyev, A.: The heat hardening of plant cells in nature. Tsitologia **3**, 702–707 (1961)

Alexandrov, V. Ya., Zavadskaya, I. G.: The protoplasm movement as a criterion of the resistance of plant cells to the adverse effects of injuring agents. Bot. Zh. **56**, 389–394 (1971)

Allen, M. B.: The dynamic nature of thermophily. J. Gen. Physiol. **33**, 205–214 (1950)

Allen, M. B.: The thermophilic aerobic spore forming bacteria. Bacteriol. Rev. **17**, 125–173 (1953)

Allen, S. D., Brock, T. D.: The adaptation of heterotrophic microcosms to different temperatures. Ecology **49**, 343–346 (1968)

Allfrey, V. G., Littau, V. C., Mirsky, A. E.: On the role of histones in regulating ribonucleic acid synthesis in the cell nucleus. Proc. Nat. Acad. Sci. USA **49**, 414–421 (1963)

Altenburg, L. C., Saunders, G. F.: Properties of hybrid ribosomes formed from the subunits of mesophilic and thermophilic bacteria. J. Mol. Biol. **55**, 487–502 (1971)

Altergot, V. F.: The action of raised temperatures upon plants. Izd. Akad. Nauk SSSR, ser. biol. **1**, 57–73 (1963)

Altergot, V. F.: Biochemical mechanisms of the death of plants and their tolerance and adaptation to high temperatures in natural conditions. In: The cell and environmental temperature. Moscow and Leningrad: Izd. Nauka 1964, pp. 185–190

Altukhov, Ju. P.: Seasonal changes in thermostability of the isolated muscle tissue of *Trachurus mediterraneus* from the Black sea. Tsitologia **2**, 241–243 (1963)

Altukhov, Ju. P., Nefyodov, G. N., Payusova, A. N.: A cytophysiological analysis of divergence between *Sebactes marinus* and *S. mentella* Travin from Northwest Atlantic. In: Izmenchivost teploustoichivocti kletok zhivotnykh v onto i filogeneze [Collection of works (Sbornik) Variability in Cellular Heat Resistance of Animals in Ontogenesis and Phylogenesis], Leningrad: Izd. Nauka 1967, pp. 82–98

Altukhov, Ju. P., Nefyodov, G. N., Payusova, A. N.: Thermostability of isolated muscles in determining the taxonomic relationship of the marinus- and mentellatypes of the redfish (Sebastes). Intern. Commiss. Northwest Atlantic Fisheries. Res. Bull. **5**, 130–136 (1968)

Altukhov, Ju. P., Ratkin, E. V.: A study of genotypic foundations of the individual variability in thermostability of isolated cells of the silkworm. Tsitologia **10**, 1546–1554 (1968)

Amaha, M., Ordal, L. F.: Effect of divalent cations in the sporulation medium on the thermal death rate of *Bacillus coagulans* var. *thermoacidurans*. J. Bacteriol. **74**, 596–604 (1957)

Amelunxen, R. E.: Some chemical and physical properties of thermostable glyceraldehyde-3-phosphate dehydrogenase from *B. stearothermophilus*. Biochim. Biophys. Acta **139**, 24–32 (1967)

Amelunxen, R. E., Clark, T.: Crystallization of thermostable glyceraldehyde-3-phosphate dehydrogenase after removal of coenzyme. Biochim. Biophys. Acta **221**, 650–652 (1970)

Amelunxen, R. E., Lins, M.: Comparative thermostability of enzymes from *Bacillus stearothermophilus* and *Bacillus cereus*. Arch. Biochem. Biophys. **125**, 765–769 (1968)

Amelunxen, R. E., Noelken, M., Singleton, R.: Studies on the subunit structure of thermostable glyceraldehyde-3-phosphate dehydrogenase from *Bacillus stearothermophilus*. Arch. Biochem. Biophys. **141**, 447–455 (1970)

Amosova, I. S.: Resistance of tissues in certain synanthropic flies *(Diptera, Calliphoridae* and *Mascidae)* to high temperatures. Rev. Entomol. URSS **41**, 816–826 (1962)

Amosova, I. S.: The influence of breeding temperature on the heat resistance of tissues of the blue fly. In: Problemy tsitoekologii zhivotnykh [Collection of works (Sbornik) Problems of Cytoecology of Animals]. Moscow, Leningrad: Izd. Nauka 1963, pp. 102–107

Amosova, I. S.: Selection of the blue fly *Calliphora erythrocephala* for thermostability of its muscle tissue. In: Izmenchivost teploustoichivosti kletok zhyvotnykh v ontogeneze i filogeneze [Collection of works (Sbornik) Variability in Cellular Heat Resistance of Animals in Ontogenesis and Phylogenesis]. Leningrad: Izd. Nauka 1967, pp. 66–70

Amosova, I. S.: Changes in thermostability of muscles from the blue meat fly larvae produced by heat treatment of the preceding generation pupa. Ekologia **6**, 74–77 (1971)

Ancel, P., Vintemberger, P.: Sur la radiosensibilité cellulaire. C. R. Acad. Sci. (Paris) **92**, 517–520 (1925)

Andreyeva, A. P.: The collagen thermostability in frogs *(Rana ridibunda)* of different age, sex and from different populations. Tsitologia **12**, 114–119 (1970)

Andreyeva, A. P.: The collagen thermostability of some species and subspecies of the ganoid fish. Tsitologia **13**, 1004–1008 (1971)

Andreyeva, A. P., Vitvitsky, V. N., Gubnitsky, L. S., Yakovleva, V. I.: Resistance to heat and tryptic digestion of mitochondrial glutamate dehydrogenase from the liver of two frog species different in their thermophily. Zh. evol. biokh. i fiziol. **11**, in press. (1975)

Andronikov, V. B.: Thermostability of the sex cells and zygotes of Sea Urchins. Tsitologia **2**, 234–237 (1963)

Andronikov, V. B.: Heat resistance of sexual cells and embryos of poikilothermal animals. In: Teploustoichivost' kletok zhivotnykh [Collection of works (Sbornik) Heat Resistance of Cells of Animals. Moscow, Leningrad: Izd. Nauka 1965, pp. 125–139

Andronikov, V. B.: The pattern of changes in thermostability of embryos of poikilotherms of early stages of ontogenesis. In: Izmenchivost teploustoichivosti kletok zhivotnykh v onto- i vilogeneze [Collection of works (Sbornik) Variability in Cellular Heat Resistance of Animals in Ontogenesis and Phylogenesis]. Leningrad: Izd. Nauka 1967, pp. 4–12

Andronikov, V. B.: Thermostability of sex cells and embryos of poikilothermal animals. Author's abstract of dissertation, Leningrad: 1968

Ansley, S. B., Campbell, L., Sypherd, P. S.: Isolation and amino acid composition of ribosomal proteins from *Bacillus stearothermophilus*. J. Bacteriol., **98**, 568–572 (1969)

Antonini, E., Wyman, J., Moretti, R., Rossi-Fanelli, A.: The interaction of bromthymol blue with hemoglobin and its effects on the oxygen equilibrium. Biochim. Biophys. Acta **71**, 124–138 (1963)

Antonov, A. S.: Gene systematics: achievements, problems and perspectives. Usp. sovr. biol. **77**, 31–47 (1974)

Antropova, T. A.: Temperature adaptation of the *Mnium affine* Bland. moss cells. Bot. Zh. **56**, 1681–1686 (1971)

Antropova, T. A.: Temperature adaptation studies on the cells of some bryophyte species. Tsitologia **16**, 38–42 (1974a)

Antropova, T. A.: The seasonal changes of cold and heat resistance of cells in two moss species. Bot. Zh. **59**, 117–122 (1974b)

Appel, P., Jang, J. T., Brown, W. D.: Denaturation of deuterated myoglobin. Stud. Biophys. **4**, 77–83 (1967)

Arronet, N. I.: The thermostability of the cells and the organism of *Rana temporaria* L. and *Unio crassus* Philipsson at various seasons. Tsitologia **1**, 443–449 (1959)

Arronet, N. I.: The repair of thermal injury in the ciliary cells of some gastropod larvae. Tsitologia **5**, 222–227 (1963)

Arronet, N. I.: Injurious effects of high hydrostatic pressure and of heating on the ciliary epithelium cells and the protective influence of glycerol. Tsitologia **6**, 432–442 (1964)

Arronet, N. I.: Studies on the action of injuring agents on the cell, employing cellular contractile models. Author's abstract of dissertation, Leningrad: 1966

Arronet, N. I.: The muscle and cell contractile (motile) models. Izd. Akad. Nauk SSSR: Leningrad: 1971

Arronet, N. I., Konstantinova, M. F.: Concerning the site of application of the injuring effects of high hydrostatic pressure on cells. Tsitologia **6**, 743–746 (1964)

Artsikhovskaya, E. B., Rubin, B. A.: Respiration of the plant as an adaptive function. Usp. sovr. biol. **37**, 136–157 (1954)

Aruga, Y.: Ecological studies of photosynthesis and matter production of phytoplankton. I. Seasonal changes in photosynthesis of natural phytoplankton. Bot. Mag. Tokyo **78**, 280–288 (1965)

Ashraf, J.: Comparative resistance of the leaf cells of diploid and autopolyploid plants to some damaging influences. Tsitologia **13**, 472–478 (1971)

Astaurov, B. L.: Ontogenetic alterations in heat resistance of silkworm eggs correlated with seasonal changes in the periods of active development and rest (diapause). In: The cell and environmental temperature. Moscow, Leningrad: Izd. Nauka 1964, pp. 253–264

Astaurov, B. L., Bednyakova, T. A., Vereiskaya, V. N., Ostryakova-Varshaver, V. L.: Deistviye vysokikh temperatur na grenu shelkovichnogo chervya (The Action of Elevated Temperatures on Silk Worm Eggs). Moscow: Izd. Akad. Nauk SSSR 1962

Atkinson, D. E.: Biological feedback control at the molecular level. Science **50**, 851–858 (1965)

Babayan, R. S.: The influence of short-term thermal treatments upon the stability of seeds to X-irradiation and to other injuring agents. Tsitologia **3**, 342–351 (1972)

Babel, W., Rosenthal, H. A., Rapoport, S.: A unified hypothesis on the causes of the cardinal temperatures on microorganisms; the temperature minimum of *Bacillus stearothermophilus*. Acta Biol. Med. Germ. **28**, 565–576 (1972)

Bachmann, K.: Temperature adaptations of amphibian embryos. Amer. Naturalist **103**, 115–131 (1969)

Badanova, K. A.: Changes in the resistance of plants to elevated temperatures as a function of age. In: Collection of Works (Sbornik) commemorating Academician N. A. Maksimov. Moscow: 1957, pp. 26–31

Baldwin, J.: Adaptation of enzymes to temperature: acetyl-cholinesterases in the central nervous system of fishes. Comp. Biochem. Physiol. **40B**, 181–187 (1971)

Baldwin, J.: Development and evolution. London: 1902

Baldwin, J., Aleksiuk, M.: Adaptation of enzymes to temperature: lactate and malate dehydrogenases from *Platypus* and *Echidna*. Comp. Biochem. and Physiol. **44**, 363–370 (1973)

Baldwin, A. N., Berg, P.: Purification and properties of isoleucyl ribonucleic acid synthetase from *Escherichia coli*. J. Biol. Chem. **241**, 831–838 (1966)

Baldwin, J., Hochachka, P. W.: Functional significance of isoenzymes in thermal acclimatization. Acetylcholinesterase from trout brain. Biochem. J. **116**, 883–887 (1970)

Ballinger, R. E., McKinney, C. O.: Developmental temperature tolerance of certain anuran species. J. Exp. Zool. **161**, 21–28 (1966)

Bandas, E. L., Bobovitch, M. A.: The influence of KCl and $CaCl_2$ on the thermostability of frog muscles. Tsitologia **3**, 100–103 (1961)

Bannikov, A. G.: Ecological conditions of the activity of anurous amphibia as a factor limiting the area of a species. Zool. Zh. **22**, 340–344 (1943)

Bannister, P.: The annual course of drought and heat resistance in heath plants from an oceanic environment. Flora **159**, 105–193 (1970)

Barabalchuk, K. A.: Response of thermolabile and thermostable functions of plant cells to the action of heat hardening. Tsitologia **11**, 1021–1032 (1969)

Barabalchuk, K. A.: Changes in the intensity of chlorophyll fluorescence in the leaves of *Tradescantia fluminensis* Vell. following heat treatment. Tsitologia **12**, 1009–1019 (1970a)

Barabalchuk, K. A.: Effects of calcium, manganese, magnesium and sodium ions on the resistance of plants cells. Tsitologia **12**, 609–621 (1970b)

Barabalchuk, K. A., Chernyavskaya, V. N.: Effect of high temperature on stomatal movement of the *Tradescantia fluminensis* Vell. leaves. Tsitologia **16**, 1481–1487 (1974)

Baranska, J., Wlodawer, P.: Influence of temperature on the composition of fatty acids and on lipogenesis in frog tissues. Comp. Biochem. Physiol. **28**, 553–570 (1969)

Barenboym, G. M., Borkhsenius, S. N., Zhirmunsky, A. V., Kosyuk, G. N., Shchelchkov, B. V.: A comparative study of thermal denaturation of deoxyribonucleoprotein complexes in the sperm of two closely related sea urchin species. Abstr. Inst. Marine Biol. (ed. A. V. Zhirmunsky) Izd. Far East Division Siberian Branch USSR Acad. Sci., Vladivostok 29–34 (1969)

Barenboym, G. M., Borkhsenius, S. N., Zhirmunsky, A. V., Shchelchkov, B. W.: Heat denaturation of DNP and DNA from sperm of two closely related species of the sea urchin. Molekuly Biol. **5**, 31–39 (1971)

Barnes, E. M., Akagi, J. M., Himes, R. H.: Properties of fructose-1.6-diphosphate aldolase from two thermophilic and a mesophilic *Clostridia*. Biochem. Biophys. Acta **227**, 199–203 (1971)

Barnes, L. D., Stellwagen, E.: Enolase from the thermophile *Thermus* X-1. Biochem. **12**, 1559–1565 (1973)

Bassel, A., Campbell, L. L.: Surface structure of *Bacillus stearothermophilus* ribosomes. J. Bacteriol. **98**, 811–815 (1969)

Battle, H.: Effects of extreme temperatures on muscle and nerve tissue in marine fishes. Trans. Roy. Soc. Can. Sect. **5**, ser. 3, 127–143 (1926)

Bauer, E. S.: Theoretical Biology. Moscow, Leningrad: Izd. Vsesoyuznogo Instituta Eksperimentalnoi Meditsiny, 1935

Bauer, H.: Hitzeresistenz und CO_2-Gaswechsel nach Hitzestress von Tanne (*Abies alba* Mill.) und Bergahorn (*Acer pseudoplatanus* L.). Diss. Leopold-Franzens-Univ. 1–100. Innsbruck: 1970

Bauer, H.: CO_2-Gaswechsel nach Hitzestress bei *Abies alba* Mill and *Acer pseudoplatanus* L. Photosynthetica **6**, 424–434 (1972)

Bauer, H., Harrasser, J., Bendetta, G., Larcher, W.: Jahresgang der Temperaturresistenz junger Holzpflanzen im Zusammenhang mit ihrer jahreszeitlichen Entwicklung. Ber. Deut. Bot. Ges. **84**, 561–570 (1971)

Bausum, H. T., Matney, T. S.: Boundary between bacterial mesophilism and thermophilism. J. Bacteriol. **90**, 50–53 (1965)

Bayley, S. T.: Composition of an extremely halophilic bacterium. J. Mol. Biol. **15**, 420–427 (1966)

Bayley, S. T., Griffiths, E.: A cell-free amino acid incorporating system from an extremely halophilic bacterium. Biochemistry **7**, 2249–2256 (1968)

Beattie, D. S., Basford, R. E., Koritz, S. B.: The turnover of the protein components of mitochondria from rat liver, kidney and brain. J. Biol. Chem. **242**, 4584–4586 (1967)

Behrisch, H. W.: Temperature and the regulation of enzyme activity in poikilotherms. Fructose diphosphatase from migrating salmon. Biochem. J. **115**, 687–696 (1969)

Behrisch, H. W., Hochachka, P. W.: Temperature and the regulation of enzyme activity in poikilotherms. Properties of rainbow-trout fructose diphosphatase. Biochem. J. **111**, 287–295 (1969a)

Behrisch, H. W., Hochachka, P. W.: Temperature and the regulation of enzyme activity in poikilotherms. Properties of lungfish fructose diphosphatase. Biochem. J. **112**, 601–607 (1969b)

Beilinsson, A.: Thermostabilisation der Eiweißlösungen mit Rohrzucker und Glycerin. Biochem. Z. **213**, 399–405 (1929)

Bělehrádek, J.: Temperature and living matter. Protoplasms Monographien **8**, 187–189. Berlin 1935

Bělehrádek, J., Melichar, J.: L'action différente des températures élevées et des températures normales sur la survie de la cellule végétale (*Helodea canadensis* Rich.). Biol. Generalis **6**, 109–124 (1930)

Belozersky, A. N., Antonov, A. S. (eds.): In: The Structure of DNA and Positions of Organisms in the System. Moscow: Izd. Moscow Univ. 1972, p. 454

Bendall, J. R.: Muscles, molecules and movement. London: Heinemann, 1969

Benesch, R. E., Benesch, R.: The influence of oxygenation on the reactivity of the SH-groups of hemoglobin. Biochemistry **1**, 735–738 (1962)

Berger, L., Slein, M. W., Colowick, S. P., Cori, C. F.: Isolation of hexokinase from baker's yeast. J. Gen. Physiol. **29**, 379–391 (1946)

Berkholz, G.: Über die Temperaturadaptation des Nerflings (*Idus idus* L., *Pisces*) nach inkonstanter Vorbehandlung. Z. wiss. Zool. **174**, 377–399 (1966)

Bernheim, F., Neurath, H., Erickson, I. O.: The denaturation of proteins and its apparent reversal. IV. Enzymatic hydrolysis of native, denatured, and apparently reversibly denatured proteins. J. Biol. Chem. **144**, 259–264 (1942)

Bianchi, C. P.: Cell calcium. London: Butterworth, 1968

Biebl, R.: Über die Temperaturresistenz von Meeresalgen verschiedener Klimazonen und verschieden tiefer Standorte. Jahrb. wiss. Bot. **88**, 389–420 (1939)

Biebl, R.: Ökologie und Zellforschung. Schriften d. Vereines z. Verbreit. naturwissensch. Kenntnisse in Wien: 1950

Biebl, R.: Temperatur und osmotische Resistenz von Meeresalgen der bretonischen Küste. Protoplasma **50**, 217–242 (1958)

Biebl, R.: Temperaturresistenz tropischer Meeresalgen. Botanica Marina **4**, 241–254 (1962a)

Biebl, R.: Protoplasmatisch-ökologische Untersuchungen an Mangrovealgen von Puerto Rico. Protoplasma **55**, 572–606 (1962b)

Biebl, R.: Protoplasmatische Ökologie der Pflanzen. Wasser und Temperatur. Wien: Springer-Verlag, 1962c

Biebl, R.: Temperaturresistenz tropischer Pflanzen auf Puerto Rico. Protoplasma **59**, 133–156 (1964)

Biebl, R.: Influence of short days on arctic plants during the Arctic long days. In: Photochemistry and Photobiology in Plant Physiol. Hvar Yugosl.: Acad. Sci. et Art. Slav. Meridianalium, 1967a, pp. 1–2

Biebl, R.: Protoplasmatische Ökologie. Naturwiss. Rundschau **20**, 248–252 (1967b)

Biebl, R.: Kurztag-Einflüsse auf arktische Pflanzen während der arktischen Langtage. Planta **75**, 77–84 (1967c)

Biebl, R.: Temperaturresistenz einiger Grünalgen warmer Bäche aus Island. Le Botaniste **50**, 33–42 (1967d)

Biebl, R.: Temperaturresistenz tropischer Urwaldmoose. Flora **157**, 25–30 (1967e)

Biebl, R.: Über Wärmehaushalt und Temperaturresistenz arktischer Pflanzen in Westgrönland. Flora B, **157**, 327–354 (1968)

Biebl, R.: Studien zur Hitzeresistenz der Gezeitenalge *Chaetomorpha cannabina* (Aresch.) Kjellm. Protoplasma **67**, 451–472 (1969a)

Biebl, R.: Untersuchungen zur Temperaturresistenz arktischer Süßwasseralgen im Raum von Barrow, Alaska. Mikroskopie **25**, 3–6 (1969b)

Biebl, R.: Resistance adaptation of marine algae at the cellular level. Tsitologia **1**, 3–13 (1969c)

Biebl, R.: Vergleichende Untersuchungen zur Temperaturresistenz von Meeresalgen entlang der pazifischen Küste Nordamerikas. Protoplasma **69**, 61–83 (1970)

Biebl, R.: Studien zur Temperaturresistenz der Gezeitenalge *Ulva pertusa* Kjellmann. Botanica Marina **15**, 139–143 (1972)

Biebl, R., Maier, R.: Tageslänge und Temperaturresistenz. Österr. Bot. Z. **117**, 176–194 (1969)

Bigelow, C.C.: On the average hydrophobicity of proteins and the relation between it and protein structure. J. Theor. Biol. **16**, 187–211 (1967)

Billings, W.D., Mooney, H.A.: The ecology of arctic and alpine plants. Biol. Rev. **43**, 481–529 (1968)

Björkman, O., Florell, C., Holmgren, P.: Studies of climatic ecotypes in higher plants. The temperature dependence of apparent photosynthesis in different populations of *Solidago virga-aurea*. Ann. Roy. Agr. Coll. Sweden **26**, 1–10 (1960)

Björkman, O., Pearcy, R.W., Harrison, A.T., Mooney, H.: Photosynthetic adaptation to high temperatures: a field study in Death Valley, California. Science **175**, 786–789 (1972)

Black, S.H., Hashimoto, T., Gerhardt, Ph: Calcium reversal of the heat susceptibility and dipicolinate deficiency of spores formed "endotrophically" in water. Can. J. Microbiol. **6**, 213–224 (1960)

Blagoveshchensky, A.V.: Biochemical Basis of the Evolutionary Process in Plants. Moscow, Leningrad: Izd. Akad. Nauk SSSR 1950

Blagoveshchensky, A.V.: Biochemical Evolution of the Flowering Plants. Moscow: Izd. Nauka 1966

Blake, C.C.F., Jonson, L.N., Mair, G.A., North, A.C.T., Phillips, D.C., Sarma, V.R.: Crystallographic studies of the activity of hen egg-white lysozyme. Proc. Roy. Soc. Ser. B. **167**, 378–388 (1967)

Blumenfeld, L.A.: Problems of Biological Physics. Moscow: Izd. Nauka 1974

Boernke, W.E.: Adaptations of amphibian arginase. II. Response to temperature. Comp. Biochem. Physiol. **44B**, 1035–1042 (1973)

Bogen, H.J.: Untersuchungen über Hitzetod und Hitzeresistenz pflanzlicher Protoplaste. Planta **36**, 298–340 (1948)

Bohley, P.: Intrazelluläre Proteolyse. Naturwissenschaften **55**, 211–217 (1968)

Bohr, N.: Essays 1958–1962 on Atomic Physics and Human Knowledge. New York, London: Interscience 1963

Bojarski, T. B., Hiatt, H. H.: Stabilization of thymidylate kinase activity by thymidylate and by thymidine. Nature **188**, 1112–1114 (1960)

Bolotina, I. A., Markovich, D. S., Volkenstein, M. V., Zavodsky, P.: Investigation of the conformation of α-glyceraldehyde-3-phosphate dehydrogenase. Biochim. Biophys. Acta **132**, 260–270 (1967)

Bolton, W., Cox, J. M., Perutz, M. F.: Structure and function of haemoglobin. IV. A three dimensional Fourier synthesis of horse deoxyhaemoglobin at 5.5 Å resolution. J. Mol. Biol. **33**, 283–297 (1968)

Bond, J. S.: A comparison of the proteolytic susceptibility of several rat liver enzymes. Biochem. and Biophys. Res. Commun. **43**, 333–339 (1971)

Bonner, J.: The chemical cure of climatic lesions. Engineering and Sci. Mag. **20**, 28–30 (1957)

Bonner, J., Huang, R. C. C.: Properties of chromosomal nucleohistone. J. Mol. Biol. **6**, 169–174 (1963)

Borasky, R.: Collagens—phylogenetic consideration. Physiol. Bohemoslovenica **14**, 206–213 (1965)

Boudet, A., Humphrey, Th. J., Davies, D. D.: The measurement of protein turnover by density labelling. Biochem. J. **152**, 409–416 (1975)

Bowler, K.: A study of the factors involved in acclimatization to temperature and death at high temperatures in *Astacus pallipes*. I. Experiments on intact animals. J. Cell. Comp. Physiol. **62**, 119–132 (1963)

Boyer, P. D.: The prevention by caprylate of urea and guanidine denaturation of serum albumin. J. Biol. Chem. **158**, 715–716 (1945)

Boyer, P. D., Ballou, G. A., Luck, J. M.: The combination of fatty acids and related compounds with serum albumin. II. Stabilization against urea and guanidine denaturation. J. Biol. Chem. **162**, 199–208 (1946a)

Boyer, P. D., Lum, F. G., Ballou, G. A., Luck, J. M., Rice, R. G.: The combination of fatty acids and related compounds with serum albumin. I. Stabilization against heat denaturation. J. Biol. Chem. **162**, 181–198 (1946b)

Brahms, J., Kay, C. M.: Molecular and enzymatic properties of cardiac myosin A as compared with those of skeletal myosin A. J. Biol. Chem. **238**, 198–205 (1963)

Brandt, J., Andersson, L.-O.: Heat denaturation of human serum albumin. Migration of bound fatty acids. Int. J. Peptide Protein Res., **8**, 33–37 (1976)

Brandts, J. F. (ed.): Heat effects on proteins and enzymes. In: Thermobiology. London, New York: 1967, pp. 25–72

Brandts, J. F.: Conformational transitions of proteins in water and in aqueous mixtures. In: Structure and Stability of Biological Macromolecules (eds. S. N. Timasheff, G. D. Fasman) New York: 1969, pp. 213–290

Brattstrom, B. H., Lawrence, P.: The rate of thermal acclimation in anuran amphibians. Physiol. Zool. **35**, 148–156 (1962)

Braun, A. D., Fizhenko, N. V.: On the thermostability of erythrocytes and their proteins—ATP-ase and haemoglobin—in two species of the frog. Tsitologia **5**, 249–253 (1963)

Braun, A. D., Nesvetayeva, N. M., Fizhenko, N. V.: On the resistance of actomyosin from *Rana temporaria* and *Rana ridibunda* to the denaturing action of heat and alcohol. Tsitologia **1**, 86–93 (1959)

Braun, A. D., Nesvetayeva, N. M., Fizhenko, N. V.: On the stability of myocardium and skeletal muscle actomyosins to the denaturative treatment with heat, ethyl alcohol and urea. Tsitologia **5**, 335–338 (1963)

Bresler, S. E., Talmud, D. L.: On the nature of globular proteins. Dokl. Akad. Nauk SSSR **43**, 326–330 (1944)

Briscoe, A. M., Loring, W. E., McClement, J. H.: Changes in human lung collagen and lipid with age. Proc. Soc. Exp. Biol. Med. **101**, 71–74 (1959)

Brock, T. D.: Life at high temperature. Science **158**, 1012–1019 (1967)

Brock, T. D.: Microbial growth under extreme conditions. In: Microbial Growth. Cambridge: Cambridge Univ. Press 1969, pp. 15–41

Brock, T. D.: High temperature systems. Ann. Rev. Ecol. Systematics **1**, 191–220 (1970)
Brock, T. D., Brock, M. L.: Relationship between environmental temperature and optimum temperature of bacteria along a hot spring thermal gradient. J. Appl. Bact. **31**, 54–58 (1968)
Brock, T. D., Brock, M. L.: Temperature optimum of nonsulphur bacteria from a spring at 90° C. Nature **233**, 494 (1971)
Brock, T. D., Brock, M. L., Bobt, T. L., Edwards, M. R.: Microbial life at 90° C: the sulfur bacteria of Boulder spring. J. Bacteriol. **107**, 303–314 (1971)
Brostrom, C. O., Jeffay, H: Protein catabolism in rat liver nuclei. Biochim. Biophys. Acta **278**, 15–27 (1972)
Brown, C. M., Rose, A. H.: Fatty acid composition of *Candida utilis* as affected by growth temperature and dissolved oxygen tension. J. Bacteriol. **99**, 371–378 (1969)
Brown, D. K., Militzer, W., Georgi, C. E.: The effect of growth temperature on the heat stability of a bacterial pyrophosphatase. Arch. Biochem. Biophys. **70**, 248–256 (1957)
Brunori, M., Engel, J., Schuster, T. M.: The effect of ligand binding on the optical rotatory dispersion of myoglobin, hemoglobin, and isolated hemoglobin subunits. J. Biol. Chem. **242**, 773–776 (1967)
Buchanon, D. L.: Total carbon turnover measured by feeding a uniformly labeled diet. Arch. Biochem. Biophys. **94**, 500–511 (1961)
Bukharin, P. D.: On the leaf temperature and heat resistance of some cultivated plants. Plant Physiol. **5**, 123–131 (1958)
Bull, H. B., Breese, K.: Thermal stability of proteins. Arch. Biochem. Biophys. **158**, 681–686 (1973)
Bullock, Th. H.: Compensation for temperature in metabolism and activity of poikilotherms. Biol. Rev. **30**, 311–342 (1955)
Bünning, E., Herdtle, H.: Physiologische Untersuchungen an thermophilen Blaualgen. Z. Naturforsch. **1**, 93–99 (1946)
Bunt, G. S., Owens, O., van H., Hoch, G.: Exploratory studies on the physiology and ecology of a psychrophilic marine diatom. J. Phycol. **2**, 96–100 (1966)
Burton, S. D., Morita, R. Y.: Denaturation and renaturation of malic dehydrogenase in a cell-free extract from a marine psychrophile. J. Bacteriol. **86**, 1019–1024 (1963)
Butt, W. D., Keilin, D.: Adsorption spectra and some other properties of cytochrome *c* and of its compounds with ligands. Proc. Roy. Soc. Ser. B **156**, 429–458 (1962)
Byrne, A. F., Burton, T. H., Koch, R. B.: Relation of dipicolinoic acid content of anaerobic bacterial endospores to their heat resistance. J. Bacteriol. **80**, 139–140 (1960)
Byrne, P., Chapman, D.: Liquid crystalline nature of phospholipids. Nature **202**, 987–988 (1964)
Caldwell, R. S., Vernberg, F. J.: The influence of acclimation temperature on the lipid composition of fish gill mitochondria. Comp. Biochem. Physiol. **34**, 179–191 (1970)
Campbell, L. L.: Crystallization of alpha-amylase from a thermophilic bacterium. J. Amer. Chem. Soc. **76** (1954)
Campbell, L. L.: Purification and properties of an α-amylase from facultative thermophilic bacteria. Arch. Biochem. Biophys. **54**, 154–161 (1955)
Campbell, L. L., Manning, G. B.: Thermostable α-amylase of *Bacillus stearothermophilus*. III. Amino acid composition. J. Biol. Chem. **236**, 2962–2965 (1961)
Candlish, J. K., Tristram, G. R.: Salts and amino acids as stabilising agents for reconstituted collagen fibres. Biochim. Biophys. Acta **88**, 553–563 (1964)
Capecchi, M. R., Capecchi, N. E., Hughes, S. H., Geoffrey, M. W.: Selective degradation of abnormal proteins in mammalian tissue culture cells. Proc. Nat. Acad. Sci. USA **71**, 4732–4736 (1974)
Carlisle, D. B., Cloudsley-Thompson, T. L.: Respiratory function and thermal acclimation in tropical invertebrates. Nature **218**, 684–685 (1968)
Carter, G. S.: On the control of the level of activity of the animal body. I. The endocrine control of seasonal variations of activity in the frog. J. Exp. Biol. **10**, 256–273 (1933)
Černý, M., Baudyšová, M., Holečková, E.: Adaptation of mammalian cells to cold. II. Cold-induced endoreduplication and polyploidy. Exp. Cell. Res. **40**, 673–677 (1965)
Chance, B., Ravilly, A., Rumen, N.: Reaction kinetics of a crystalline hemoprotein: an effect of crystal structure on reactivity of ferrimyoglobin. J. Mol. Biol. **17**, 525–534 (1966)

Chapman, D.: The effect of heat on membranes and membrane constituents. In: Thermobiology (ed. A. H. Rose) London, New York: Academic Press 1967, pp. 123–145
Charles-Edwards, D. A., Charles-Edwards, J., Cooper, J. P.: The influence of temperature on photosynthesis and transpiration in ten temperature grass varieties grown in four different environments. J. Exp. Bot. **22**, 650–662 (1971)
Chatfield, P. O., Lyman, C. P., Irving, L.: Physiological adaptation to cold of peripheral nerve in the leg of the herring gull *(Larus augentatus)*. Amer. J. Physiol. **172**, 639–644 (1953)
Chatterton, N. J., McKell, C. M., Strain, B. R.: Intraspecific differences in temperature—induced respiratory acclimation of desert saltbush. Ecology **51**, 545–547 (1970)
Chernitsky, E. A.: Luminescence and Structural Lability of Proteins in Solution and in the Cell. Minsk: Izd. Nauka i tekhnika 1972
Chernokozheva, I. S.: Changes in thermostability of isolated tissues of frogs as a result of preliminary heating. In: Teploustoichivost kletok zhivotnykh (Collection of works (Sbornik) Heat Resistance of Cells of Animals) Moscow, Leningrad: Izd. Nauka 1965, pp. 171–
Chernokozheva, I. S.: A study of thermostability of muscles and muscle models in connection with the growth of frogs. In: Izmenchivost teploustoichivosti kletok zhivotnykh v onto- i filogeneze (Collection of works (Sbornik) Variability in Cellular Heat Resistance of Animals in Ontogenesis and Phylogenesis) Leningrad: Izd. Nauka 1967a, pp. 13–19
Chernokozheva, I. S.: A study on thermostability of muscles and muscle model of juvenile frogs at different seasons of the year. In: Izmenchivost teploustoichivosti kletok zhivotnykh v onto- i filogeneze (Collection of works (Sbornik) Variability in Cellular Heat Resistance of Animals in Ontogenesis and Phylogenesis) Leningrad: Izd. Nauka 1967b, pp. 20–26
Chernokozheva, I. S.: Variability of the heat resistance of muscles and their contractile proteins in the ontogeny of *Rana temporaria*. Author's abstract of dissertation Leningrad 1970
Chernokozheva, I. S., Shlyakhter, N. A.: Increase in heat resistance of muscles of *Rana temporaria* L. as a result of preliminary heating in connection with seasonal changes of their heat resistance. In: Problemy tsytoekologii zhivotnykh (Collection of works (Sbornik) Problems of Cytoecology of Animals) Moscow, Leningrad: Izd. Akad. Nauk SSSR 1963, pp. 62–67
Child, Ch. M.: Senescence and Rejuvenescence. Chicago: Univ. Chicago Press, 1915
Child, C. M.: The physiological gradients. Protoplasma **5**, 447–476 (1928)
Child, C. M.: Patterns and Problems of Development. Chicago: Univ. Chicago Press, 1941
Christian, J. H. B., Waltho, J. A.: Solute concentrations within cells of halophilic and non-halophilic bacteria. Biochim. Biophys. Acta **65**, 506–508 (1962)
Christophersen, J.: Adaptive temperature responses of microorganisms. In: Molecular Mechanisms of Temperature Adaptation (ed. C. L. Prosser) Washington: 1967, pp. 327–348
Christophersen, J., Precht, H.: Fermentative Temperaturadaptation. Biol. Zbl. **69**, 240–256 (1950a)
Christophersen, J., Precht, H.: Untersuchungen über die Temperaturabhängigkeit von Lebensprozessen bei Hefen. Biol. Zbl. **69**, 300–323 (1960b)
Christophersen, J., Precht, H.: Über den Umkehrpunkt der Atmungskurven und die Fermentwirkungs-maxima bei Einwirkung steigender Temperaturen auf Hefen. Biol. Zbl. **70**, 261–274 (1951)
Christophersen, J., Precht, H.: Untersuchungen zum Problem der Hitzeresistenz. II. Untersuchungen an Hefezellen. Biol. Zbl. **71**, 585–601 (1952)
Chuang, H. Y. K., Atherly, A. G., Bell, F. E.: Protection of the proline and valine-activating enzymes by their amino acid substrates against thermal inactivation. Biochem. Biophys. Res. Commun. **28**, 1013–1018 (1967)
Church, B. D., Halvorson, H.: Dependence of the heat resistance of bacterial endospores on their dipicolinic acid content. Nature **183**, 124–125 (1959)
Citri, N.: Conformational adaptability in enzymes. Adv. Enzymol. **37**, 397–649 (1973)
Clark, A. Y.: The Mode of Action of Drugs on Cells. London E. Arnold: 1933
Clausen, J., Øvlisen, B.: Lactate dehydrogenase isoenzymes of human semen. Biochem. J. **97**, 513–517 (1965)
Cobon, G. S., Haslam, T. M.: The effect of altered membrane sterol composition on the temperature dependence of yeast mitochondrial ATPase. Biochem. Biophys. Res. Commun. **52**, 320–326 (1973)

Coffey, J. W., de Duve, C.: Digestive activity of lysosome. I. The digestion of proteins by extracts of rat liver lysosomes. J. Biol. Chem. **243**, 3255–3263 (1968)

Coffman, F. A.: Factors influencing heat resistance in oats. Agron. J. **49**, 368–373 (1957)

Comings, D. E., Riggs, A. D.: Molecular mechanisms of chromosome pairing, folding and function. Nature **233**, 48–50 (1971)

Connell, J. J.: The relative stabilities of the skeletal-muscle myosins of some animals. Biochem. J. **80**, 503–509 (1961)

Cooper, J. P., Tainton, N. M.: Light and temperature requirements for the growth of tropical and temperate grasses. Herb. Abstr. **38**, 167–176 (1968)

Costes, C., Deroche, M., Ferrou, F., Chaitrer, Ph.: L'utilisation du gaz carbonique et de la lumière dans les serres. Bull. Tech. Inform. **217**, 1–9 (1967)

Cowey, C. B.: Comparative studies on the activity of α-glyceraldehyde-3-phosphate dehydrogenase from cold- and warm-blooded animals with reference to temperature. Comp. Biochem. Physiol. **23**, 969–976 (1967)

Crabb, J. W., Murdock, A. L., Amelunxen, R. E.: A proposed mechanism of thermopily in facultative thermophiles. Biochem. Biophys. Res. Commun. **62**, 627–633 (1975)

Crippa-Franceschi, T.: Dauermodifications. J. Gen. Biol. **31**, 572–577 (1970)

Cronan, J. E.: Regulation of the fatty acid composition of the membrane phospholipids of *Escherichia coli*. Proc. Nat. Acad. Sci. USA **71**, 3758–3762 (1974)

Cullen, J., Phillips, M. C., Shipley, G. G.: The effects of temperature on the composition and physical properties of the lipids of *Pseudomonas fluorescens*. Biochem. J. **125**, 733–742 (1971)

Daron, H. H.: Fatty acid composition of lipid extracts of a thermophilic *Bacillus* species, J. Bacteriol. **101**, 145–151 (1970)

Das, A. B.: Biochemical changes in tissues of goldfish acclimated to high and low temperatures. II. Synthesis of protein and RNA of subcellular fraction and tissue composition. Comp. Biochem. Physiol. **21**, 469–485 (1967)

Davidson, B. E., Hird, F. J. R.: The reactivity of the disulphide bonds of purified proteins in relationship to primary structure. Biochem. J. **104**, 473–479 (1967)

Davis, F. L. Y., Williams, O. B.: Studies on heat resistance. I. Increasing resistance to heat of bacterial spores by selection. J. Bacteriol. **56**, 555–559 (1948)

Decicco, B. T., Noon, K. F.: Thermophilic mutants of *Pseudomonas fluorescens*. Arch. Mikrobiol. **90**, 297–304 (1973)

Dehnel, P. A.: Rates of growth of Gastropods as a function of latitude. Physiol. Zool. **28**, 115–144 (1955)

Denko, E. I.: The influence of cultivation temperature on cellular resistance of *Cabomba aquatica* Aubl. to various agents. In: The Cell and Environmental Temperature (ed. A. S. Troshin) Moscow, Leningrad: Izd. Nauka 1964, pp. 111–115

Denko, E. I.: A study of plant cell adaptation to heavy water (D_2O). Tsitologia **9**, 940–948 (1967)

Devanathan, T., Akagi, J. M., Hersh, R. T., Himes, R. H.: Ferredoxin from two thermophilic *Clostridia*. J. Biol. Chem. **244**, 2846–2853 (1969)

De Yong, D. W., Olson, A. C., Hawker, K. M., Jansen, E. F.: Effect of cultivation temperature on peroxidase isozymes of plant cells grown in suspension. Plant Physiology **43**, 841–844 (1968)

Dianoux, A. C., Jollès, P.: Etude d'un lysozyme pauvre en cystine et en tryptophane: le lysozyme de blanc d'oeuf et d'oie. Biochim. Biophys. Acta **133**, 472–479 (1967)

Dice, J. E., Schimke, R. T.: Turnover of chromosomal proteins from rat liver. Arch. Biochem. Biophys. **158**, 97–105 (1973)

Dickerson, R. E.: The structure and history of an ancient protein. Scientific American **226**, 58–72 (1972)

Dietrich, F. M.: Inactivation of egg- white lisozyme by ultrasonic waves and protective effect of amino acids. Nature **195**, 146–148 (1962)

Dircksen, A.: Vergleichende Untersuchungen zur Frost-, Hitze- und Austrocknungsresistenz einheimischer Laub- und Lebermoose unter besonderer Berücksichtigung jahreszeitlicher Veränderungen. Dissert. Georg-August-Univ. Göttingen: 1964

Dowben, R. M., Weidemüller, R.: Adaptation of mesophilic bacteria to growth at elevated temperatures. Biochim. Biophys. Acta **158**, 255–261 (1968)

Dregolskaya, I. N.: The influence of the sea water salinity on the thermostability of ciliary epithelium of actinia. Tsitologia **3**, 471–473 (1961)

Dregolskaya, I. N.: Thermostability of the ciliary epithelium cells of the Black Sea actinias in different seasons. Tsitologia **5**, 538–544 (1962)

Dregolskaya, I. N.: Effect of the rearing temperature on the heat stability of hydras and their cells. Tsitologia **5**, 194–203 (1963a)

Dregolskaya, I. N.: Heat resistance of ciliated epithelium of gills of *Mytilus galloprovincialis* L. from the Black sea. In: Problemy tsitoekologii zhivotnykh (Collection of works (Sbornik) Problems of Cytoecology of Animals) Moscow, Leningrad: Izd. Acad. Sci. SSSR 1963b, pp. 43–50

Dregolskaya, I. N., Chernokozheva, I. S.: Thermostability of the *Rana temporaria* muscles and cells of ciliary epithelium and of their glycerinated models in different seasons. Tsitologia **12**, 51–58 (1970)

Drysdale, J. W., Munro, H. N.: Regulation of synthesis and turnover of ferritin in rat liver. J. Biol. Chem. **241**, 3630–3637 (1966)

Dua, R. D., Burris, R. H.: Stability of nitrogen fixing enzymes and the reactivation of a cold labile enzyme. Proc. Nat. Acad. Sci. USA **50**, 169–175 (1963)

Dundas, J. E. D.: Purification of ornithine carbamoyl transferase from *Halobacterium solinarium*. Eur. J. Biochem. **16**, 393–398 (1970)

Dzhamusova, T. A.: Thermostability of the marine mollusc muscle tissue. Tsitologia **2**, 274–286 (1960a)

Dzhamusova, T. A.: Thermostability of the muscle tissue as a cytophysiological indicator of terrestrial *Mollusc* species. Tsitologia **2**, 561–572 (1960b)

Dzhamusova, T. A.: Thermostability of the muscles from some fresh water molluscs. In: Voprosy tsilologii i protistologii (Collection of works (Sbornik) Problems of Cytology and Protistology) Moscow and Leningrad: Izd. Akad. Nauk SSSR 1960c, pp. 100–107

Dzhamusova, T. A.: Thermostability of the muscle tissue from molluscs as a cytophysiological characteristic of a species. Author's abstract of dissertation Leningrad: 1961

Dzhamusova, T. A.: Heat resistance of the muscle tissue from molluscs with respect to the problem of species. In: Problemy tsitoekologii zhivotnykh (Collection of works (Sbornik) Problems of Cytoecology of Animals) Moscow, Leningrad: Izd. Akad. Nauk SSSR 1963, pp. 108–133

Dzhamusova, T. A.: Thermostability of the muscle tissue as a physiological characteristic of a mollusc species. All-Union Inst. Sci. Tech. Inform., pp. 53–61 (1965a)

Dzhamusova, T. A.: Heat contracture and irreversible loss of muscle excitability in connection with thermostability of muscle tissue. In: Teploustoichivost kletok zhivotnykh (Collection of works (Sbornik) Heat Resistance of Cells of Animals) Moscow, Leningrad: Izd. Nauka 1965b, pp. 61–69

Dzhamusova, T. A.: Changes in the level of thermostability of different muscles from the frog produced by thyroxin injections. Ekologia **5**, 53–58 (1971)

Dzhamusova, T. A., Pashkova, I. M.: Some physiological indices of the heat injury of the muscles in hibernating and spawning frogs. In: Izmenchivost teploustoichivosti kletok zhivotnykh v onto- i filogeneze (Collection of works (Sbornik) Variability in Cellular Heat Resistance of Animals in Ontogenesis and Phylogenesis) Leningrad: Izd. Nauka 1967, pp. 27–36

Dzhamusova, T. A., Shapiro, E. A.: Heat resistance of the muscle tissue in various species and populations of the fresh water molluscs. Zh. obshch. biol. **21**, 447–454 (1960)

Eastoe, J. E.: The amino acid composition of fish collagen and gelatin. Biochem. J. **65**, 363–368 (1957)

Ebashi, S., Endo, M., Ohtsuki, J.: Control of muscle contraction. Quart. Rev. Biophys. **2**, 351–384 (1969)

Eisen, A. Z., Jeffrey, J. J., Gross, J.: Human skin collagenase. Isolation and mechanism of attack on the collagen molecule. Biochim. Biophys. Acta **151**, 637–645 (1968)

Ekberg, D. R.: Respiration in tissues of goldfish adapted to high and low temperatures. Biol. Bull. **114** 308–316 (1958)

Elödi, P., Szabolcsi, G.: Role of the coenzyme in the stabilization of glyceraldehyde-3-phosphate dehydrogenase. Nature **184**, 56 (1959)

Endo, S.: The protease produced by thermophilic bacteria. J. Ferment. Technol. **40**, 346–353 (1962)

Engelbrecht, L., Mothes, K.: Weitere Untersuchungen zur experimentellen Beeinflussung der Hitzewirkung bei Blättern von *Nicotiana rustica*. Flora **154**, 279–298 (1964)

Engelman, D.M.: X-ray diffraction studies of phase transition in the membrane of *Mycoplasma laidlawii*. J. Mol. Biol. **47**, 115–117 (1970)

Epstein, J., Grossowicz, N.: Intracellular protein breakdown in a thermophile. J. Bacteriol. **99**, 418–421 (1969)

Epstein, S.J., Possick, P.A.: Inhibition of enzymic hydrolysis of plasma albumin by detergents. Arch. Biochem. Biophys. **93**, 538–541 (1961)

Esfahani, M., Barnes, E.M. Jr., Wakil, S.: Control of fatty acid composition in phospholipids of *Escherichia coli*: response to fatty acid supplements in a fatty acid auxotroph. Proc. Nat. Acad. Sci. USA **64**, 1057–1064 (1969)

Esfahani, M., Limbrick, A.R., Knutton, S., Oka, T., Wakil, S.J.: The molecular organization of lipids in the membrane of *Escherichia coli*: phase transitions. Proc. Nat. Acad. Sci. USA **68**, 3180–3184 (1971)

Esipova, N.G.: On the characteristics of certain hydrogen bonds in collagen. Biofizika **2**, 461–464 (1957)

Esser, A.F., Souza, K.A.: Correlation between thermal death and membrane fluidity in *Bacillus stearothermophilus*. Proc. Nat. Acad. Sci. USA **71**, 4111–4115 (1974)

Ettienne, E.M.: Control of contractility in *Spirostomum* by dissociated calcium ions. J. Gen. Physiol. **56**, 168–179 (1970)

Etingof, R.N.: On the biological basis of light reception. Usp. sovr. biol. **64**, 425–443 (1967)

Ewald, A.: Beiträge zur Kenntnis des Collogens. Hoppe-Seylers Z. physiol. Chem. **105**, 115–134 (1919)

Ewart, A.: On the physics and physiology of protoplasmic streaming in plants. Oxford: Claredon Press 1903

Falkova, T.V.: Seasonal changes of thermoresistance of higher plants cells under the conditions of the mediterranean type subtropics. Bot. Z. **58**, 1424–1438 (1973)

Falkova, T.V.: Heat hardening of higher plant cells under the conditions of the semi-arid subtropics. Ekologia **1**, 90–98 (1975)

Falkova, T.V., Galushko, R.V.: Seasonal variations of the heat stability of the protoplasm of *Daphne laureola* L. on the southern Crimean coast. Ekologia **6**, 10–15 (1974)

Farrell, Y., Rose, A.H.: Temperature effects on microorganisms. In: Thermobiology (ed. A.H. Rose) London, New York: Academic Press 1967, pp. 147–218

Fatt, H.V., Dougherty, E.C.: Genetic control of differential heat tolerance in two strains of the nematode *Caenorhabditis elegans*. Science **141**, 266–267 (1963)

Favard, A.: Contributions à l'étude vitale de l'adaptation des Droser aux températures élevées. Rev. gén. bot. **70**, 661–678 (1963)

Feder, J., Garrett, L.R., Wildi, B.S.: Studies on the role of calcium in thermolysin. Biochemistry **10**, 4552 (1971)

Feeney, R.E., Miller, H.T., Komatsa, S.K.: Properties of proteins from cold-adapted Antarctic fish. 7th Intern. Congr. Biochem. Abstr. **5**, 993 (1967)

Feierabend, Y., Berger, Ch., Meyer, A.: Spezifische Störung von Entwicklung und Enzymbildung des Plastiden höherer Pflanzen durch hohe Wachstumstemperaturen. Z. Naturforsch. **248**, 1641–1647 (1969)

Feigelson, P., Dashman, T., Margolis, F.: The halflife time of induced tryptophan peroxidase *in vivo*. Arch. Biochem. Biophys. **85**, 478–482 (1959)

Feldberg, R.S. and Datta, P.: Cold inactivation of l-threonine desaminase from *Rhodospirillum rubrum*. Involvement of hydrophobic interactions. Eur. J. Biochem. **21**, 447–454 (1971)

Feldman, N.L.: The effect of heat hardening on the heat resistance of some enzymes from plant leaves. Planta **78**, 213–225 (1968)

Feldman, N.L.: Effect of wound irritation on the sensitivity of plant cells. In: Voprosy tsitologii i protistologii (Collection of works (Sbornik) Problems of Cytology and Protistology) Moscow, Leningrad: Izd. Akad. Nauk SSSR 1960, pp. 216–223

Feldman, N.L.: The influence of sugars on the cell stability of some higher plants to heating and high hydrostatic pressure. Tsitologia **4**, 633–643 (1962)

Feldman, N. L.: Heat resistance of the leaf cells of prevernal ephemeroids with different periods of vegetations. In: Tsitologicheskiye osnovy prisposobleniya rastenii k faktoram sredy (Collection of works (Sbornik) Cytological Aspects of Adaptation of Plants to the Environmental Factors) Moscow, Leningrad: Izd. Nauka 1964, pp. 82–87
Feldman, N. L.: The increase of urease heat stability resistance of after heat treatment of the leaves. Dokl. Akad. Nauk SSSR **167**, 946–949 (1966)
Feldman, N. L.: Thermostability of acid phosphatase from the leaves of two species of *Leucojum* with different periods of vegetation. Tsitologia **11**, 592–599 (1969)
Feldman, N. L.: The temperature dependence of the enzymatic activity and the K_m of acid phosphatase from the leaves of spring and summer species of *Leucojum*. Tsitologia **15**, 170–176 (1973)
Feldman, N. L., Agapova, N. D., Kamentseva, I. E., Shukhtina, H. G.: Thermostability of various cellular functions of leaves and petals in the genus *Ornithogalum*. Bot. Zh. **60**, 356–364 (1975a)
Feldmann, N. L., Ageeva, O. G., Lutova, M. I.: Heat resistance of ferredoxin from heat hardened leaves of *Pisum sativum* L. Dokl. Akad. Nauk SSSR **208**, 479–483 (1973)
Feldmann, N. L., Artyushenko, Z. T., Shukhtina, H. G.: Cellular heat resistance in certain species of different genera of *Amaryllidaceae*. Bot. Zh. **55**, 1678–1683 (1970)
Feldman, N. L., Kamentseva, I. E.: Thermostability at different stages of development. Bot. Zh. **48**, 414–419 (1963)
Feldmann, N. L., Kamentseva, I. E.: The urease thermostability in leaf extracts of two *Leucojum species* with different periods of vegetation. Tsitologia **9**, 886–889 (1967)
Feldman, N. L., Kamentseva, I. E.: Thermostability of cells and of intracellular proteins in the spring and summer snowflakes *(Leucojum)*. Tsitologia **13**, 479–483 (1971)
Feldman, N. L., Kamentseva, I. E.: Thermostability of cells in the middle Asian *Allium* species with ephemeroid and long-term vegetation cycles. Ekologia **4**, 48–52 (1974)
Feldmann, N. L., Kamentseva, I. E., Shukhtina, H. G.: Thermostability of the cells of the spring and summer leaves of the lungwort *(Pulmonaria obscura* Dumort.). Bot. Zh. **51**, 828–830 (1966a)
Feldman, N. L., Kamentseva, I. E., Yurashevskaya, K. N.: Acid phosphatase thermostability in the extracts of the cucumber and wheat seedling leaves after heat hardening. Tsitologia **8**, 755–759 (1966b)
Feldman, N. L., Lutova, M. I.: Studies on the heat stability of the cells of some marine grasses. Bot. Zh. **47**, 542–546 (1962)
Feldman, N. L., Lutova, M. I.: Variations de la thermostabilité cellulaire des algues en fonction des changements de la température du milieu. Cahiers biol. marine **4** 435–458 (1963)
Feldman, N. L., Lutova, M. I., Shcherbakova, A. M.: Resistance of some proteins of *Pisum sativum* L. leaves after heat hardening to elevated temperature, proteolysis and shifts in pH. Thermal Biology **1**, 47–51 (1975b)
Feldman, N. L., Zavadskaya, I. G., Lutova, M. I.: A study of the temperature resistance of some marine algae in nature and under experimental conditions. Tsitologia **5**, 125–134 (1963)
Felföldy, L. Y. M.: Effect of temperature on photosynthesis in three unicellular green algae strains. Acta biol. Acad. Sci. Hung. **12**, 153–159 (1961)
Fenstein, R. N., Howard, J. B., Savol, R.: Heat and urea stability of blood catalase of catalase—mutant mouse strain. Experientia **27**, 1152–1153 (1971)
Filner, P., Wray, J. L., Varner, J. E.: Enzyme induction in higher plants. Science **165**, 358–367 (1969)
Fincham, J. R. S.: A modified glutamic dehydrogenase as a result of gene mutation in *Neurospora crassa*. Biochem. J. **65**, 721 (1957)
Fincham, J. R. S.: On the nature of the glutamic dehydrogenase produced by inter-allele complementation at the *amlocus* of *Neurospora crassa*. J. Gen. Microbiol. **21**, 600–611 (1959)
Fincham, J. R. S.: Genetically controlled differences in enzyme activity. Adv. Enzymol. **22**, 1–43 (1960)
Fischer, E.: Einfluß der Configuration auf die Wirkung der Enzyme. Ber. Deut. Ges. **27**, 2985–2993 (1894)
Fisher, H. F.: A limiting law relating the size and shape of protein molecules to their composition. Proc. Nat. Acad. Sci. USA **51**, 1285–1291 (1964)

Fischman, L., Lery, M.: A comparison of soluble dogfish skin collagens. Biol. Bull. **127**, 369–370 (1964)
Fox, A. S., Burnett, J. B.: Tyrosinases of diverse thermostabilities and their interconversion in *Neurospora crasse* Ncr. Biochim. biophys. Acta **61**, 108–120 (1962)
Fox, H. M.: The activity and metabolism of poikilothermal animals in different latitudes. Proc. Zool. Soc. **A 109**, 141–156 (1939)
Fraenkel, G., Hopp, H. S.: Temperature adaptation and degree of saturation of the phosphatides. Biochem. J. **34**, 1085–1092 (1940)
Frank-Kamenetskii, M. D., Lazurkin, Yu. S.: Conformational changes in DNA molecules. Ann. Rev. Biophys. Bioengin. **3**, 127–150 (1974)
Freed, J.: Changes in activity of cytochrome oxidase during adaptation of goldfish to different temperatures. Comp. Biochem. Physiol. **14**, 651–659 (1965)
Freed, T. M.: Temperature effects on muscle phosphofructokinase of the Alaskan king crab, *Paralithodes camtschatica*. Comp. Biochem. Physiol. **39 B**, 765–774 (1971)
Freeze, H., Brock, Th. D.: Thermostable aldolase from *Thermus aquaticus*. J. Bacteriol. **101**, 541–550 (1970)
Frenkel, S. Ya.: Some problems of the statistics and morphology of polymers. Author's abstract of dissertation. Leningrad: 1962
Frieden, C.: Protein-protein interaction and enzymatic activity. Ann. Rev. Biochem. **40**, 653–696 (1971)
Friedman, S. M.: Protein-synthesizing machinery of thermophilic bacteria. Bacteriol. Rev. **32**, 27–28 (1968)
Friedman, S. M.: Heat stabilities of ribosomal subunits and reassociated ribosomes from *Bacillus stearothermophilus*. J. Bacteriol. **108**, 589–591 (1971)
Friedman, S. M., Axel, R., Weinstein, I. B.: Stability of ribosomes and ribosomal ribonucleic acid from *Bacillus stearothermophilus*. J. Bacteriol. **93**, 1521–1526 (1967)
Friedman, S. M., Weinstein, I. B.: Polyribonucleotide stimulation of amino acid in a subcellular system derived from a thermophilic bacterium. Fed. Proc. **23**, 1–164 (1964)
Friedman, S. M., Weinstein, J. B.: Protein synthesis in a subcellular system from *Bacillus stearothermophilus*. Biochim. Biophys. Acta **114**, 593–605 (1966)
Friedrich, L.: Experimentelle Untersuchungen zum Problem zellulärer nichtgenetischer Resistenzänderungen bei der Miesmuschel *Mytilus edulis* L. Kieler Meeresforschung **23**, 105–126 (1967)
Fries, N.: Induced thermosensitivity in *Ophiostoma* and *Rhodotorula*. Physiol. Plantarum **16**, 415–422 (1963)
Fry, F. E. J.: The lethal temperature as a tool in taxonomy. L'année biolog. 3 ser. **33**, 205–219 (1957)
Fry, F. E. J.: Temperature compensation. Ann. Rev. Physiol. **20**, 207–224 (1958)
Fukushi, T., Imanishi, A., Isemuro, T.: Changes in enzymatic activity and conformation during regeneration of native bacterial amylase from denatured form. J. Biochem. (Tokyo) **63**, 409–416 (1968)
Fulco, A. J.: The effect of temperature on the formation of 5-unsaturated fatty acids by bacilli. Biochim. biophys. Acta **144**, 701–703 (1967)
Fulco, A. J.: The biosynthesis of unsaturated fatty acid by bacilli. J. Biol. Chem. **247**, 3511–3519 (1972)
Furch, K.: Der Einfluß einer Vorbehandlung mit konstanten und wechselnden Temperaturen auf die Hitzeresistenz von *Gammarus salinus* und *Idafea balthica*. Marine Biol. **15**, 12–34 (1972)
Galston, A. W.: Adenine and plant growth. Science, **129**, 57 (1959)
Galston, A. W., Hand, M. E.: Adenine as a growth factor for etiolated peas and its relation to the thermal inactivation of growth. Arch. Biochem. **22**, 434–443 (1949)
Gates, D. M.: Transpiration and leaf temperature. Ann. Rev. Plant Physiology **19**, 211–238 (1968)
Gaughran, E. R. L.: The thermophilic microorganisms. Bacteriol. Rev. **11**, 189–225 (1947 a)
Gaughran, E. R. L.: The saturation of bacterial lipids as a function of temperature. J. Bacteriol. **53**, 506 (1947 b)
Générmont, J.: The problem of long-lasting modifications in *Protozoa*. Zh. obshch. biol. **31**, 661–671 (1970)

George, H., McMahan, J., Bowler, K., Elliott, M.: Stabilization of lactate and malate dehydrogenase by organic solvents. Biochim. biophys. Acta **191**, 466–468 (1969)

Georgiev, G.P., Samarina, O.P.: The problem of transport of messenger RNA in the animal cell. Usp. sovr. biol. khimii **10**, 5–35 (1969)

Gerez de Burgos, N.M., Burgos, C., Gutierrez, M., Blanco, A.: Effect of temperature upon catalytic properties of lactate dehydrogenase isozymes from a poikilotherm. Biochem. Biophys. Acta **315**, 250–258 (1973)

Gerloff, E.D., Richardson, T., Stahmann, M.A.: Changes in fatty acids of alfalfa roots during cold hardening. Plant Physiology **41**, 1280–1284 (1966)

Givol, D., Goldberger, R.F., Anfinsen, C.B.: Oxidation and disulfide interchange in the reactivation of reduced ribonuclease. J. Biol. Chem. **239**, 3114–3116 (1964)

Glasziou, K.T., Waldron, J.C., Bull, T.A.: Control of invertase synthesis in sugar cane. Loci of auxin and glucose effects. Plant Physiology **41**, 282–288 (1966)

Glushankova, M.A.: Evaluation of the heat stability of actomyosin preparations and the iodine number of isolated lipids in *Trachurus mediterraneus* from the Black Sea. Tsitologia **5**, 244–246 (1963a)

Glushankova, M.A.: The effect of a lipid diet on the iodine number and heat resistance of the muscle tissue. In: Problemy tsitoekologii zhivotnykh (Collection of works (Sbornik) Problems of Cytoecology of Animals) Moscow, Leningrad: Izd. Akad. Nauk SSSR 1963b, pp. 208–212

Glushankova, M.A.: The environmental temperature and thermostability of actomyosin, alkaline phosphatase and adenylate kinase from poikilotherms. In: Izmenchivost teploustoichivosti kletok zhivotnykh v onto- i filogeneze (Collection of works (Sbornik) Variability in Cellular Heat Resistance of Animals in Ontogenesis and Phylogenesis) Leningrad: Izd. Nauka 1967a, pp. 126–141

Glushankova, M.A.: Thermostability of actin from the lake and grass frog *Rana ridibunda* Pall. and *R. temporaria* L. In: Izmenchivost teploustoichivosti kletok zhivotnykh v onto- i filogeneze (Collection of works (Sbornik) Variability in Cellular Heat Resistance of Animals in Ontogenesis and Phylogenesis) Leningrad: Izd. Nauka 1967b, pp. 153–157

Glushankova, M.A., Chernokozheva, I.S.: The dependence of thermostability of different muscles from *Rana ridibunda* on the duration of keeping the animals in a warm spring. Ekologia **4**, 51–56 (1971)

Glushankova, M.A., Dregolskaya, I.N., Kusakina, A.A., Pashkova, I.M.: Thermostability of protein preparations and cells of ciliary epithelium in European and Asian groups of the lake frog *Rana ridibunda* Pall. In: Izmenchivost teploustoichivosti kletok zhivotnykh v onto- i filogeneze (Collection of works (Sbornik) Variability in Cellular Heat Resistance of Animals in Ontogenesis and Phylogenesis) Leningrad: Izd. Nauka 1967, pp. 99–106

Glushankova, M.A., Kusakina, A.A.: Thermostability of some proteins from the representatives of various populations of the lizard *Phrynocephalus helioscopus*. In: Izmenchivost teploustoichivosti kletok zhivotnykh v onto- i filogeneze (Collection of works (Sbornik) Variability of Cellular Heat Resistance of Animals in Ontogenesis and Phylogenesis) Leningrad: Izd. Nauka 1967, pp. 114–118

Goldberg, A.C., Dice, J.E.: Intracellular protein degradation in mammalion and bacterial cells. Ann. Rev. Biochem. **43**, 835–869 (1974)

Goldberg, A.L.: Degradation of abnormal proteins in *Escherichia coli*. Proc. Nat. Acad. Sci. USA **69**, 422–426 (1972a)

Goldberg, A.L.: Correlation between rates of degradation of bacterial proteins *in vivo* and their sensitivity to protease. Proc. Nat. Acad. Sci. USA **69**, 2640–2644 (1972b)

Goldsack, D.E.: Relation of the hydrophobicity index to thermal stability of homologous proteins. Biopolymers. **9**, 247–252 (1970)

Goldschmidt, R.: *In vivo* degradation of nonsense fragments in *E. coli*. Nature **228**, 1151–1154 (1970)

Gorban, I.S.: On the correlation between the growth and thermostability of plant cells. Tsitologia **4**, 182–192 (1962)

Gorban, I.S.: The repair of heat injury in plant cells of different age, Tsitologia **5**, 169–174 (1963)

Gorban, I.S.: Repair of heat injury in cells of different age. In: The Cell and Environmental Temperature, (ed. A.S. Troshin) Moscow, Leningrad: Izd. Nauka 1964, pp. 197–200

Gorban, I. S.: Effects of indol-3-acetic acid and maleic hydrozide on the elongation and thermostability of wheat coleoptiles. Tsitologia **10**, 76–87 (1968)

Gorban, I. S.: Thermostability of urease from the cucumber cotyledon seedlings assayed in intact organs and in homogenates at elevated temperatures. Tsitologia **14**, 1504–1512 (1972)

Gorban, I. S.: Alteration of the reparatory properties of plant cells under the influence of heating to superoptimal temperatures. Tsitologia **16**, 1111–1116 (1974)

Gordon, J.: The protective action of some amino acids against the effect of heat on complement. J. Hygiene **51**, 140–144 (1953)

Gorini, L., Andrian, L.: Relations entre degré de dénaturation et sensivilité à la trypsine de la sérumalbumine. Influence de Ca^{++} et de Mn^{++} et rôle des ponts disulfure. Biochim. biophys. Acta **19**, 289–296 (1956)

Gorke, H.: Über chemische Vorgänge beim Erfrieren der Pflanzen. Landwirtsch. Versuchs. Stat. **65**, 149–160 (1907)

Gorodilov, Yu. N.: Thermal stability of the cells of *Arenicola marina* as a function of ambient temperature. Tsitologia **3**, 469–471 (1961)

Gorodilov, Yu. N.: A study of the fish sensitivity to high temperature during embryogenesis. II. The patterns on the sensitivity changes in the developing spawn from the viewpoint of different methods of its estimation and possible ecological significance on the changes. Tsitologia **3**, 366–376 (1969)

Gorovsky, M. A.: Heat denaturation studies on *Arbacia punctulata* nucleoproteins. Biol. Bull. **127**, 2–371 (1964)

Graevsky, E. I.: The temperature preference and the temperature optimum of the fresh-water molluscs and arthropoda. Zh. obshch. biol. **7**, 455–472 (1946)

Green, D. E.: The conformational basis of energy transductions in biological systems. Proc. Nat. Acad. Sci. USA **67**, 544–549 (1970)

Green, D. E., Asai, J., Harris, R. A., Penniston, J. T.: Conformational basis of energy transformations in membrane systems. III. Configurational changes in the mitochondrial inner membrane induced by changes in functional states. Arch. Biochem. Biophys. **125**, 684–705 (1968)

Green, D. E., Baum, H.: Energy and Mitochondrion. New York: 1970

Greengard, O., Smith, M. A., Aes, G.: Relation of cortisone and synthesis of ribonucleic acid to induced and developmental enzyme formation. Biol. Chem. **238**, 1548–1551 (1963)

Greer, J.: Three-dimensional structure of abnormal human haemoglobins Kansas and Richmond. J. Mol. Biol. **59**, 99–105 (1971a)

Greer, J.: Three-dimensional structure of abnormal human haemoglobins M Hyde Park and M Iwate. J. Mol. Biol. **59**, 107–126 (1971b)

Greshnykh, K. P., Grigorian, A. N., Dikanskaya, E. M., Dyatlovskaya, E. V., Bergelson, L. D.: Effect of temperature and nitrogen sources on lipid biosynthesis in yeasts grown on n-alkanes. Mikrobiologia **37**, 251–254 (1968)

Grigoryan, G. A.: Relation between thermal stability of *Paramecium caudatum* and Ca^{++} ion concentration in the surrounding medium. Tsitologia **1**, 105–109 (1964)

Grisolia, A.: The catalytic environment and its biological implications. Physiol. Rev. **44**, 657–712 (1964)

Grisolia, S., Fernandez, M., Amelunxen, R., Quijada, C. L.: Glutamate-dehydrogenase inactivation by reduced nicotinamide-adenine dinucleotide phosphate. Biochem. J. **85**, 568–576 (1962)

Gromysz-Kołkowska, K.: Minimum, maximum and perferential temperature in *Orthomorpha gracilis* C. L. Koch. (Diphopoda). Folia Biol. **15**, 101–115 (1967)

Grossbard, L., Schimke, R. T.: Multiple hexokinases of rat tissues. Purification and comparison of soluble forms. J. Biol. Chem. **241**, 3546–3560 (1966)

Grossman, A., Mavrides, C.: Regulation of tyrosine-α-keto-glutarate aminotransferase induction and subsequent inactivation. Fed. Proc. **25**, 285 (1966)

Grossman, A., Mavrides, C.: Studies on the regulation of tyrosine aminotransferase in rats. J. Biol. Chem. **242**, 1398–1405 (1967)

Gurley, L. R., Hardin, J. M.: The metabolism of histone fractions. II. Conservation and turnover of histone fractions in mamalian cells. Arch. Biochem. Biophys. **130**, 1–2 (1969)

Gustavson, K. H.: Behaviour of collagen in combination with tanning agents towards trypsin. Särtryck ur Svensk Kem. Tidskr. **54**, 249–256 (1942)

Gustavson, K. H.: Hydrothermal stability and intermolecular organization of collagens from mammalion and teleost skins. Svensk Kem. Tidskr. **65**, 70–76 (1953)

Gustavson, K. H.: The function of hydroxyproline in collagens. Nature **175**, 70–74 (1955)

Gustavson, K. H.: The Chemistry and Reactivity of Collagen. New York: Academic Press 1956

Hachimori, A., Muramatsu, N., Nosoh, Y.: Studies on an ATPase of thermophilic bacteria. I. Purification and properties. Biochim. biophys. Acta **206**, 426–437 (1970)

Hagen, P. O., Kushner, D. J., Gibbons, N. E.: Temperature induced death analysis in a psychrophilic bacterium. Can. J. Microbiol. **10**, 813–822 (1964)

Hagen, P. O., Rose, A. H.: Studies on the biochemical basis of the low maximum temperature in a psychrophilic *Cryptococcus*. J. Gen. Microbiol. **72**, 89–99 (1962)

Hagerup, O.: Zytoökologische Bicornesstudien. Planta **32**, 6–14 (1941)

Hagihara, B., Nakayama, T., Matsubara, H., Okunuki, K.: Denaturation and inactivation of enzyme proteins. II. Denaturation and inactivation of bacterial amylase. J. Biochem. (Tokyo) **43**, 469–481 (1956a)

Hagihara, B., Nakayama, T., Matsubara, H., Okunuki, K.: Denaturation and inactivation of enzyme proteins. III. Denaturation and inactivation of taka-α-amylase. J. Biochem. (Tokyo) **43**, 483–494 (1956b)

Hajdu, S.: Observations on the temperature dependence of tension developed by the frog muscle. Arch. Intern. Physiol. **59**, 58–61 (1951)

Hakala, M. T., Glaid, A. J., Schwert, G. W.: Lactic dehydrogenase. II. Variation of kinetic and equilibrium constants with temperature. J. Biol. Chem. **221**, 191–209 (1956)

Hammouda, M., Lange, O. L.: Zur Hitzeresistenz der Blätter höherer Pflanzen in Abhängigkeit von ihrem Wassergehalt. Naturwissenschaften **49**, 1–2 (1962)

Harder, R.: Über die Assimilation von Kälte- und Wärmeindividuen der gleichen Pflanzenspezies. Jahrb. wiss. Bot. **64**, 169–200 (1925)

Harris, P., James, A. T.: The effect of low temperatures on fatty acid biosynthesis in plants. Biochem. J. **112**, 325–330 (1969)

Hartwell, L. H., McLaughlin, C. S.: Mutants of yeast with temperature-sensitive isoleusyl tRNA synthetases. Proc. Nat. Acad. Sci. USA **59**, 422–428 (1968a)

Hartwell, L. H., McLaughlin, C. S.: Temperature sensitive mutants of yeast exhibiting a rapid inhibition of protein synthesis. J. Bacteriol. **96**, 1664–1671 (1968b)

Hatfaludi, F., Strashilov, T., Straub, F. B.: Effect of urea and calcium ions on pancreatic amylase. Acta biochim. biophys. Acad. Sci. Hung. **1**, 39–43 (1966)

Hattori, A., Crespi, H. L., Katz, J. J.: Effect of side chain denaturation on protein stability. Biochem. **4**, 1213–1225 (1965)

Hazel, J. R.: The effect of temperature acclimation upon succinic dehydrogenase activity from the epaxial muscle of the common goldfish (*Carassius auratus* L.). I. Properties of the enzyme and the effect of lipid extraction. Comp. Biochem. Physiol. **43B**, 837–861 (1972a)

Hazel, J. R.: The effect of temperature acclimation upon succinic dehydrogenase activity from the epaxial muscle of the common goldfish (*Carassius auratus* L.). II. Lipid reactivation of the soluble enzyme. Comp. Biochem. Physiol. **43B**, 863–882 (1972b)

Hazel, J. R., Prosser, C. L.: Interpretation of inverse acclimation to temperature. Z. vergl. Physiol. **67**, 217–228 (1970)

Hazel, J. R., Prosser, C. L.: Molecular mechanisms of temperature compensation in poikilotherms. Physiolog. Rev. **54**, 620–677 (1974)

Hebb, C., Stephens, T. C., Smith, M. W.: Effect of environmental temperature on the kinetic properties of goldfish brain choline acetyltransferase. Biochem. J. **129**, 1013–1021 (1972)

Heber, U.: Ursachen der Frostresistenz bei Winterweizen. I. Die Bedeutung der Zucker für die Frostresistenz. Planta **52**, 144–172 (1958a)

Heber, U.: Ursachen der Frostresistenz bei Winterweizen. II. Die Bedeutung von Aminosäuren und Peptiden für die Frostresistenz. Planta **52**, 431–446 (1958b)

Heber, U.: Ursachen der Frostresistenz bei Winterweizen. III. Die Bedeutung von Proteinen für die Frostresistenz. Plante **54**, 34–67 (1959)

Heilbrunn, L. V.: The Dynamics of Living Protoplasm. New York: Academic Press, 1956

Hellmuth, E.: Eco-physiological studies on plants in arid and semi-arid regions in Western Australia. J. Ecol. **59**, 365–374 (1971)

Henderson, R., Wright, C.S., Hess, G.P., Blow, D.M.: α-chymotrypsin. What can we learn about catalysis from X-ray diffraction? Cold Spr. Harb. Symp. Quant. Biol. **36**, 63–70 (1971)

Henning, U., Yanofsky, C.: An alteration in the primary structure of a protein predicted on the basis of genetic recombination data. Proc. Nat. Acad. Sci. USA **48**, 183–190 (1962)

Hermans, J., Jr., Acampora, G.: Reversible denaturation of sperm whale myoglobin. II. Thermodynamic analysis. J. Amer. Chem. Soc. **89**, 1547–1552 (1967)

Hermans, J., Jr., Scheraga, H.A.: The thermally induced configurational change of ribonuclease in H_2O and D_2O. Biochim. biophys. Acta **36**, 534–535 (1959)

Herner, A.E., Frieden, E.: Biochemistry of *Anuran metamorphosis*. VII. Changes in serum proteins during spontaneous and induced metamorphosis. Biol. Chem. **235**, 2845–2851 (1960)

Herrlinger, F., Kiermeier, F.: Inaktivierung und Regeneration der Peroxydase in wärmebehandelten Pflanzengeweben. Biochem. Z. **318**, 413–424 (1948)

Herter, K.: Die Beziehungen zwischen der Ökologie und der Thermotaxis der Tiere. Biol. Gen. **17**, 243–309 (1943)

Hiesey, W.M., Milner, H.W.: Physiology of ecological races and species. Ann. Rev. Plant Physiol. **16**, 203–216 (1965)

Hippel, P.H., Schleicht, T.: The effects of neutral salts on the structure and conformational stability of macromolecules in solution. In: Structure and Stability of Biological Macromolecules (eds. S.M. Timasheff, G.D. Fasman) New York: Marcel Dekker 1969

Hoar, W.S., Cottle, M.K.: Some effects of temperature acclimatization on the chemical constitution of goldfish tissues. Can. J. Zool. **30**, 49–54 (1952)

Hochachka, P.W.: Action of temperature on branch points in glucose and acetate metabolism. Comp. Biochem. Physiol. **25**, 107–118 (1968a)

Hochachka, P.W.: The nature of the thermal optimum for lungfish lactate dehydrogenase. Comp. Biochem. Physiol. **27**, 609–611 (1968b)

Hochachka, P.W., Clayton-Hochachka, B.: Glucose-6-phosphate dehydrogenase and thermal acclimation in the mulletfish. Marine Biol. **18**, 251–259 (1973)

Hochachka, P.W., Hayes, F.R.: The effect of temperature acclimation on pathways of glucose metabolism in the trout. Can. J. Zool. **40**, 261–270 (1962)

Hochachka, P.W., Lewis, J.K.: Interacting effects of pH and temperature on the K_m values for fish tissues lactate dehydrogenases. Comp. Biochem. Physiol. **B 39**, 925–933 (1971)

Hochachka, P.W., Somero, G.N.: The adaptation of enzymes to temperature. Comp. Biochem. Physiol. **27**, 659–668 (1968)

Hochachka, P.W., Somero, G.N.: Biochemical adaptation to the environment. Fish Physiol. **6**, 99–156 (1971)

Hochachka, P.W., Somero, G.N.: Strategies of Biochemical Adaptation. Philadelphia, London, Toronto: Saunders 1973

Hocking, J.D., Harris, J.I.: Purification by affinity chromatography of thermostable glyceraldehyde-3-phosphate dehydrogenase from *Thermus aquaticus*. Biochem. J. **130**, 240–250 (1972)

Hogness, D.S., Cohn, M., Monod, J.: Studies on the induced synthesis of β-galactosidase in *Echerichia coli*: the kinetics and mechanism of sulfur incorporation. Biochim. biophys. Acta **16**, 99–116 (1955)

Holmes, P.K., Halvorson, H.O.: The inactivation and reactivation of salt-requiring enzymes from an extreme obligate halophile. Can. J. Microbiol. **9**, 904–906 (1963)

Holmes, P.K., Halvorson, H.O.: Purification of a salt-requiring enzyme from an obligately halophilic bacterium. J. Bacteriol. **90**, 312–315 (1965)

Hopkins, D.L.: The relation between temperature and locomotion in the marine *Amoeba, Flobellula mira* Schaeffer, with special reference to adaptation to temperature. Protoplasma **28**, 161–174 (1937)

Horowitz, N.H.: Progress in developing chemical concepts of genetic phenomena. Fed. Proc. **15**, 818–822 (1956)

Horowitz, N. H., Fling, M.: The role of the genus in the synthesis of enzymes. In: Enzymes: Units of Biological Structure and Function. New York, London: Academic Press, 1956, pp. 139–145

Horowitz, N. H., Fling, M., MacLeod, H., Suloka, N.: A genetic study of two new structural forms of tyrosinase in *Neurospora*. Genetics **46**, 1015–1024 (1961)

House, H. L., Riordan, D. F., Barlow, J. S.: Effects of thermal conditioning and of degree of saturation of dietary lipids on resistance of an insect to a high temperature. Can. J. Zool. **36**, 629–632 (1958)

Howard, R. L., Becker, R. R.: Isolation and some properties of the triphosphopyridine nucleotide isocitrate dehydrogenase from *Bacillus stearothermophilus*. J. Biol. Chem. **245**, 3186–3194 (1970)

Hsia, J. C., Wong, P. T. S., MacLennan, D. H.: Salt-dependent conformational changes in the cell membrane of *Halobacterium salinarium*. Biochem. Biophys. Res. Commun. **43**, 88–93 (1971)

Hsiu, F., Fischer, E. H., Stein, E. A.: Alpha-amylases as calcium-metalloenzymes. II. Calcium and the catalytic activity. Biochemistry **3**, 61–66 (1964)

Huang, R. C. C., Bonner, J.: Histone, a suppressor of chromosomal RNA synthesis. Proc. Nat. Acad. Sci. USA **48**, 1216–1222 (1962)

Huang, R. C. C., Bonner, J., Murray, K.: Physical and biological properties of soluble nucleohistones. J. Mol. Biol. **8**, 54–64 (1964)

Hubbard, J. S., Miller, A. B.: Purification and reversible inactivation of the isocitrate dehydrogenase from an obligate halophile. J. Bacteriol. **99**, 161–168 (1969)

Huber, B.: Der Wärmehaushalt der Pflanzen. Naturwiss. und Landwirt. **17**, (1935)

Hunter, A. S.: Effects of temperature on *Drosophila*. I. Respiration of *D. melanogaster* grown at different temperatures. Comp. Biochem. Physiol. **11**, 411–417 (1964)

Huttunen, M., Johansson, B. W.: The influence of the dietary fat on the lethal temperature in the hypothermic rat. Acta Physiol. Scand. **59**, 7–11 (1963)

Hvidt, A., Wallevik, K.: Conformational changes in human serum albumin as revealed by hydrogen-deuterium exchange studies. J. Biol. Chem. **247**, 1530–1535 (1972)

Igarashi, S.: Temperature-sensitive mutation in *Paramecium aurelia*. I. Induction and inheritance. Mutation Res. **3**, 13–24 (1966)

Ikonnikova, T. B.: Studies on proteolysis of hemoglobin by pepsin under aerobic and anaerobic conditions. Diploma, Leningrad State Univ. 1965

Illert, H.: Botanische Untersuchungen über Hitzetod und Stoffwechselgifte. Bot. Arch. **7**, 133–141 (1924)

Ilyinskaya, N. B.: The resistance of insect cells to injuring agents during diapause. In: Reaktsiya kletok i ikh belkovykh komponentov na ekstremalnye vozdeistviya (Collection of works (Sbornik) Reactions of Cells and Their Protein Components on the Action of Extremal Factors) Moscow, Leningrad: Izd. Nauka 1966, pp. 59–72

Imanishi, A.: Calcium binding by bacterial α-amylase. J. Biochem. (Tokyo) **60**, 381–390 (1966)

Inagaki, C.: Beiträge zur Kenntnis der Wärmestarre des Muskels. Z. Biol. **48**, 313–339 (1906)

Inagaki, M.: Denaturation and inactivation of enzyme proteins. XI. Inactivation and denaturation of glutamatic acid dehydrogenase by urea, and the effect of its coenzyme on these processes. J. Biochem. (Tokyo) **46**, 893–901 (1959a)

Inagaki, M.: Denaturation and inactivation of enzyme proteins. XII. Thermal inactivation and denaturation of glutamic acid dehydrogenase and the effect of its coenzyme on these processes. J. Biochem. (Tokyo) **46**, 1001–1010 (1959b)

Inesi, G., Millman, M., Eletr, S.: Temperature-induced transitions of function and structure in sarcoplasmic reticulum membranes. J. Mol. Biol. **81**, 483–504 (1973)

Ingraham, J. L.: Temperature relationships. In: The Bacteria (eds. I. C. Gunsalus, R. Y. Stainer) New York, London: Academic Press 1962, pp. 265–269

Ingraham, J. L., Bailey, G. F.: A comparative study of the effect of temperature on the metabolism of psychrophilic and mesophilic bacteria. J. Bacteriol. **77**, 609–613 (1959)

Ingraham, J. L., Neuhard, J.: Cold-sensitive mutants of *Salmonella typhimurium* defective in uridine monophosphate kinase (pyrH). J. Biol. Chem. **247**, 6259–6265 (1972)

Ingram, D. J. E.: Biological and Biochemical Applications of Electron Spin Resonance. London: Hilger, 1969

Irlina, I. S.: Changes in thermostability of some free-living *Protozoa* produced by a temperature pre-treatment. Tsitologia **2**, 227–234 (1960)
Irlina, I. S.: Some physiological and cytochemical peculiarities of *Paramecium caudatum* adapted to different temperatures. Tsitologia **5**, 183–193 (1963a)
Irlina, I. S.: Effects of various concentrations of some salts and ethyl alcohol on the ciliates adapted to different temperatures. Tsitologia **5**, 287–294 (1963b)
Irlina, I. S.: Temperature adaptations of *Paramecium caudatum* and some other infusoria and their resistance to various injuring agents. Author's abstract of dissertation, Leningrad 1964
Irlina, I. S.: Thermoresistance of *Tetrahymena pyriformis* in the cell cycle. Tsitologia **14**, 69–79 (1972)
Isono, K.: Enzymological differences of α-amylase from *Bacillus stearothermophilus* grown at 37° C and 55° C. Biochem. Biophys. Res. Commun. **41**, 852–857 (1970)
Ivanov, N. N.: Variability of the chemical composition of oil-bearing plant grains as a function of geographical factors. Tr. po prikld. botaniki i sistematiki **16**, 3–59 (1926)
Ivanov, V. I., Bocharov, A. L., Volkenstein, M. V., Karpeisky, M. Ya., Mora, S., Okina, E. I., Yudina, L. V.: Conformational properties and catalytic function of aspartate aminotransferase. Eur. J. Biochem. **40**, 519–526 (1973)
Ivleva, I. V.: Thermostability of the muscle tissue in polychaetas from the Mediterranean basin. Zool. Zh. **41**, 1798–1810 (1962)
Ivleva, I. V.: The relation of the tissue heat resistance of polychaetas to osmotic and temperature conditions of the environment. In: Kletka i temperatura sredy (Collection of works (Sbornik) The Cell and Environmental Temperature) Moscow, Leningrad: Izd. Nauka 1964, pp. 158–162
Jacobs, M. H.: Acclimatization as a factor affecting the upper thermal death points of organisms. J. Exp. Zool. **27**, 427–442 (1919)
Jacobson, K. B.: Alcohol dehydrogenase of *Drosophila:* interconversion of isoenzymes. Science **159**, 324–325 (1968)
Jankowsky, H. D.: Über die hormonale Beeinflussung der Temperaturadaptation beim Grasfrosch, *Rana temporaria.* Z. vergl. Physiol. **43**, 392–410 (1960)
Jarabak, J., Seeds, A. E., Talalay, P.: Reversible cold inactivation of a 17β-hydroxysteroid dehydrogenase of human placenta: protective effect of glycerol. Biochemistry **5**, 1269–1278 (1966)
Jick, H., Shuster, L.: The turnover of microsomal reduced nicotinamide adenine dinucleotide, phosphate cytochrome C reductase in the liver of mice treated with phenobarbital. J. Biol. Chem. **241**, 5366–5369 (1966)
Jinks, J. L.: Extrachromosomal Inheritance. Englewood Cliffs New Jersey, USA: Prentice-Hall, 1964
Jockusch, H.: Relations between temperature sensitivity, amino acid replacements and quaternary structure of mutant proteins. Biochem. Biophys. Res. Commun. **24**, 577–583 (1966)
Jockusch, H.: Stability and genetic variation of a structural protein. Naturwissenschaften **55**, 514–518 (1968)
Johnson, F. H., Shimomura, O.: The chemistry of lumin *Coelenterates.* In: Chemical Zoology (eds. M. A. Florkin, B. T. Scheer) New York, London: Academic Press 1968, pp. 233–261
Johnston, I. A., Davison, W., Goldspink, G.: Adaptations in Mg^{2+} activated myofibrillar ATPase activity induced by temperature acclimation. FEBS. Letters **50**, 293–295 (1975a)
Johnston, I. A., Frearson, N., Goldspink, G.: The effects of environmental temperature on the properties of myofibrillar adenosine triphosphatase from various species of fish. Biochem. J. **133**, 735–738 (1973)
Johnston, I. A., Walesby, N. J., Davison, W., Goldspink, G.: Temperature adaptation in myosin of Antarctic fish. Nature **254**, 74–75 (1975b)
Johnston, P. V., Roots, B. I.: Brain lipid fatty acids and temperature acclimation. Comp. Biochem. Physiol. **11**, 303–309 (1964)
Jollès, J., Dianoux, A. C., Hermann, J., Niemann, B., Jollès, P.: Relationship between the cystine and tryptophan contents of 5 different lysozymes and their heat stability and specific activity. Biochim. biophys. Acta **128**, 568–570 (1966)
Jollos, V.: Experimentelle Untersuchungen an Infusorien. Biol. Zbl. **33**, 222–236 (1913)

Jollos, V.: Experimentelle Protistenstudien. I. Untersuchungen über Variabilität und Vererbung bei Infusorien. Arch. Protistenk. **43**, 1–222 (1921)
Jollos, V.: Dauermodifikationen und Mutationen bei Protozoen. Arch. Protistenk. **83**, 197–219 (1934)
Joly, M.: A Physico-Chemical Approach to the Denaturation of Proteins. London, New York: Academic Press, 1965
Jonxis, J. H. P.: On the spreading of different haemoglobins muscle haemoglobins and cytochrome C. Biochem. J. **33**, 1743–1751 (1939)
Jorgensen, E. G.: The adaptation of plankton algae. II. Aspects of the temperature adaptation of *Skeletonema costatum*. Physiol. Plantarum **21**, 423–427 (1968)
Joseph, K. T., Bose, S. M.: Influence of biological ageing on the stability of skin collagen in albino rats. In: Collagen (ed. N. Ramanathan) New York and London: 1962, pp. 371–394
Joshi, L., Holmes, F. O.: Induced tolerance of tobacco-mosaic virus to heat. Phytopathology **58**, 60–61 (1968)
Josse, J., Harrington, W. F.: Role of pyrrolidine residues in the structure and stabilization of collagen. J. Mol. Biol. **9**, 269–287 (1964)
Julander, O.: Drought resistance in range and pasture grasses. Plant Physiology **20**, 573–599 (1945)
Kageyama, M.: Studies of a pyocin. I. Physical and Chemical properties. J. Biochem. **55**, 49–53 (1964)
Kalakutsky, L. V., Agre, N. S., Aslanijan, R. R.: Thermoresistance of spores in actinomycetes with reference to their dipicolinic acid, Ca^{++} and Mg^{++} content. Dokl. Akad. Nauk SSSR **184**, 1214–1216 (1969)
Kamentseva, I. E.: Thermostability of photosynthesis and respiration spring and summer leaves of *Pulmonaria obscura* Dumort. Dokl. Akad. Nauk SSSR, ser. biol. **186**, 968–970 (1969)
Kamentseva, I. E.: Thermostability of cells and some proteins in the spring and summer leaves of *Pulmonaria obscura* Dumort. Bot. Zh. **59**, 417–421 (1974a)
Kamentseva, I. E.: Thermostability of some functions of leaf cells in ephemeroid and long-vegetating *Allium* species. Bot. Zh. **59**, 1669–1675 (1974b)
Kamshilov, M. M.: On the "systemic" and "cellular" adaptation. Tr. Murmanskogo biologicheskogo instituta (Collection of works of the Murmansk Institute of Marine Biology) Murmansk: 1960, pp. 226–235
Kang, A. H., Nagai, Y., Piez, K. A., Gross, J.: Studies on the structure of collagen utilizing a collagenolytic enzyme from tadpole. Biochemistry **5**, 509–515 (1966)
Kappen, L.: Untersuchungen über den Jahreslauf der Frost-, Hitze- und Austrocknungsresistenz von Sporophyten einheimischer Polypodiaceen *(Filicinae)*. Flora **155**, 123–166 (1964)
Kappen, L.: Der Einfluß des Wassergehaltes auf die Widerstandsfähigkeit von Pflanzen gegenüber hohen und tiefen Temperaturen, untersucht an Blättern einiger Farne und von *Ramonda myconi*. Flora **B 156**, 427–445 (1966)
Kappen, L., Lange, O. L.: Die Hitzeresistenz angetrockneter Blätter von *Commelina africana* — ein Vergleich zwischen zwei Untersuchungsmethoden. Protoplasma **65**, 119–132 (1968)
Karimov, Kh. Kh., Popova, A. I.: On the heat stability of cellular protoplasm of different organs in ephemeroids in respect to the developmental cycles. Dokl. Akad. Nauk Tadzh. SSR **9**, 40–43 (1966)
Karpukhina, S. J., Sosfenov, N. I.: Alpha-amylase solutions investigated by the method of small angle X-ray scattering. Dokl. Akad. Nauk SSSR **175**, 723–725 (1967)
Karush, F.: Heterogeneity of binding sites of bovine serum albumin. J. Amer. Chem. Soc. **72**, 2705–2713 (1950)
Kates, M., Baxner, R. M.: Lipid composition of mesophilic and psychrophilic yeasts *(Candida species)* as influenced by environmental temperature. Can. J. Biochem. Physiol. **40**, 1213–1227 (1962)
Kates, M., Paradis, M.: Phospholipid desaturation in *Candida lipolytica* as a function of temperature and growth. Can. J. Biochem. **51**, 184–197 (1973)
Kauzmann, W.: Some factors in the interpretation of protein denaturation. Adv. Protein Chem. **14**, 1–63 (1959)

Kawashima, N., Singh, S., Wildman, S. G.: Reversible cold inactivation and heat reactivation of RuDP carboxylase activity of crystallized tobacco fraction I protein. Biochem. Biophys. Res. Commun. **42**, 664–668 (1971)

Kazakova, O. V., Orekhovich, V. N., Shpikiter, V. O.: The rate of procollagen splitting by collagenase as affected by temperature. Dokl. Akad. Nauk SSSR **120**, 359–360 (1958)

Kelly, M. T., Brock, T. D.: Warm-water strain of *Leucothrix mucor*. J. Bacteriol. **98**, 1402–1403 (1969)

Kemp, P., Smith, M. W.: Effect of temperature acclimatization on the fatty acid composition of goldfish intestinal lipids. Biochem. J. **117**, 9–15 (1970)

Kenney, F. T.: Induction of tyrosine-α-ketoglutarate transaminase in rat liver. IV. Evidence for an increase in the rate of enzyme synthesis. J. Biol. Chem. **237**, 3495–3498 (1962)

Kenney, F. T.: Turnover of rat liver tyrosine transaminase: stabilization of the inhibition of protein synthesis. Science **156**, 525–528 (1967)

Kenny, J. W., Richards, A. G.: Differences between leg and flight muscle of the giant water bug, *Lethocerus americanus*. Entomol. News **66**, 29–36 (1955)

Keradjopoulos, D., Wulff, K.: Thermophilic alanine dehydrogenase from *Halobacterium salinarium*. Can. J. Biochem. **52**, 1033–1037 (1974)

Ketellapper, H. J.: Temperature-induced chemical defects in higher plants. Plant Physiology **38**, 175–179 (1963)

Keynes, R. D.: Evidence for structural changes during nerve activity and their relation to the conduction mechanism. In: The Neurosciences: 2nd Study Program (ed. F. O. Schmitt) New York: Rockefeller Univ. Press 1970, pp. 707–714

Khesin, R. B., Astaurova, O. B., Shemyakin, M. F., Kamzolova, S. G., Manyakov, V. F.: Changes in RNA-polymerase properties during binding to DNA and initiation of RNA synthesis. Molek. biol. **1**, 736–753 (1967)

Khlebnikova, N. A.: On the heat resistance of plants. Izd. Akad. Nauk SSSR **7**, 1127–1146 (1932)

Kiknadze, G. S.: Fluorescent microscopic investigation of the chlorophyll in the leaves of *Campanula persicifolia* L. as affected by various injuring agents. Tsitologia **2**, 144–152 (1960)

Kinbacher, E. J., Sullivan, Ch. Y.: Effect of high temperature on the respiration rate of *Phaseolus sp.* Proc. Amer. Soc. Hort. Sci. **90**, 163–168 (1967)

Kinbacher, E. J., Sullivan, Ch. Y., Knull, H. R.: Thermal stability of malic dehydrogenase from heat-hardened *Phaseolus acutifolius* 'Tepary buff'. Crop. Sci. **7**, 148–151 (1967)

King, J. H., Jukes, Th. H.: Non-Darwinean evolution. Science **164**, 788–800 (1969)

Kirberger, C.: Untersuchungen über die Temperaturabhängigkeit von Lebensprozessen bei verschiedenen Wirbellosen. Z. vergl. Physiol. **35**, 175–198 (1953)

Kirkman, H. N., Hendrickson, E. M.: Glucose-6-phosphate dehydrogenase from human erythrocytes. J. Biol. Chem. **237**, 2371–2376 (1962)

Kiro, M. B.: Differences in thermostability of coronas of the rotifer *Epiphanes shenta* (Ehrb.) cultivated at various temperatures for serveral generations. In: Izmenchivost teploustoichivosti kletok zhivotnykh v onto- i filogeneze (Collection of works (Sbornik) Variability in Cellular Heat Resistance of Animals in Ontogenesis and Phylogenesis) Leningrad: Izd. Nauka 1967, pp. 63–65

Kirpichnikov, V. S.: The role of non-hereditary variability in the process of natural selection (a hypothesis of indirect selection). Biol. Zh. **4**, 775–800 (1935)

Kirpichnikov, V. S.: The problem of non-hereditary adaptive modifications (coincident of organic selection). J. Genetics **48**, 164–175 (1947)

Kirpichnikov, V. S.: Biochemical polymorphism and the problem of the so-called Non-Darwinean evolution. Usp. sovr. biol. **74**, 231–246 (1972)

Kislyuk, I. M.: Increase in the heat resistance of young specimes of cereals by heat and cold hardening. Bot. Zh. **47**, 713–715 (1962)

Klee, W. A.: Intermediate stages in the thermally induced transconformation reactions of bovine pancreatic ribonuclease A. Biochemistry **6**, 3736–3742 (1967)

Kleinschmidt, M. G., McMahon, V. A.: Effect of growth temperature on the lipid composition of *Cyanidium caldarum*. Plant Physiology **46**, 286–289 (1970a)

Kleinschmidt, M. G., McMahon, V. A.: Effect of growth temperature on the lipid composition of *Cyanidium caldarium*. II. Glycolipid and phospholipid components. Plant Physiology **46**, 290–293 (1970b)

Klikoff, L. G.: Temperature dependence of the oxidative rates of mitochondria in *Danthonia intermedia*, *Pentstemon davidsonii* and *Sitanion hystrix*. Nature **212**, 529–530 (1966)

Klyachko, N. L., Kulaeva, O. N.: Thermostability of protein synthesis in leaves of different age. Dokl. Akad. Nauk SSSR **188**, 230–232 (1969)

Koffler, H.: Protoplasmic differences between mesophiles and thermophiles. Bacteriol. Rev. **21**, 227–240 (1957)

Koffler, H.: Relative stability of thermophilic organisms. 5th Intern. Biochem. Congr. 1, Abstr. **88.** Moscow: 1961

Koffler, H., Gale, G. O.: The relative thermostability of cytoplasmic proteins from thermophilic bacteria. Arch. Biochem. Biophys. **67**, 249–251 (1957)

Koffler, H., Mallett, G. E., Adye, J.: Molecular basis of biological stability to high temperatures. Proc. Nat. Acad. Sci. USA **43**, 464–477 (1957)

Kohonen, J., Tirri, R., Lagerspetz, K. J. H.: Temperature dependence of the ATPase activities in brain homogenates from a cold-water fish and warm-water fish. Comp. Biochem. Physiol. **44**, 819–822 (1973)

Komatsu, S. K., Feeney, R. E.: A heat labile fructose diphosphate aldolase from cold-adapted antarctic fishes. Biochim. biophys. Acta **206**, 305–315 (1970)

Komissarchik, Ya. Yu., Levin, S. V., Rosental, D. L., Troshin, A. S.: On the structural changes in the nerve fiber membranes during excitation. Ukr. biokhim. Zh. **43**, 166–172 (1971)

Komkova, A. I., Ushakov, B. P.: Thermal inactivation of adenosine triphosphatase from *Rana ridibunda* and *Rana temporaria* muscles. Dokl. Akad. Nauk SSSR **102**, 1185–1188 (1955)

Kondo, M.: Effect of fatty acids on the proteolysis of proteins. II. Effect of C_8–C_{14} fatty acids on the tryptic digestion of serum albumin. J. Biochem. (Tokyo) **46**, 705–709 (1959)

Konev, S. V., Aksenov, S. P., Chernitsky, E. A.: Cooperative Transitions of Proteins in the Cell. Minsk: Izd. Nauka i tekhnika 1970a

Konev, A. D., Burtseva, V. M.: Changes in the heat resistance of entire organisms of molluscs and their cells in response to changes in ambient temperature. Ekologia **6**, 80–88 (1970)

Konev, S. V., Chernitsky, E. A., Lin, E. I.: Luminescence and specific features of the structural state of cell proteins. Biofizika **13**, 1040–1047 (1968)

Konev, S. V., Chernitsky, E. A., Mazhul, V. M., Yaskevich, V. P.: On the conformational transitions of proteins in intact cells. Dokl. Akad. Nauk BSSR **14**, 68–70 (1970b)

Konev, S. V., Okun, I. M., Aksentsev, S. L., Nisenbaum, G. D., Adzerikho, R. D.: Conformation transitions in erythrocyte membranes at physiological moderate temperatures. Dokl. Akad. Nauk SSSR **205**, 979–982 (1972)

Konis, E.: The resistance of maquis plants to supramaximal temperatures. Ecology **30**, 425–429 (1949)

Konstantinova, M. F.: Resistance of adenylate kinase from two species of frogs (differing in their thermophily) to heating and trypsin. Dokl. Akad. Nauk SSSR **203**, 1204–1206 (1972)

Konstantinova, M. F.: Comparative valuation of myosin stability to heat and proteolysis in the frogs *Rana temperaria* and *Rana ridibunda*. In: Fiziologiya i biokhimiya nizshikh pozvonochnykh (Collection of works (Sbornik) Physiology and Biochemistry of Lower Vertebrates) (ed. E. M. Kreps) Leningrad: Izd. Nauka 1974, pp. 140–147

Konstantinova, M. F., Grigoryeva, G. M.: Thermostability of acetylcholinesterase in frogs of two species. Dokl. Akad. Nauk SSSR **184**, 972–974 (1969)

Kornberg, T., Malcolm, L. G.: Deoxyribonucleic acid synthesis in cell-free extracts. IV. Purification and catalytic properties of deoxyribonucleic acid polymerase III. J. Biol. Chem. **247**, 5369–5375 (1972)

Korsgaard, C. L.: Trypsin splitting and denaturation of β-lactoglobulin. Nature **163**, 1003–1004 (1949)

Korzhuev, P.: Effects of elevated temperature on trypsin from the warm-blooded and cold-blooded vertebrates. Fiziol. Zh. SSSR **21**, 433–437 (1936)

Koshland, D. E.: Application of a theory of enzyme specificity to protein synthesis. Proc. Nat. Acad. Sci. USA **44**, 98–104 (1958)

Koshland, D. E.: Mechanisms of transfer enzymes. In: The Enzymes I (eds. P. D. Boyer, H. Lardy, K. Myrbäck) New York: Academic Press 1959, pp. 305–346

Koshland, D. E.: The active site and enzyme action. Adv. Enzymol. **22**, 45–97 (1960)

Koshland, D. E.: Catalysis in life and in the test tube. In: Horizons in Biochemistry (eds. M. Kasha, B. Pullmann) New York, London: Academic Press 1962, pp. 265–283

Koshland, D. E.: Conformation changes at the active site during enzyme action. Fed. Proc. **23**, part I, 719–726 (1964a)

Koshland, D. E.: Catalysis in live nature and *in vitro*. In: Collection of works "Gorizonty v biokhimii" (Horizons in Biochemistry) Moscow: Izd. Mir 1964b, pp. 202–217

Koshland, D. E.: Conformational aspects of enzyme regulation. In: Current Topics in Cellular Regulation Vol. I, New York, London: Academic Press 1969, pp. 1–27

Koshland, D. E.: Molecular basis for enzymatic catalysis and enzyme regulation. Zh. Vsesouysn. khim. obshch. im. D.I. Mendeleeva **16**, 158–164 (1971)

Koshland, D. E., Kirtley, M. E.: Protein structure in relation to cell dynamics and differentiation. In: Major Problems in Developmental Biology (ed. M. Jocke) New York, London: Academic Press 1966, pp. 217–249

Koshland, D. E., Neet, K. E.: The catalytic and regulatory properties of enzymes. Ann. Rev. Biochem. **37**, 359–410 (1968)

Koshland, D. E., Yankeelov, J. A., Thoma, J. A.: Specificity and catalytic power in enzyme action. Fed. Proc. **21**, 1031–1038 (1962)

Kraut, J., Wright, H. T., Kellerman, M., Freer, S. T.: π-, δ-, and γ-chymotrypsin: Three dimensional electron density and difference maps at 5 Å resolution, and comparison with chymotrypsinogen. Proc. Nat. Acad. Sci. USA **58**, 304–311 (1967)

Krolenko, S. A., Nikolsky, N. N.: Thermostability of the muscle tissue of some tropical amphibians of south-eastern Asia. Tsitologia **5**, 230–234 (1963)

Krylov, V. N., Zapadnaya, A. A.; Temperature-sensitive r-mutations (r^{ts}) of bacteriophage T4B. Genetika **4**, 7–11 (1965)

Kuczenski, R. T., Suelter, C. H.: Effect of temperature and effectors on the conformations of yeast pyruvate kinase. Biochemistry **9**, 939–945 (1970)

Kuehl, L., Sumsion, E. N.: Turnover of several glycolytic enzymes in rat liver. J. Biol. Chem. **245**, 6619–6623 (1970)

Kuiper, P. J. C.: Lipids in alfalfa leaves in relation to cold hardiness. Plant. Physiol. **45**, 684–686 (1970)

Kumamoto, J., Raison, J. K., Lyons, J. M.: Temperature "breaks" in Arrhenius plots: a thermodynamic consequence of phase change. J. Theor. Biol. **31**, 47–51 (1971)

Kunina, O. V., Levdikova, G. A., Shpikiter, B. O.: Studies of collagenase action on tropocollagen. In: Aktualnye voprosy med. biokhimii (Collection of works (Sbornik) Current Problems in Medical Biochemistry) (ed. V. N. Orekhovich), 1 Izd. Inst. Med. Biokhimii AMN SSSR, 1962, Vol II. pp. 92–98

Kunitz, M., McDonald, M. R.: Crystalline hexokinase (heterophosphatase). J. Gen. Physiol. **29**, 393–412 (1946)

Künnemann, H.: Der Einfluss der Temperatur auf Thermostabilität, Isoenzym-Muster und Reaktionskinetik der Laktat-Dehydrogenase aus Fischen. Marine Biol. **18**, 37–45 (1973)

Künnemann, H., Passia, D.: $NADP^+$-isocitraten-dehydrogenase aus *Idus idus* (Pisces: Cyprinidae). II. Einfluß der Temperatur auf Substrat und Cosubstrataffinität. Marine Biol. **23**, 205–211 (1973)

Kurtz, E.: Chemical basis for adaptation in plants. Science **128**, 1115–1117 (1958)

Kusakina, A. A.: Thermostability of the muscles of some representatives of the Atlantic salmon and Sea trout. In: Problemy tsitologii i protistologii (Collection of works (Sbornik) Problems of Cytology and Protistology) Moscow, Leningrad: Izd. Akad. Nauk SSSR 1960, pp. 107–111

Kusakina, A. A.: Rate of decrease in cholinesterase activity in liver homogenates of *Rana temporaria* and *Rana ridibunda* as a function of temperature. Dokl. Akad. Nauk SSSR **139**, 1258–1261 (1961)

Kusakina, A. A.: Thermostability of muscles and of cholinesterase in homogenates of muscles in *Carassius* fishes from thermal springs or ordinary water bodies. Dokl. Akad. Nauk SSSR **144**, 1160–1162 (1962a)

Kusakina, A. A.: On the correspondence between the muscle and cholinesterase thermostability and the temperature environment in some fishes. Tsitologia **4**, 68–71 (1962b)

Kusakina, A. A.: Specific differences in heat resistance of protoplasmic proteins. In: Problemy tsitoekologii zhivotnykh (Collection of works (Sbornik) Problems of Cytoecology of Animals) Moscow, Leningrad: Izd. Akad. Nauk SSSR 1963a, pp. 169–188

Kusakina, A. A.: Heat resistance of acetylcholinesterase in homogenates from muscles and brain of rats whose tissues differ in lipid content. In: Problemy tsitoekologii zhivotnykh (Collection of works (Sbornik) Problems of Cytoecology of Animals) Moscow, Leningrad: Izd. Akad. Nauk SSSR 1963b, pp. 213–216

Kusakina, A. A.: Haemoglobin thermostability of five species of lizards from the Kara-Kum desert. In: Teploustoichivost kletok zhivotnykh (Collection of works (Sbornik) Heat Resistance of Cells of Animals) Moscow, Leningrad. Izd. Nauka 1965a, pp. 212–215

Kusakina, A. A.: Thermostability of haemoglobin and cholinesterase of muscles and liver in three subspecies of *Bufo bufo* L. In: Teploustoichivost kletok zhivotnykh (Collection of works (Sbornik) Heat Resistance of Cells of Animals) Moscow, Leningrad: Izd. Nauka 1965b, pp. 208–211

Kusakina, A. A.: Thermostability of aldolase and cholinesterase in closely related species of poikilotherms. In: Izmenchivost teploustoichivosti kletok zhivotnykh v onto- i filogeneze (Collection of works (Sbornik) Variability in Cellular Heat Resistance of Animals in Ontogenesis and Phylogenesis) Leningrad: Izd. Nauka 1967, pp. 142–148

Kusakina, A. A., Glushankova, M. A., Vasyanin, S. I.: Thermostability of some proteins in specimens of intraspecific groups of *Coregonus autumnalis* and *Thymalys arcticus* from the Baikal Lake. Tsitologia **13**, 994–1003 (1971)

Kusakina, A. A., Skholl, E. D.: Thermostability of erythrocytes and haemoglobin derivatives of four *Mocaca* species. Tsitologia **5**, 253–257 (1963)

Lagerpetz, K. J. H., Kohonen, J., Tirri, R.: Temperature acclimation of the ATPase activities in the nerve cord of the earthworm *Lumbricus terrestris* L. Comp. Biochem. Physiol. **44**, 823–828 (1973)

Laidler, K. J.: The Chemical Kinetics of Enzyme Action. Oxford: Clarendon Press 1958

Lange, O. L.: Hitze- und Trockenresistenz der Flechten in Beziehung zu ihrer Verbreitung. Flora **140**, 39–97 (1953)

Lange, O. L.: Untersuchungen über die Hitzeresistenz der Moose in Beziehung zu ihrer Verbreitung. Flora **142**, 381–399 (1955)

Lange, O. L.: Untersuchungen über Wärmehaushalt und Hitzeresistenz mauretanischer Wüsten- und Savannenpflanzen. Flora **147**, 595–651 (1959)

Lange, O. L.: Die Hitzeresistenz einheimischer immer- und wintergrüner Pflanzen im Jahreslauf. Planta **56**, 666–683 (1961)

Lange, O. L.: Versuche zur Hitzeresistenz—Adaptation bei höheren Pflanzen. Naturwissenschaften **49**, 20–21 (1962a)

Lange, O. L.: Über die Beziehungen zwischen Wasser- und Wärmehaushalt von Wüstenpflanzen. Veröffentl. Geobot. Inst. Eidg. Techn. Hochschule Zürich **37**, 155–168 (1962b)

Lange, O. L.: Investigations of the variability of heat resistance in higher plants. In: Kletka i temperatura sredy (Collection of works (Sbornik) The Cell and Environmental Temperature) Moscow, Leningrad: Izd. Nauka 1964, pp. 91–97

Lange, O. L.: Der CO_2-Gaswechsel von Flechten bei tiefen Temperaturen. Planta **64**, 1–19 (1965)

Lange, O. L.: Die funktionellen Anpassungen der Flechten an die ökologischen Bedingungen arider Gebiete. Ber. Deut. Bot. Ges. **82**, 3–22 (1969)

Lange, O. L., Lange, R.: Die Hitzeresistenz einiger mediterraner Pflanzen in Abhängigkeit von der Höhenlage ihrer Standorte. Flora **152**, 707–710 (1962)

Lange, O. L., Lange, R.: Untersuchungen über Blattemperaturen, Transpiration und Hitzeresistenz an Pflanzen mediterraner Standorte (Costa brava, Spanien). Flora **153**, 387–425 (1963)

Lange, O. L., Schulze, E.-D., Koch, W.: Experimentell-ökologische Untersuchungen an Flechten der Negev-Wüste. II. CO_2-Gaswechsel und Wasserhaushalt von *Ramalina maciformis* (Del.) Bory am natürlichen Standort während der sommerlichen Trockenperiode. Flora **159**, 38–62 (1970a)

Lange, O. L., Schulze, E.-D., Koch, W.: Experimentell-ökologische Untersuchungen an Flechten der Negev-Wüste. III. CO_2-Gaswechsel und Wasserhaushalt von Krusten- und Blattflechten am natürlichen Standort während der sommerlichen Trockenperiode. Flora **159**, 525–528 (1970b)

Langridge, J.: The genetic basis of climatic response. In: Environmental Control of Plant Growth. New York, London: Academic Press 1963a, pp. 367–379

Langridge, J.: Biochemical aspects of temperature response. Ann. Rev. Plant Physiol. **14**, 441–462 (1963b)

Langridge, J.: Temperature-sensitive, vitamin-requiring mutants of *Arabidopsis thaliana*. Austral. J. Biol. Sci. **18**, 311–321 (1965)

Langridge, J., Griffing, B.: A study of high temperature lesions in *Arabidopsis thaliana*. Austral. J. Biol. Sci. **12**, 117–135 (1959)

Lanyi, J. K., Stevenson, J.: Studies of the electron transport chain of extremely halophilic bacteria. IV. Role of hydrophobic forces in the structure of menadione reductase. J. Biol. Chem. **245**, 4074–4080 (1970)

Larkin, J. M., Stokes, J. L.: Growth of psychrophilic microorganisms at subzero temperatures. Can. J. Microbiol. **14**, 97–101 (1968)

Larsen, H.: Biochemical aspects of extreme halophilism. Adv. Microbiol. Physiol. **1**, 97–132 (1967)

Laude, H. H.: Diurnal cycle of heat resistance in plants. Science **89**, 556–557 (1939)

Laude, H. M., Chaugule, B. A.: Effect of stage of seedling development upon heat tolerance in bromegrasses. J. Range Management **6**, 320–324 (1953)

Lazarev, Yu. A., Esipova, N. G., Lobashov, V. M., Grishkovsky, B. A.: The study of the collagen-like triple-helical structure by the spectroscopic methods. Abstract 5th Internat. Biophys. Congr. Copenhagen: 1975, p. 123

Lepeschkin, W.: The constancy of the living substance. Studies from the Plant Physiol. Labor. Charles Univ. Prague **1**, 6–44 (1923)

Levinson, H. S., Hyatt, M. T., Moore, F. E.: Dependence of the heat resistance of bacterial spores on the calcium: dipicolinic acid ratio. Biochem. Biophys. Res. Commun. **5**, 417–421 (1961)

Levitt, J.: Frost, drought, and heat resistance. Ann. Rev. Plant Physiol. **2**, 245–268 (1951)

Levitt, J.: The hardiness of plants. New York, London: Academic Press, 1956a

Levitt, J.: Temperature (heat and cold resistance, frost hardening). Handb. Pflanzenphysiol. **2**, 632–638 (1956b)

Levitt, J.: Frost, drought and heat resistance. Protoplasmatologia **8**, 87 (1958)

Levitt, J.: Winter hardiness in plants. In: Cryobiology, London, New York: 1966, pp. 495–563

Levitt, J.: Growth and survival of plants at extremes of temperature—a unified concept. Symp. Soc. Exper. Biol. **23**, 395–448 (1969)

Levitt, J.: Responses of plants to environmental stresses. New York, London: Academic Press, 1972

Levitt, J., Dear, J.: The role of membrane proteins in freezing injury and resistance. In: Ciba Found. Symp. on the Frozen Cell, London: 1970, pp. 149–174

Licht, P.: The temperature dependence of myosin adenosine triphosphatase and alkaline phosphatase in lizards. Comp. Biochem. Physiol. **12**, 331–340 (1964a)

Licht, P.: A comparative study of the thermal dependence of contractility in saurian skeletal muscle. Comp. Biochem. Physiol. **13**, 27–31 (1964b)

Licht, P.: Thermal adaptation in the enzymes of lizards in relation to preferred body temperatures. In: Molecular Mechanisms of Temperature Adaptation (ed. C. L. Prosser) Washington: Am. Ass. Advan. Sci. 1967, pp. 131–145

Lieberman, M. M., Lanyi, J. K.: Threonine deaminase from extremely halophilic bacteria. Cooperative substrate kinetics and salt dependence. Biochemistry **11**, 211–216 (1972)

Likhtenshtein, G. I.: Studies of the structure and function of proteins using the method of paramagnetic labels. Usp. biol. khim. **12**, 3–27 (1971)

Lim, S. T., Botts, J.: Temperature and aging effects on the fluorescence intensity of myosin-ANS complex. Arch. Biochem. Biophys. **122**, 153–156 (1967)

Lim, V. I., Ptitsyn, O. B.: On the constancy of the hydrophobic core volume in the molecules of myoglobins and hemoglobins. Mol. Biol. **4**, 372–382 (1970)

Linderstrøm-Lang, K.: Structure and enzymatic breakdown of proteins. Cold Spr. Harb. Symp. Quant. Biol. **14**, 117–126 (1950)

Linderstrøm-Lang, K., Hotchkiss, R. D., Johansen, G.: Peptide bonds in globular proteins. Nature **142**, 996 (1938)

Linderstrøm-Lang, K., Jacobsen, C. F.: The contraction accompanying the enzymatic breakdown of proteins. C. R. Lab. Carlsberg, ser. chem. **24**, 1–48 (1941)

Linderstrøm-Lang, K. U., Schellman, J. A.: Protein structure and enzyme activity. (ed. P. D. Boyer, H. Lardy, K. Myrbäck). In: The Enzymes, New York, London: Academic Press, 1959, pp. 443–510

Lindsay, J. A., Creaser, E. H.: Enzyme thermostability is a transformable property between *Bacillus* ssp. Nature **255**, 650–652 (1975)

Lineweaver, H., Hoover, S. R.: A comparison of the action of crystalline papain on native and urea-denatured proteins. J. Biol. Chem. **137**, 325–335 (1941)

Ljunger, C.: On the nature of the heat resistance of thermophilic bacteria. Physiol. Plant. **23**, 351–364 (1970)

Ljunger, C.: Further investigation on the nature of the heat resistance of thermophilic bacteria. Physiol. Plant **28**, 415–418 (1973)

Lobyshev, V. I., Pisachenko, A. I., Khromova, T. B.: The nature of biological solvent isotope effects. I. Investigation of collagen thermostability in light and heavy water. Studia Biophysica **46**, 173–182 (1974)

Loginov, M. A., Akhmedov, A. Ya.: On the temperature adaptation of plant photosynthesis. Dokl. Akad. Nauk Tadzh. SSR **13**, 53–56 (1970)

Loginov, M. A., Nasyrov, U. S.: Ecologo-physiological analysis of photosynthesis of *Astragalus* species indigenous to high altitudes of the Western Pamiro-Alay. Bot. Zh. **55**, 1171–1176 (1970)

Loginova, L. G.: Adaptation in the yeast. Short-term exposure to external factors. Mikrobiologia **14**, 310–319 (1945)

Loginova, L. G., Golovacheva, R. S., Egorova, L. A.: Life of Mikroorganisms at High Temperatures. Moscow: Izd. Nauka 1966

Loginova, L. G., Karpukhina, S. J.: On certain peculiar features of α-amylase in thermophilic bacteria. Dokl. Akad. Nauk SSSR **180**, 225–228 (1968)

Lomagin, A. G.: Changes in the resistance of plant cells after a short action of high temperature. Tsitologia **3**, 426–436 (1961)

Lomagin, A. G.: The effect of thermal hardening on the development of injury produced by ultraviolet rays in vegetable cells. Dokl. Akad. Nauk SSSR **157**, 1477–1479 (1964)

Lomagin, A. G., Antropova, T. A.: A study of the capacity of *Physarum polycephalum* for temperature adaptation. Tsitologia **10**, 1094–1104 (1968)

Lomagin, A. G., Antropova, T. A., Ilmete, A.: The influence of heat hardening on the resistance of plant cells to different injuring agents. Tsitologia **5**, 142–150 (1963)

Lomagin, A. G., Antropova, T. A., Semenikhina, L. V.: Phototaxis of chloroplasts as a criterion of viability of leaf parenchyma. Planta **71**, 119–124 (1966)

Lomagin, A. G., Zubova, M. E., Antropova, T. A.: Heat hardening of *Vicia faba* L. roots. Bot. Zh. **55**, 1327–1330 (1970)

Loomis, W. F.: Temperature—sensitive mutants of *Dictyostelium discoideum*. J. Bacteriol. **99**, 65–69 (1969)

Low, Ph. S., Somero, G. N.: Temperature adaptation of enzymes; a proposed molecular basis for the different catalytic efficiencies of enzymes from ectotherms and endotherms. Comp. Biochem. Physiol. **49**, 307–312 (1974)

Low, Ph. S., Bada, J. L., Somero, G. N.: Temperature adaptation of enzymes: roles of the free energy, the enthalpy, and the entropy of activation. Proc. Nat. Acad. Sci. **70**, 430–432 (1973)

Lozina-Lozinsky, L. K.: The stability of the *Paramecia* adapted to life in warm radioactive springs to various extremal agents. Tsitologia **3**, 154–166 (1961)

Luknitskaya, A. F.: Effect of the rearing temperature on thermostability of some algae. Tsitologia **5**, 135–141 (1963)

Luknitskaya, A. F.: Do *Chlamydomonas* have a capacity for heat hardening? Tsitologia **9**, 800–803 (1967)

Lukyanenko, V.I., Nikolayev, L.E.: A comparative study of thermostability of myocardial and skeletal muscles in the fingerlings of two sturgeon species. In: Izmenchivost teploustoichivosti kletok zhivotnykh v onto- i filogeneze (Collection of works (Sbornik) Variability in Cellular Heat Resistance of Animals in Ontogenesis and Phylogenesis. Leningrad: Izd. Nauka 1967, pp. 149–152

Lumry, R., Biltonen, R.: Thermodynamic and kinetic aspects of protein conformations in relation to physiological function. In: Structure and Stability of Biological Macromolecules (eds. S.N. Timasheff, G.D. Fasman) New York: Marcel Dekker 1969, pp. 65–212

Lundgren, H.P.: The catalytic effect of active crystalline papain on the denaturation of thyroglobulin. J. Biol. Chem. **138**, 293–303 (1941)

Lundgren, L.: Effect of variation of sporulation time and temperature on thermostability of *Bacillus cereus* spores. Physiol. Plantarum **20**, 392–399 (1967)

Lutova, M.I.: Investigation of photosynthesis in the cells with experimentally increased resistance. Bot. Zh. **43**, 283–287 (1958)

Lutova, M.I.: The effect of heat hardening on photosynthesis and respiration of leaves. Bot. Zh. **47**, 1761–1774 (1962)

Lutova, M.I.: The strength of the bond between chlorophyll and protein in plants with high heat resistance. Dokl. Akad. Nauk SSSR **149**, 1206–1208 (1963a)

Lutova, M.I.: Temperature dependence of photosynthesis in *Tradescantia* leaves heat-hardened by exposure to high temperature. Bot. Zh. **48**, 890–893 (1963b)

Lutova, M.I., Feldman, N.L.: A study of the ability of temperature adaptation in some marine algae. Tsitologia **2**, 699–710 (1960)

Lutova, M.I., Feldman, N.L., Drobyshev, V.P.: Changes in the cellular thermoresistance of marine algae under the influence of environmental temperature. Tsitologia **10**, 1538–1545 (1968)

Lutova, M.I., Zavadskaya, I.G.: Effects of the exposure of plants to different temperatures on the cell heat resistance. Tsitologia **8**, 484–493 (1966)

Luzikov, V.N.: Stabilization of the enzymatic systems of the inner mitochondrial membrane and related problems. Sub. Cell. Biochem. **2**, 1–31 (1973)

Luzikov, V.N., Rakhimov, M.M., Berezin, I.V.: A study of thermal inactivation of $NADH_2$ oxidase. Dokl. Akad. Nauk SSSR **174**, 1211–1214 (1967)

Luzikov, V.N., Rakhimov, M.M., Sacs, V.A., Berezin, I.V.: Heat inactivation of succinate oxidase and of its fragments. Biokhimia **32**, 1032–1035 (1967a)

Luzikov, V.N., Rakhimov, M.M., Sacs, V.A., Berezin, I.V.: The protective effect of substrates to inactivation of NADH oxidase and succinate oxidase by the cobra venom and trypsin. Biokhimia **32**, 1234–1241 (1967b)

Luzzati, V., Husson, F.: The structure of liquid crystalline phases of lipid-water systems. J. Cell. Biol. **12**, 207–219 (1962)

Lyons, J.M., Asmundson, C.M.: Solidification of unsaturated/saturated fatty acid mixtures and its relationship to chilling sensitivity in plants. J. Amer. Oil. Chem. Soc. **42**, 1056–1058 (1965)

Lyons, J.M., Raison, J.K.: Oxidative activity of mitochondria isolated from plant tissues sensitive and resistant to chilling injury. Plant Physiology **45**, 386–389 (1970a)

Lyons, J.M., Raison, J.K.: A temperature—induced transition in mitochondrial oxidation: contrasts between cold and warm-blooded animals. Comp. Biochem. Physiol. **37**, 405–411 (1970b)

Maas, W.K., Davis, B.D.: Production of an altered pantothenate-synthesizing enzyme by a temperature sensitive mutant of *Escherichia coli*. Proc. Nat. Acad. Sci. USA **38**, 785–797 (1952)

MacLennan, D.H., Wong, P.T.S.: Isolation of a calcium-sequestring protein from sarcoplasmic reticulum. Proc. Nat. Acad. Sci. USA **68**, 1231–1235 (1971)

MacLennan, D.H., Yip, C.C., Iles, G.H., Seeman, P.: Isolation of sarcoplasmic reticulum proteins. Cold Spr. Harb. Symp. Quant. Biol. **37**, 469–477 (1973)

Maier, R.: Einfluß von Photoperiode und Einstrahlungsstärke auf die Temperaturrestistenz einiger Samenpflanzen. Österr. Bot. Z. **119**, 306–322 (1971)

Makarewicz, W.: Thermal inactivation of AMP-aminohydrolase in muscle extracts of homeo- and poikilothermic animals. Bull. Acad. Pol. Sci. Cl. II, **13**, 447–450 (1965)

Makarov, A. A., Ivanov, V. I., Volkenshtein, M. V.: The kinetics of proteolysis of aspartate aminotransferase as a function of the state of the active site. Molek. biol. **8**, 433–442 (1974)
Makhlin, E. E.: Thermostability of pereopodes of the Baikal Lake gammarids. Dokl. Akad. Nauk SSSR **139**, 980–983 (1961)
Makhlin, E. E.: Heat resistance of protein complexes from corneas of *Rana temporaria* L. and *Rana ridibunda* Pall. In: Problemy tsitoekologii zhivotnykh (Collection of works (Sbornik) Problems of Cytoecology of Animals) Moscow, Leningrad: Izd. Akad. Nauk SSSR 1963, pp. 199–207
Makhlin, E. E.: Studies on succinic dehydrogenase thermostability in muscle homogenates of *Rana temporaria* L. at different levels of muscle thermostability. In: Teploustoichivost kletok zhivotnykh (Collection of works (Sbornik) "Heat Resistance of Cells of Animals") Moscow, Leningrad: Izd. Nauka 1965, pp. 119–124
Maksimov, N. A.: On the freeze-out and cold resistance of plants. Selected works **2**, 27–241. Moscow: 1952 (Year of original publication 1913)
Maksimov, V. I.: Adaptation of enzymes by their stability. Usp. sovr. biol. **76**, 21–33 (1973)
Malcolm, N. L.: Synthesis of protein and ribonucleic acid in a psychrophile at normal and restrictive growth temperatures. J. Bacteriol. **95**, 1388–1399 (1968a)
Malcolm, N. L.: A temperature—induced lesion in amino acid transfer ribonucleic acid attachment in a psychrophile. Biochim. biophys. Acta **157**, 493–503 (1968b)
Malcolm, N. L.: Enzymic bases of physiological changes in a mutant of the psychrophile *Micrococcus cryophilus*. Biochim. biophys. Acta **190**, 337–346 (1969a)
Malcolm, N. L.: Molecular determinants of obligate psychrophily. Nature **221**, 1031–1033 (1969b)
Mallett, G. E., Koffler, H.: Hypotheses concerning the relative stability of flagella from thermophilic bacteria. Arch. Biochem. Biophys. **67**, 254–256 (1957)
Mandelstam, J.: Turnover of protein in growing and nongrowing population of *Escherichia coli*. Biochem. J. **69**, 110–119 (1958)
Mangiantini, M. T., Tecce, G., Toschi, G., Trentalance, A.: A study of ribosomes and of RNA from a thermophilic organism. Biochim. biophys. Acta **103**, 252–274 (1965)
Manning, G. B., Campbell, L. L.: Thermostable α-amylase of *Bacillus stearothermophilus*. I. Crystallization and some general properties. J. Biol. Chem. **236**, 2952–2957 (1961)
Manning, G. B., Campbell, L. L., Foster, K. J.: Thermostable α-amylase of *Bacillus stearothermophilus*. II. Physical properties and molecular weight. J. Biol. Chem. **236**, 2958–2961 (1961)
Margoliash, E., Schejter, A.: Cytochrome C. Adv. Protein. Chem. **21**, 113–286 (1966)
Markus, G.: Protein substrate conformation and proteolysis. Proc. Nat. Acad. Sci. USA **54**, 253–266 (1965)
Markus, G., Barnard, E., Castellani, B. A., Saunders, D.: Ligand-induced conformational changes in ribonuclease. J. Biol. Chem. **243**, 4070–4076 (1968)
Markus, G., McClintock, D. K., Castellani, B. A.: Ligand-stabilized conformations in serum albumin. J. Biol. Chem. **242**, 4402–4408 (1967)
Marmur, J.: Thermal denaturation of deoxyribosenucleic acid isolated from a thermophile. Biochim. biophys. Acta **38**, 342–343 (1960)
Marmur, J., Doty, P.: Determination of the base composition of deoxyribonucleic acid from its thermal denaturation temperature. J. Mol. Biol. **5**, 109–118 (1962)
Marquez, E., Brodie, A. F.: The effect of cations on the heat stability of a halophilic nitrate reductase. Biochim. biophys. Acta **321**, 84–89 (1973)
Marr, A. G., Ingraham, J. L.: Effect of temperature on the composition of fatty acids in *Escherichia coli*. J. Bacteriol. **84**, 1260–1267 (1962)
Marre, E. E.: Temperature. In: Physiology and Biochemistry of Algae, New York, London: Academic Press 1962, pp. 541–550
Marre, E. E., Albertario, M., Vaccari, E.: Ricerche sull 'adattamento proteico in organismi termoresistenti. III. Relativa insensibilita di enzimi di cianoficee termali a denaturanti che agiscono rompendo i ponti di idrogeno. Atti Accad. nazl. Lincei Rend. Classe Sci. fis. mat. e nat. **24**, 349–353 (1958)
Marre, E., Servettaz, O.: Ricerche sull'adattamento proteico in organismi termoresistenti. II. Sulla termoresistenza in vitro del sistema citrocromo riduttasico di cianoficee termali. Atti Accad. nazl. Lincei Rend. Classe Sci. fis. mat. e nat. **22**, 91–98 (1957)

Marsh, C., Militzer, W.: Thermal enzymes. VIII. Properties of a heat stable inorganic pyrophosphatase. Arch. Biochem. Biophys. **60**, 439–451 (1956)
Marx, K., Engels, F.: Collection of works. Second edition Moscow: State Publ. House Polit. Lit., 1961, **20**, pp. 534–635
Mason, M., Gullekson, E. H.: Estrogen-enzyme interaction: inhibition and protection of kynurenine transaminase by the sulfate esters of diethylstilbestrol, estradiol, and estrone. J. Biol. Chem. **235**, 1312–1316 (1960)
Massey, V., Curti, B., Ganther, H.: A temperature—dependent conformational change in α-amino acid oxidase and its effect on catalysis. J. Biol. Chem. **241**, 2347–2357 (1966)
Mathias, M. M., Kemp, R. G.: Allosteric properties of muscle phosphofructokinase. III. Thiol reactivity as an indicator of conformational state. Biochem. **11**, 578–584 (1972)
Matsubara, H.: Some properties of thermolysin. In: Molecular Mechanisms of Temperature Adaptation (ed. C. L. Prosser) Washington: Am. Ass. Advan. Sci. 1967, pp. 283–294
Maurer, P. H.: Modified bovine serum albumin. VI. Immunochemical and physicochemical properties of bovine serum albumin denatured by various agents. Arch. Biochem. Biophys. **79**, 13–26 (1959)
Maybury, R. H., Katz, J. J.: Protein denaturation in heavy water. Nature **177**, 629–630 (1956)
Mazhul, V. M., Chernitsky, E. A., Konev, S. V.: Conformation transitions of native proteins in solution and in the cell. Biofizika **15**, 5–11 (1970)
McClintock, D. K., Markus, G.: Conformational changes in asparate transcarbamylase. I. Proteolysis of the intact enzyme. J. Biol. Chem. **243**, 2855–2862 (1968)
McDonald, W. C., Matney, T. S.: Genetic transfer of the ability to grow at 55° C in *Bacillus subtilis*. J. Bacteriol. **85**, 218–220 (1963)
McMurchie, E. J., Raison, J. K., Cairncross, K. D.: Temperature-induced phase changes in membranes of heart: a contrast between the thermal response of poikilotherms and homotherms. Comp. Biochem. Physiol. **44**, 1017–1026 (1973)
McMurrough, J., Rose, A. H.: Effects of temperature variation on the fatty acid composition of *Candida utilis*. J. Bacteriol. **107**, 753–758 (1971)
McNaughton, S. J.: Differential enzymatic activity in ecological races of *Typha latifolia* L. Science **150**, 1829–1830 (1965)
McNaughton, S. J.: Thermal inactivation properties of enzymes from *Typha latifolia* L. ecotypes. Plant Physiol. **41**, 1736–1738 (1966)
McNaughton, S. J.: Enzymic thermal adaptations: the evolution of homeostasis in plants. Amer. Naturalist **106**, 165–172 (1972)
Meeks, J. C., Castenholz, R. W.: Growth and photosynthesis in an extreme thermophile *Synechococcus lividus* (Cyanophyta). Arch. Mikrobiol. **78**, 25–41 (1971)
Meier, M. N., Orlov, V. N., Skholl, E. D.: A new *Rodentia* species isolated by application of the data of karyological, physiological, and cytophysiological analyses. Dokl. Akad. Nauk SSSR **188**, 1411–1414 (1969)
Meier, M. N., Orlov, V. N., Skholl, E. D.: Sibling species in the group *Microtus arvalis* (Rodentia, Cricetidae) Zool. Zh. **51**, 724–737 (1972)
Melchior, D. L., Morowitz, H. I., Sturtevant, I. M., Tsong, T. Y.: Characterization of the plasma membrane of *Mycoplasma laidlawii*. VII. Phase transition of membrane lipids. Biochim. biophys. Acta **219**, 114–122 (1970)
Mertens, R., Müller, L.: Liste der Amphibien und Reptilien Europas. Abhandl. Senckenberg. Naturforsch. Ges. **41**, 1–62 (1928)
Mews, H. H.: Über die Temperaturadaptation der Sekretion von Verdauungsfermenten und deren Hitzeresistenz. Z. vergl. Physiol. **40**, 345–355 (1957)
Meyer, G. H., Morrow, M. B., Wyss, O., Berg, S. E., Littlepage, T. L.: Antarctica: the microbiology of an saline pond. Science **138**, 1103–1104 (1962)
Mikhalchenko, T. V.: Heat resistance of ciliary epithelium and its variations in *Rana temporaria* L. Dokl. Akad. Nauk SSSR **111**, 1352–1355 (1956)
Mikhalchenko, T. V.: Temperature sensitivity of *Rana temporaria* and some of her parasites. Uch. zapiski Len. gos. ped. inst. im. A. I. Hertsena **143**, 261–287 (1958)
Mikhalchenko, T. V.: Sex differences in sensitivity to elevated temperature in males and females of *Rana temporaria*. Uch. zapiski Len. gos. ped. inst. im. A. I. Hertsena **176**, 103–110 (1959)

Militzer,W., Burns,L.: Thermal enzymes. VI. Heat stability of pyruvic oxidase. Arch. Biochem. Biophys. **52**, 66–73 (1954)

Militzer,W., Sonderegger,T.B., Tuttle,L.C., Georgi,C.E.: Thermal enzymes. Arch. Biochem. **24**, 75–82 (1949)

Miller,L.K.: Activity in mammalian peripheral nerves during supercooling. Science **149**, 74–75 (1965)

Miller,V.J.: Temperature effect on the rate of apparent photosynthesis of seaside bent and Bermuda grass. Proc. Amer. Soc. Hort. Sci. **75**, 700–703 (1960)

Mishiro,Y., Ochi,M.: Effect of dipicolinate on the heat denaturation of proteins. Nature **211**, 1190 (1966)

Mizushima,H., Nozaki,M., Horio,T., Okunuki,K.: Digestion of baker's yeast cytochrome *C* in oxidized form by bacterial proteinase under aerobic and anaerobic conditions. J. Biochem. (Tokyo) **45**, 845–846 (1958)

Moewus,F.: Die Analyse von 42 erblichen Eigenschaften der *Chlamydomonas eugametos-Gruppe*. II. Teil: Zellresistenz, Sexualität, Zygote, Besprechung der Ergebnisse. Z. ind. Abst. Vererbungslehre **78**, 462–518 (1940)

Monfort,C., Ried,A., Ried,J.: Die Wirkung kurzfristiger warmer Bäder auf Atmung und Photosynthese im Vergleich von eurythermen und kaltstenothermen Meeresalgen. Beitr. Biol. Pflanz. **31**, 349–375 (1955)

Monfort,C., Ried,A., Ried,J.: Abstufungen der funktionellen Wärmeresistenz bei Meeresalgen in ihren Beziehungen zu Umwelt und Erbgut. Biol. Zbl. **76**, 257–289 (1957)

Monod,J., Wyman,J., Changeux,J.P.: On the nature of allosteric transitions: a plausible model. J. Mol. Biol. **12**, 88–118 (1965)

Moon,T.W., Hochachka,P.W.: Temperature and enzyme activity in poikilotherm. Isocitrate dehydrogenases in rainbow-trout liver. Biochem. J. **123**, 695–705 (1971)

Moon,T.W., Hochachka,P.W.: Temperature and the kinetic analysis of trout isocitrate dehydrogenases. Comp. Biochem. and Physiol. **42**, 725–730 (1972)

Mooney,H.A., Billings,W.D.: Comparative physiological ecology of arctic and alpine populations of *Oxyria digyna*. Ecological Monographs. **31**, 1–29 (1961)

Mooney,H.A., West,M.: Photosynthetic acclimation of plants of diverse origin. Amer. J. Bot. **51**, 825–827 (1964)

Mooney,H.A., Wright,R.D., Strain,B.R.: The gas exchange capacity of plants in relation to vegetation zonation in the White Mountains of California. Amer. Midland Naturalist **72**, 281–297 (1964)

Moore,J.A.: Temperature tolerance and rates of development in the eggs of *Amphibia*. Ecology **20**, 459–478 (1939)

Moore,J.A.: The role of temperature in speciation of frogs. Biol. Symp. **6**, 189–213 (1942)

Morgan,L.: Habit and Instinct. London: 1896

Morin,J.G., Hastings,J.W.: Biochemistry of the bioluminescence of colonial hydroids and other Coelenterates. J. Cell. Physiol. **77**, 305–311 (1971)

Morita,R.Y.: Marine psychrophilic bacteria. Oceanogr. Mar. Biol. Ann. Rev. **4**, 105–121 (1966)

Morita,R.Y., Burton,S.D.: Influence of moderate temperature on growth and malic dehydrogenase activity of a marine psychrophile. J. Bacteriol. **86**, 1025–1029 (1963)

Mothes,K.: Über das Altern der Blätter und die Möglichkeit ihrer Wiederverjüngung. Naturwissenschaften **47**, 337–351 (1960)

Mueller,H., Theiner,M., Olson,R.E.: Macromolecular fragments of canine cardiac myosin obtained by tryptic digestion. J. Biol. Chem. **239**, 2153–2159 (1964)

Muirhead,H., Cox,J.M., Mazzarella,L., Perutz,M.F.: Structure and function of haemoglobin. III. A three-dimensional Fourier synthesis of human deoxyhaemoglobin at 5.5 Å resolution. J. Mol. Biol. **28**, 117–156 (1967)

Murata,Y., Iyama,J.: Studies on the photosynthesis of forage crops. II. Influence of air-temperature upon the photosynthesis of some forage and grain crops. Proc. Crop. Sci. Soc. Japan **31**, 315–322 (1963)

Mutchmor,J.A.: Temperature adaptation in insects. In: Molecular Mechanisms of Temperature Adaptation, Washington: 1967, pp. 165–176

Mutchmor, J. A., Richards, A. G.: Low temperature tolerance of insects in relation to the influence of temperature on muscle apyrase activity. J. Insect Physiol. **7**, 141–158 (1961)

Myalo, E. G.: Peculiarities of distribution of reed grass within an area. Bull. Mosk. obshch. ispytat. prirody **67**, 83–95 (1962)

Nakajima, M., Mizusawa, K., Yoshida, F.: Purification and properties of an extracellular proteinase of psychrophilic *Escherichia freundii*. Eur. J. Biochem. **44**, 87–96 (1974)

Nash, C. H., Grant, D. W.: Thermal stability of ribosomes from a psychrophilic and a mesophilic yeast. Can. J. Microbiol. **15**, 1116–1118 (1969)

Nash, C. H., Grant, D. W., Sinclair, N. A.: Thermolability of protein synthesis in a cell—free system from the obligately psychrophilic yeast *Candida gelida*. Can. J. Microbiol. **15**, 339–343 (1969)

Nasonov, D. N.: Local reaction of protoplasm and the spreading irritation. Moscow, Leningrad: Izd. Akad. Nauk SSSR, 1959

Nasonov, D. N., Alexandrov, V. Ya.: Reaction of the live matter to external influences. Moscow, Leningrad: Izd. Akad. Nauk SSSR, 1940

Nath, K., Koch, A. L.: Protein degradation in *Escherichia coli*. II. Strain differences in the degradation of protein and nucleic acid resulting from starvation. J. Biol. Chem. **246**, 6956–6967 (1971)

Neifakh, S. A.: On the sources of animal warmth at the cellular level. In: Fiziologiya teploobmena i gigiena promyshlennogo mikroklimata (Collection of works (Sbornik) Physiology of Heat-Exchange and the Hygiene of Industrial Microclimate) Moscow: Inst. gigieny truda i profzabolevanii, 1961, pp. 35–51

Neurath, H.: Mechanism of zymogen activation. Fed. Proc. **23**, part I, 1–17 (1964)

Neurath, H., Greenstein, J. P., Putnam, F. W., Erickson, J. O.: The chemistry of protein denaturation. Chem. Rev. **34**, 157–265 (1944)

Neurath, H., Rupley, J. A., Dreyer, J.: Structural changes in the activation of chymotrypsinogen and trypsinogen. Effect of urea on chymotrypsinogen and chymotrypsin. Arch. biochem. Biophys. **65**, 243–259 (1956)

Newell, R. C., Northcroft, H. R.: A re-interpretation of the effect of temperature on the metabolism of certain marine vertebrates. J. Zool. **151**, 277–298 (1967)

Newell, R. C., Pye, V. I.: Seasonal variations in the effect of temperature on the respiration of certain intertidal algae. J. Mar. Biol. **48**, 341–348 (1968)

Newell, R. C., Pye, V. I.: Temperature—induced variations in the respiration of mitochondria from the winkle *Littorina littorea* (L.). Comp. Biochem. Physiol. **40 B**, 249–261 (1971)

Nickerson, K. W.: Biological functions of multistable proteins. J. Theor. Biol. **40**, 507–515 (1973)

Nikiforov, V. G.: RNA-polymerase from a thermophilic *Bacillus megaterium* strain. Studies of initiation of RNA synthesis. Molek. biol. **4**, 159–174 (1970)

Nomoto, M., Narahashi, Y., Murakami, M.: A proteolytic of *Streptomyces griseus*. V. Protective effect of calcium ion on the stability of protease. J. Biochem. (Tokyo) **48**, 453–463 (1960)

Novoselova, A. N., Sevrova, O. K., Volgina, K. P.: Enzymes from plants with experimentally elevated heat stability. In: Fiziologicheskiye mekhanismy regulyatsii, prisposobleniya i ustoichivosti u rastenii (Collection of works (Sbornik) Physiological Mechanisms of Regulation, Adaptation and Resistance in Plants) Novosibirsk: Izd. Nauka, 1966, pp. 22–32

Novoselova, A. N., Sevrova, O. K., Kiselev, V. E.: Peculiarities of proteins from wheat seedlings in relation to their adaptation to high temperature. Izv. Sib. otd. Akad. Nauk SSSR **2**, 64–71 (1971)

Nowell, N., Akagi, J., Himes, R. H.: Thermostability of glycolytic enzymes from thermophilic *Clostridia*. Can. J. Microbiol. **15**, 461–464 (1969)

Nozaki, M., Kagamigama, H., Hayashi, O.: Metapyrocatechase. I. Purification, crystallization and some properties. Bioch. Z. **338**, 582–590 (1963)

Nozaki, M., Mizushima, H., Horio T., Okunuki, K.: Studies on cytochrome C. II. Further study on proteinase digestion of baker's yeast cytochrome C. J. Biochem. (Tokyo) **45**, 815–823 (1958)

Nyns, E. J.: Reactivation by organic solvents of an alcohol dehydrogenase from *Candida lipolytica* grown on n-hexadecane. Biochim. biophys. Acta **212**, 351–352 (1970)

O'Brien, W.S., Brewer, J.M., Ljungdahl, L.Y.: Purification and characterization of thermostable 5.10-methylenetetra hydrofolate dehydrogenase from *Clostridium thermoaceticum*. J. Biol. Chem. **248**, 403–408 (1973)

Ochi, M.: Studies on the effect of dipicolinate on the heat denaturation of serum albumin. Odontology **55**, 93–109 (1967)

Ogasahara, K., Imanishi, A., Isemura, T.: Studies on thermophilic α-amylase from *Bacillus stearothermophilus*. I. Some general and physico-chemical properties of thermophilic α-amylase. J. Biochem. **67**, 65–75 (1970a)

Ogasahara, K., Imanishi, A., Isemura, T.: Studies on thermophilic α-amylase from *Bacillus stearothermophilus*. II. Thermal stability of thermophilic α-amylase. J. Biochem. **67**, 77–82 (1970b)

Ohta, Y.: Thermostable protease from thermophilic bacteria. II. Studies on the stability of the protease. J. Biol. Chem. **242**, 509–515 (1967)

Ohta, Y., Ogura, Y., Wada, A.: Thermostable protease from thermophilic bacteria. I. Thermostability, physico-chemical properties and amino acid composition. J. Biol. Chem. **241**, 5919–5925 (1966)

Okunuki, K.: Denaturation and inactivation of enzyme proteins. Adv. Enzymol. **23**, 29–82 (1961)

Okunuki, K., Hagihara, B., Matsubara, H., Nakayama, T.: Denaturation and inactivation of enzyme proteins. I. Bacterial proteinase method for the determination of the ration of denaturation of globular proteins. J. Biochem. (Tokyo) **43**, 453–467 (1956)

Oleynikova, T.V.: Effects of high temperature and light on the heat stability of cells in different crop varieties of plants. In: Kletka i temperatura sredy (Collection of works (Sbornik) The Cell and Environmental Temperature) Moscow, Leningrad: Izd. Nauka, 1964a, pp. 119–122

Oleynikova, T.V.: High temperature and light effects on the permeability of cells of a spring grass leaves. In: Tsitologicheskiye osnovy prisposobleniya rastenii k faktoram sredy (Collection of works (Sbornik) Cytological Aspects of Adaptation of Plants to the Environmental Factors) Moscow, Leningrad: Izd. Nauka, 1964b, pp. 70–81

Oleynikova, T.V., Uglov, P.D.: Thermostability of the cell protoplasm in some varieties of a spring wheat. Bot. Zh. **47**, 337–343 (1962)

Olsen, R.H., Metcalf, E.S.: Conversion of mesophilic to psychrophilic bacteria. Science **162**, 1288–1289 (1968)

Olsson, S.-O.R.: Comparative studies on the temperature dependence of lactic and malic dehydrogenase from a homeotherm, guinea pig *(Cavia porcellus)*; two hibernators, hedgehog *(Erinaceus europaeus)* and bat *(Nystalus noctula)*; and two poikilotherms, frog *(Rana temporaria)* and cod *(Gadus callarias)* Comp. Biochem. Physiol. **51B**, 5–18 (1975)

Omura, T., Siekevitz, P., Palade, G.E.: Turnover of constituents of the endoplasmic reticulum membranes of rat hepatocytes. J. Biol. Chem. **242**, 2389–2396 (1967)

Ono, K., Nozaki, M., Hayashi, O.: Purification and some properties of protocatecholate 4,5-dioxygenase. Biochim. biophys. Acta **220**, 224–238 (1970)

Orgel, J.E.: Adaptation to wide-spred disturbance of enzyme function. J. Mol. Biol. **9**, 208–212 (1964)

Oshima, T., Imahori, K.: Physicochemical properties of deoxyribonucleic acid from an extreme thermophile. J. Biochem. **75**, 179–183 (1974)

Osipov, D.V.: An analysis of hereditary mechanisms responsible for the heat resistance of *Paramecium caudatum* Ehrbg. Genetika **1**, 119–131 (1966a)

Osipov, D.V.: Methods for isolating homozygous *Paramecium caudatum* clones. Genetika **2**, 41–48 (1966b)

Osipov, D.V.: Thermostability of the *Paramecium caudatum* clones isolated from various natural populations. Vestn. Len. Univ. (ser. biol.) **3**, 107–112 and 129 (1966c)

Ostner, U., Hultin, T.: The use of proteolytic enzymes in the study of ribosomal structure. Biochim. biophys. Acta **154**, 376–387 (1968)

O'Sullivan, S.A., Wedding, R.T.: Malate dehydrogenase isoenzymes in cotton leaves of different ages. Physiol. Plantarum **26**, 34–38 (1972)

O'Sullivan, W.J., Cohn, M.: Nucleotide specificity and conformation of the active site of creatine kinase. J. Biol. Chem. **241**, 3116–3125 (1966)

Pace,B., Campbell,L,L.: Correlation of maximal growth temperature and ribosome heat stability. Proc. Nat. Acad. Sci. USA **57** 1110–1116 (1967)

Pace,N.C., Tanford,C.: Thermodynamics of the unfolding of β-lactoglobulin A in aqueous urea solutions between 5 and 55°. Biochemistry **7**, 198–208 (1968)

Padayatty,J.D., Hensley,M.D., Kley,H. van: Stabilization of nucleic acids by proteins toward enzymic digestion. Biochim. biophys. Acta **161**, 51–55 (1968)

Palm,D., Katzendobler,H.: Effect of allosteric ligands on the mechanism and stability of nicotinamide-adenine dinucleotide specific isocitrate dehydrogenase from yeast. Biochemistry **11**, 1283–1289 (1972)

Pandey,K.K.: Isozyme specificity to temperature. Nature New Biol. **239**, 27–28 (1972)

Parker,J.: Heat resistance and respiratory response in twigs of some common tree species. Bot. Gaz. **132**, 268–273 (1971)

Pashkova,I.M.: On the analysis of seasonal changes in common frog cells. Zh. obshch. biol. **23**, 313–317 (1962a)

Pashkova,I.M.: Variation in thermostability of *Rana temporaria* L. muscles as a “fright” reaction. Dokl. Akad. Nauk SSSR **144**, 1425–1428 (1962b)

Pashkova,I.M.: Physiological analysis of seasonal changes in heat resistance of the muscle tissue of *Rana temporaria* L. In: Problemy tsitoekologii zhivotnykh (Collection of works (Sbornik) Problems of Cytoecology of Animals) Moscow, Leningrad: Izd. Akad. Nauk SSSR, 1963a, pp.62–68

Pashkova,I.M.: Changes in muscle thermostability of *Rana temporaria* L. after a short storage period under conditions of increased temperatures at different seasons of the year. In: Problemy tsitoekologii zhivotnykh (Collection of works (Sbornik) Problems of Cytoecology of Animals) Moscow, Leningrad: Izd. Akad. Nauk SSSR, 1963b, pp.87–92

Pashkova,I.M.: Relationships between muscle thermostability and activity of thyroid gland of *Rana temporaria* L. in different seasons of the year. In: Teploustoichivost kletok zhivotnykh (Collection of works (Sbornik) Heat Resistance of Cells of Animals) Moscow, Leningrad: Izd. Nauka 1965, pp.82–89

Patzel,H.: Vergleichende Untersuchungen über die Wärmekontraktur und Wärmelähmung der quergestreiften Muskeln von Eidechsen und Fröschen. Pflüg. Arch. ges. Physiol. **231**, 90–101 (1933)

Pauli,Wo., Valkó,E.: Kolloidchemie der Eiweißkörper. Dresden, Leipzig: Steinkopf, 1933

Payusova,A.N., Koreshkova,N.D.: Cytophysiological analysis of divergence in the Issyk-Kul Lake *Leuciscus*. Tsitologia **14**, 961–967 (1972)

Pearson,L.K., Raper,H.S.: The influence of temperature on the nature of the fat formes by living organisms. Bioch. J. **21**, 875–879 (1927)

Peary,J.A., Castenholz,R.W.: Temperature strains of a thermophilic blue-green alga. Nature **202**, 720–721 (1964)

Permogorov,V.I., Prozorov,A.A., Shemyakin,M.F., Lazurkin,Yu.S., Khesin,R.B.: On the mechanism of suppression of biological activity of DNA by Actinomycin. In: Molekulyarnaya biofizika (Collection of works (Sbornik) Molecular Biophysics) Moscow: Izd. Nauka 1965, pp.162–180

Perutz,M.F.: Stereochemistry of cooperative effects in haemoglobin. Nature **228**, 726–734 (1970)

Pfueller,S.L., Elliott,N.H.: The extracellular α-amylase of *Bacillus stearothermophilus*. J. Biol. Chem. **244**, 48–54 (1969)

Pigulevsky,G.V.: On the investigations of the influence of climatic conditions on the composition of plant oils. Zh. Russk. Khimich. obshch. **46**, 324–341 (1916)

Pine,M.J.: Turnover of intracellular proteins. Ann. Rev. Microbiol. **26**, 103–126 (1972)

Pisek,A., Larcher,W., Moser,W., Pack,T.: Kardinale Temperaturbereiche der Photosynthese und Grenztemperaturen des Lebens der Blätter verschiedener Spermatophyten. III. Temperaturabhängigkeit und optimaler Temperaturbereich der Netto-Photosynthese. Flora **B, 158**, 608–630 (1969)

Platt,T., Miller,J.H., Weber,K.: *In vivo* degradation of mutant *lac*-repressor. Nature **228**, 1154–1156 (1970)

Polgar,L.: The mechanism of action of glyceraldehyde-3-phosphate dehydrogenase. Experientia **20**, 1–14 (1964)

Polyanovsky, O. L.: The quaternary structure of enzymes. Usp. sovr. biol. khimii **7**, 34–60 (1967)

Polyansky, Yu. I. (with the participation of Orlova, A. F.): Temperature adaptation in *Infusoria*. I. Relation of heat resistance of *Paramecium caudatum* to the temperature conditions of existence. Zool. Zh. **36**, 1630–1646 (1957)

Polyansky, Yu. I.: Temperature adaptations of Infusoria. II. Changes in resistance to high and low temperatures in *Paramecium caudatum* at low cultivation temperatures. Tsitologia **1**, 714–727 (1959)

Polyansky, Yu. I.: On the capability of *Paramecium caudatum* to survive subzero temperatures. Acta Protozool. **1**, 165–175 (1963)

Polyansky, Yu. I.: The problem of physiological adaptation with regard to the forms of variability in the free living *Protozoa* (some results and perspectives). In: Progress in Protozoology Abst. 4th Intern. Congr. Protozool. Clermont-Ferrant: 1973, pp. 40–53

Polyansky, Yu. I., Irlina, I. S.: On "heat hardening" in ciliates. Tsitologia **9**, 791–799 (1967)

Polyansky, Yu. I., Irlina, I. S.: On the variability in the norm of reaction of *Paramecium caudatum* to various cultivation temperatures. Acta Protozool. **12**, 85–95 (1973)

Polyansky, Yu. I., Orlova, A. F.: On the adaptive changes and longlasting modifications in the infusorium *Paramecium caudatum* produced by the action of high and low temperatures. Dokl. Akad. Nauk SSSR **59**, 1025–1028 (1948)

Polyansky, Yu. I., Poznanskaya, T. M.: Prolonged cultivation of *Paramecium caudatum* at 0°. Acta protozool. **2**, 271–278 (1964)

Polyansky, Yu. I., Sukhanova, K. M.: Some peculiarities in temperature adaptations of *Protozoa* as compared to multicellular poikilotherms. In: The Cell and Environmental Temperature. Moscow, Leningrad: Izd. Nauka, 1964, pp. 135–142

Polyansky, Yu. I., Sukhanova, K. M., Sopina, V. A., Yudin, A. L.: Resistance of *Amoeba proteus* to the effect of lethal temperatures and ethyl alcohol. In: Izmenchivost teploustoichivosti kletok zhivotnykh v onto- i filogeneze (Collection of works (Sbornik) Variability in Cellular Heat Resistance of Animals in Ontogenesis and Phylogenesis) Leningrad: Izd. Nauka, 1967, pp. 43–62

Popova, A. I.: Thermostability of protoplasm in maize leaves in relation to the sowing time. Izv. otd. biol. nauk Akad. Nauk Tadzh. SSR **3**, 37–41 (1964)

Popova, A. I.: Cellular thermostability of leaves from different layers of maize plants. Izv. otd. biol. nauk Akad. Nauk Tadzh. SSR **3**, 83–88 (1968)

Port, J.: Die Wirkung der Neutralsälze auf die Koagulation des Protoplasmas bei *Paramecium caudatum*. Protoplasma **2**, 401–419 (1927)

Prat, S.: Contribution to the physiology of the vegetation of thermal and mineral waters. VIII. Congress. Internat. de Botanique, Sect. 17. Rapports et Communication 42–44. Paris: 1954

Pravdina, K. I.: The thermostability of aldolase in taxonomically related species of amphibians. In: Teploustoichivost kletok zhivotnykh (Collection of works (Sbornik) Heat Resistance of Cells of Animals) Moscow, Leningrad: Izd. Nauka, 1965, pp. 193–199

Pravdina, K. I.: Determination of heterogeneity of sarcoplasmic proteins in two species of frogs by the method of electrophoresis in agar. Tsitologia **9**, 61–67 (1967)

Pravdina, K. I.: Thermostability of water-soluble esterases of poikilothermal animals. Tsitologia **12**, 1541–1549 (1970)

Precht, H.: Die Temperaturabhängigkeit von Lebensprozessen. Z. Naturforsch. **46**, 26–35 (1949)

Precht, H.: Concepts of the temperature adaptation of unchanging reaction systems of cold-blooded animals. In: Physiological Adaptation (ed. L. Prosser). Washington: Am. Phys. Soc. 1958, pp. 50–78

Precht, H.: Über die Resistenzadaptation wechselwarmer Tiere an extreme Temperaturen und ihre Ursachen. Helgol. wiss. Meeresuntersuch. **9**, 392–411 (1964a)

Precht, H.: Über die Bedeutung des Blutes für die Temperaturadaptation von Fischen. Zool. Jahrb. Physiol. **71**, 313–327 (1964b)

Precht, H.: Der Einfluß „normaler" Temperaturen auf Lebensprozesse bei wechselwarmen Tieren unter Ausschluß der Wachstums- und Entwicklungsprozesse. Helgol. wiss. Meeresuntersuch. **18**, 487–548 (1968)

Precht, H., Basedow, T., Bereck, R., Lange, F., Thiede, W., Wilke, L.: Reaktionen und Adaptationen wechselwarmer Tiere nach einer Änderung der Anpassungstemperatur und der zeitliche Verlauf. Helgol. wiss. Meeresuntersuch. **13**, 369–401 (1966)
Precht, H., Christophersen, J., Hensel, H.: Temperatur und Leben. Berlin, Göttingen, Heidelberg: Springer 1955
Preer, J.R., Jr.: A gene determining temperature sensitivity in *Paramecium*. J. Genetics **55**, 375–378 (1957)
Preobrazhenskaya, T.A., Shnol, S.E.: The kinetics of changes of growth to the *Rhizopus nigricans* mycelium as a function of temperature variations. In: Biofizika kletki (Collection of works (Sbornik) The Biophysics of the Cell) Moscow: Izd. Nauka, 1965, pp. 68–72
Price, V.E., Sterling, W.R., Tarantola, V.A., Hartley, R.W., Rechcige, M.: The kinetics of catalase synthesis and destruction *in vivo*. J. Biol. Chem. **237**, 3468–3475 (1962)
Privalov, P.L.: Study of thermal transconformation of tropocollagen. I. Denaturation enthalpy of tropocollagens with different amino acid content. Biofizika **13**, 955–963 (1968)
Privalov, P.L.: Thermodynamical investigations of biological macromolecules. In press
Privalov, P.L., Khechinashvili, N.N.: A thermodynamical approach to the problem of stabilization of globular protein structure. A calorimetric study. J. Mol. Biol. **86**, 665–684 (1974)
Privalov, P.L., Khechinashvili, N.N., Atanasov, B.P.: Thermodynamic analysis of thermal transitions in globular proteins. I. Calorimetric study of chymotrypsinogen, ribonuclease and myoglobin. Biopolymers **10**, 1865–1890 (1971)
Privalov, P.L., Monaselidze, D.R.: Investigation of thermal denaturation of serum albumin. Biofizika **8**, 420–426 (1963)
Prosser, L., Brown, F.: Comparative Animal Physiology. Philadelphia, London: W.B. Saunders Co. 1962
Ptitsyn, O.B.: The nature of the forces determining native spatial structures of globular proteins. Usp. sovr. biol. **63**, 3–27 (1967
Purkinje, J.E., Valentin, G.: Phaenomeno generali et fundamentali motus vibratorii continui. Wratislawiae: 1835
Purohit, K., Stokes, J.L.: Heat-labile enzymes in a psychrophilic bacterium. J. Bacteriol. **93**, 199–206 (1967)
Putnam, F.W.: Protein denaturation. In: The Proteins (eds. H. Neurath, K. Bailey) Vol. I. B, New York: Academic Press, 1953, pp. 807–892
Putnam, F.W., Erickson, J.O., Volkin, E., Neurath, H.: Native and regenerated bovine albumin. I. Preparation and physicochemical properties. J. Gen. Physiol. **26**, 513–531 (1943)
Quist, R.G., Stokes, J.L.: Comparative effect of temperature on the induced synthesis of hydrogenase and enzymes of the benzoate exidation system in psychrophilic and mesophilic bacteria. Can. J. Microbiol. **18**, 1233–1239 (1972)
Radda, G.K.: Enzyme and membrane conformation in biochemical control. The seventh Colworth medal lecture. Biochem. J. **122**, 385–396 (1971)
Radouco-Thomas, S.: Cellular and molecular aspects of transmitter release: calcium/monoamine dynamics. In: Advances in Cytopharmacology, I Intern. Symp. Cell. Biol. Cytopharmacol. New York: Raven Press 1971, pp. 457–475
Raison, J.K., Lyons, J.M.: Hibernation: alteration of mitochondrial membranes as a requisite for metabolism at low temperature. Proc. Nat. Acad. Sci. USA **68**, 2092–2094 (1971)
Raison, J.K., Lyons, J.M., Mehlhorn, R.J., Keith, A.D.: Temperature-induced phase changes in mitochondrial membranes detected by spin labeling. J. Biol. Chem. **246**, 4036–4040 (1971 a)
Raison, J.K., Lyons, J.M., Thomson, W.W.: The influence of membranes on the temperature-induced changes in the kinetics of some respiratory enzymes of mitochondria. Arch. Biochem. Biophys. **142**, 83–90 (1971 b)
Ram, J.S., Maurer, P.H.: Modified bovine serum albumin. I. Immunochemical and other studies of bovine serum albumin after precipitation with trichloroacetic acid and solution in ethanol. Arch. Biochem. Biophys. **76**, 28–31 (1958)
Rao, P.K., Bullock, T.H.: Q_{10} as a function of size and habitat temperature in poikilotherms. Amer. Naturalist **88**, 33–44 (1954)
Ray, P.H., White, D.S., Brock, T.D.: Effect of temperature on the fatty acid composition of *Thermus aquaticus*. J. Bacteriol. **106**, 25–30 (1971)

Read, K. R. H.: Respiration of the bivalved molluscs *Mytilus edulis* L. and *Brachidontes Demissus plicatulus* Lamarck as a function of size and temperature. Comp. Biochem. Physiol. **7**, 89–101 (1962)

Read, K. R. H.: Thermal inactivation of preparations of aspartic/glutamic transaminase from species of bivalved molluscs from the sublittoral and intertidal zones. Comp. Biochem. Physiol. **9**, 161–180 (1963)

Read, K. R. H.: Comparative biochemistry of adaptations of poikilotherms to the thermal environment. Proc. Symp. Exp. Marine Ecol. Grad. School of Oceanogr. **2**, 39–47 (1964a)

Read, K. R. H.: The temperature-coefficients of ribonucleases from two species of gastropod molluscs from different thermal environments. Biol. Bull. **127**, 489–498 (1964b)

Read, K. R. H.: Thermostability of proteins in poikilotherms. In: Molecular Mechanisms of Temperature Adaptation, Washington: 1967, pp. 93–106

Reeke, G. N., Hartsuck, J. A., Ludwig, M. L., Quiocho, F. A., Steitz, T. A., Lipscomb, W. N.: The structure of carboxypeptidase A. VI. Some results at 2.0 Å-resolution and the complex with glycyltyrosine at 2.8 Å-resolution. Proc. Nat. Acad. Sci. USA **58**, 2220–2226 (1967)

Reistand, R.: On the composition and nature of the bulk protein of extremely halophilic bacteria. Arch. Microbiol. **71**, 353–360 (1970)

Remold-O'Donnel, E., Zillig, W.: Purification and properties of DNA-dependent RNA-polymerase from *Bacillus stearothermophilus*. Eur. J. Biochem. **7**, 318–323 (1969)

Reshöft, K.: Untersuchungen zur zellulären osmotischen und thermischen Resistenz verschiedener Lamellibranchier der deutschen Küstengewässer. Kieler Meeresforschung **17**, 65–84 (1961)

Rice, R. G., Ballou, G. A., Boyer, P. D., Luck, J. M., Lum, F. G.: The papain digestion of native, denatured and "stabilized" human serum albumin. J. Biol. Chem. **158**, 609–617 (1945)

Riegel, J. A.: Energy, Life and Animal Organisation. London: English Univ. Press, 1965

Rigby, B. J.: Thermal transitions in normal and denaturated rat tail tendon, human skin and tunafish skin. Biochim. biophys. Acta **62**, 183–185 (1962)

Rigby, B. J.: Correlation between serine and thermal stability of collagen. Nature **214**, 87–88 (1967a)

Rigby, B. J.: Relation between the shrinkage of native collagen in acid solution and the melting temperature of the tropocollagen molecule. Biochim. Biophys. Acta **133**, 272–277 (1967b)

Rigby, B. J.: Amino acid composition and thermal stability of the skin collagen of the antarctic ice-fish. Nature **219**, 166–167 (1968a)

Rigby, B. J.: Thermal transitions in some invertebrate collagens and their relation to amino acid content and environmental temperature. In: Symposium on Fibrous Proteins (ed. W. G. Grewther) Sydney: Butterworths, 1968b, pp. 217–225

Rigby, B. J.: Temperature relationship of poikilotherms and the melting temperature of molecular collagen. Biol. Bull. **135**, 223–229 (1968c)

Rigby, B. J.: The thermal stability of collagen: its significance in biology and physiology. In: Chemical Dynamic (ed. Hirschfelder) Sydney: John Wiley and Sons 1971, pp. 537–555

Rigby, B. J., Hafey, M.: Thermal properties of the collagen of jellyfish (*Aurelia coerulea*) and their relation to its thermal behaviour. Austral. J. Biol. Sci. **25**, 1361–1363 (1972)

Rigby, B. J., Mason, P.: Thermal transitions in gastropod collagen and their correlation with environmental temperature. Austral. J. Biol. Sci. **20**, 265–271 (1967)

Rigby, B. J., Prosser, C. L.: Thermal transition of collagen from fish recovered from different depths. Comp. Biochem. Biophys. **50B**, 1975

Rigby, B. J., Robinson, M. S.: Thermal transitions in collagen and the preferred temperature range of animals. Nature **253**, 277–279 (1975)

Rigby, B. J., Spikes, J. D.: Hydroxyproline and the shrinkage temperature of collagen. Nature **187**, 150–151 (1960)

Roberts, J. L.: Thermal acclimation of metabolism in the crab *Pachygrapsus crassipes* Randall. II. Mechanisms and the influence of season and latitude. Physiol. Zool. **30**, 242–255 (1957)

Roberts, D. W. A.: Temperature coefficient of invertase from the leaves of cold hardened and cold susceptible wheat plants. Can. J. Bot. **45**, 1347–1357 (1967)

Ron, E. Z., Davis, B. D.: Growth rate of *Escherichia coli* at elevated temperatures: limitation by methionine. J. Bacteriol. **107**, 391–396 (1971 a)

Ron, E. Z., Shani, M.: Growth rate of *Escherichia coli* at elevated temperatures: reversible inhibition of homoserine succinylase. J. Bacteriol. **107**, 397–400 (1971 b)

Roots, B. I.: Phospholipids of goldfish (*Carassius auratus* L.) brain: the influence of environmental temperature. Comp. Biochem. Physiol. **25**, 457–466 (1968)

Roots, B. I., Johnston, P. V.: Plasmalogens of the nervous system and environmental temperature. Comp. Biochem. Physiol. **26**, 553–560 (1968)

Rose, A. H., Evison, L. M.: Studies on the biochemical basis of the minimum temperatures for growth of certain psychrophilic and mesophilic microorganisms. J. Gen. Microbiol. **38**, 131–145 (1965)

Rosenbloom, J., Harsch, M., Jimenez, S.: Hydroxyproline content determines the denaturation temperature of chick tendon collagen. Arch. Biochem. Biophys. **158**, 478–484 (1973)

Rossi, G.: Sulla temperatura e sul tempo di coagulazione delle proteine del siero di sangue in rapporto con la viscosita di questa. Arch. di Fisiol. 2 fasc. **5**, 599–608 (1905)

Rottem, S., Hubbell, W. L., Hayflick, L., McConnel, H. M.: Motion of fatty acid spin labels in the plasma membrane of *Mycoplasma*. Biochim. Biophys. Acta **219**, 104–113 (1970)

Rudenok, A. N., Konev, S. V.: On the phenomenon of cell self-defence against thermal injury. Dokl. Akad. Nauk SSSR **208**, 977–980 (1973)

Rumyantsev, P. P.: Thermostability of the myocardium and its explants. Tsitologia **2**, 547–560 (1960)

Runnström, S.: Über die Thermopathie der Fortpflanzung und Entwicklung mariner Tiere in Beziehung zu ihrer geographischen Verbreitung. Bergens Museums Årbok **2**, 1–67 (1927)

Runnström, S.: Weitere Studien über die Temperaturanpassung der Fortpflanzung und Entwicklung mariner Tiere. Bergens Museums Årbok **10**, 1–46 (1930)

Runnström, S.: Die Anpassung der Fortpflanzung und Entwicklung mariner Tiere an die Temperaturverhältnisse verschiedener Verbreitungsgebiete. Bergens Museums Årbok **3**, 1–36 (1936)

Ryrie, I. J., Jagendorf, A. T.: An energy-linked conformational change in the coupling factor protein in chloroplasts. Studies with hydrogen exchange. J. Biol. Chem. **246**, 3771–3774 (1971)

Ryumin, A. V.: Temperature sensitivity of vertebrates and the biological pathway of the origin of the warm-blooded forms. In: Sbornik stud. nauch. rabot MGU (Collection of Stud. Sci. Works, Moscow State Univ.) Moscow: Moscow State Univ. Press, 1939, pp. 55–85

Saburova, E. A.: Thermal and proteolytic stability of proteins. Author's abstract of dissertation. Pushchino: 1973

Saburova, E. A., Markovich, D. S.: Conformational changes of tropomyosin in the process of tryptic hydrolysis. In: Konformatsionnyie izmeneniya biopolimerov v rastvorakh (Collection of works (Sbornik) Conformational Changes of Biopolymers in Solution) Moscow: Izd. Nauka 1973, pp. 157–161

Sachs, T.: Über die obere Temperaturgrenze der Vegetation. Flora **47**, 5–12; 33–39; 65–75 (1864)

Sakai, A., Yoshida, S.: The role of sugar and related compounds in variations of freezing resistance. Cryobiology **5**, 160–174 (1968)

Sakai, T., Gross, J.: Some properties of the products of reaction of tadpole collagenase with collagen. Biochemistry **6**, 518–528 (1967)

Salawry, O. S., McCormick, T.: The influence of elevated temperature on the growth and morphology of HeLa cultures. Anat. Rec. **124**, 498 (1956)

Salcheva, G., Samygin, G.: Microscopical observations of freezing in winter wheat tissues. Fiziol. rast. **10**, 65–72 (1963)

Samejima, T., Takamiya, A.: Comparative studies on heat stability of proteins of thermophilic and mesophilic bacteria. Cytologia **23**, 509–519 (1958)

Sando, G. N., Hogenkamp, P. C.: Ribonucleotide reductase from *Thermus X*—1, a thermophilic organism. Biochemistry **12**, 3316–3322 (1973)

Santarius, K. A.: Ursachen der Frostschäden und Frostadaptation bei Pflanzen. Ber. Deut. Bot. Ges. **84**, 425–436 (1971)

Santarius, K.A.: The protective effect of sugars on chloroplast membranes during temperature and water stress and its relationship to frost, desiccation and heat resistance. Planta **113**, 105–114 (1973)
Sapper, I.: Versuche zur Hitzeresistenz der Pflanzen. Planta **23**, 518–556 (1935)
Sauer, H.W., Babcock, K.L., Rusch, H.P.: Changes in nucleic acid and protein synthesis during starvation and spherule formation in *Physarum polycephalum*. W.Roux' Archiv **165**, 110–123 (1970)
Saunders, G.F., Campbell, L.L.: Ribonucleic acid and ribosomes of *Bacillus stearothermophilus*. J. Bacteriol. **91**, 332–339 (1966)
Saunders, P.P., Wilson, B.A., Saunders, G.F.: Purification and comparative properties of a pyrimidine nucleoside phosphorylase from *Bacillus stearothermophilus*. J. Biol. Chem. **244**, 3691–3697 (1969)
Sauvan, R.L., Mira, O.J., Amelunxen, R.E.: Thermostable glyceraldehyde-3-phosphate dehydrogenase from *Bacillus stearothermophilus*. I. Immunochemical studies. Biochim. biophys. Acta **263**, 794–804 (1972)
Schechter, A., Moravek, L., Anfinsen, C.B.: Suppression of hydrogen exchange in staphylococcal nuclease by ligands. Proc. Nat. Acad. Sci. USA **61**, 1478–1485 (1968)
Scheibmair, G.: Hitzeresistenz-Studien an Mooszellen. Protoplasma **29**, 394–424 (1938)
Scheraga, H.A.: Protein structure. New York, London: Academic Press, 1961
Scheraga, H.A., Nemethy, G., Steinberg, J.Z.: The contribution of hydrophobic bonds to the thermal stability of protein conformations. J. Biol. Chem. **237**, 2506–2508 (1962)
Schimke, R.T.: The importance of both synthesis and degradation in the control of arginase levels in rat liver. J. Biol. Chem. **239**, 3808–3817 (1964)
Schimke, R.T.: Studies on the roles of synthesis and degradation in the control of enzyme levels in animal tissues. Bull. Soc. Chem. Biol. **48**, 1009–1030 (1966a)
Schimke, R.T.: Protein turnover and the regulation of enzyme levels in rat liver. In: Intern. Symp. Enzymatic Aspects of Metabolic Regulation, Mexico City: 1966,b, pp. 301–314
Schimke, R.T.: The role of protein turnover in the regulation of enzyme levels in animal tissues. In: Abstract 7th Intern. Congr. Biochem. Tokyo: 1967, pp. 263–264
Schimke, R.T.: On the roles of synthesis and degradation in regulation of enzyme levels in mammalian tissues. In: Current Topics in Cellular Regulation (eds. B.L. Horecker, E.R.-Stadtman) **1**, New York, London: Academic Press 1969, pp. 77–124
Schimke, R.T.: Control of enzyme levels in mammalian tissues. Adv. Enzymol. **37**, 135–187 (1973)
Schimke, R.T., Doyle, D.: Control of enzyme levels in animal tissues. Ann. Rev. Biochem. **39**, 929–976 (1970)
Schimke, R.T., Ganschow, R., Doyle, D., Arias, T.M.: Regulation of protein turnover in mammalian tissues. Fed. Proc. **27**, 1223–1230 (1968)
Schimke, R.T., Sweeney, E.W., Berlin, C.M.: The roles of synthesis and degradation in the control of rat liver tryptophan pyrrolase. J. Biol. Chem. **240**, 322–331 (1965a)
Schimke, R.T., Sweeney, E.W., Berlin, C.M.: Studies of the stability *in vivo* and *in vitro* of rat liver tryptophan pyrrolase. J. Biol. Chem. **240**, 4609–4620 (1965b)
Schlieper, C.: Genotypische und phaenotypische Temperatur- und Salzgehalts-Adaptationen bei marinen Bodenvertebraten der Nord- und Ostsee. Kieler Meeresforschung **16**, 180–185 (1960)
Schlieper, C.: Cellular ecological adaptations and reactions demonstrated in surviving isolated gill tissues of *Bivalves*. In: The Cell and Environmental Temperature, Moscow, Leningrad: Izd. Nauka 1964, pp. 129–135
Schlieper, C., Flügel, H., Rudolf, J.: Temperature and salinity relationships in marine bottom invertebrates. Experientia **16**, 1–8 (1960)
Schlieper, C., Kowalski, R.: Über den Einfluß des Mediums auf die thermische und osmotische Resistenz des Kiemengewebes der Miesmuschel *Mytitus edulis* L. Kieler Meeresforschung **12**, 37–45 (1956a)
Schlieper, C., Kowalski, R.: Quantitative Beobachtungen über physiologische Ionenwirkungen im Brackwasser. Kieler Meeresforschung **12**, 154–165 (1956b)
Schmalhausen, I.I.: Pathways and Regularities of the Evolutionary Process. Moscow, Leningrad: Izd. Akad. Nauk SSSR 1940

Schmalhausen, I. I.: Factors of the Evolution. Moscow, Leningrad: Izd. Akad. Nauk SSSR 1946

Scholander, P. F., Flagg, W., Walters, V., Irving, L.: Respiration in some arctic and tropical lichens in relation to temperature. Amer. J. Bot. **39**, 707–713 (1952)

Scholander, P. F., Flagg, W., Walters, W., Irving, L.: Climatic adaptation in arctic and tropical poikilotherms. Physiol. Zool. **26**, 67–92 (1953)

Schölm, H. E.: Untersuchungen zur Hitze- und Frostresistenz einheimischer Süßwasseralgen. Protoplasma **65**, 97–118 (1968)

Schroeder, C. A.: Induced temperature tolerance of plant tissue *in vitro*. Nature **200**, 1301–1302 (1963)

Schroeder, C. A.: Induction of temperature tolerance in excised plant tissue. In: Molecular Mechanism of Temperature Adaptations, (ed. C. L. Prosser) Washington: Am. Ass. Advan. Sci. 1967, pp. 61–72

Schuckmann, W., Piekarski, G.: Beiträge zum Problem der Dauermodifikation bei Protozoen. Arch. Protistenk. **93**, 355–416 (1940)

Schuster, L., Jick, H.: The turnover of microsomal protein in the livers of phenobarbital-treated mice. J. Biol. Chem. **241**, 5361–5365 (1966)

Schwarz, W.: Der Einfluß der Tageslänge auf die Frosthärte, die Hitzeresistenz und das Photosynthesevermögen von Zirben und Alpenrosen. Diss. Innsbruck: 1968

Schwarz, W.: Der Einfluß der Photoperiode auf das Austreiben, die Frosthärte und die Hitzeresistenz von Zirben und Alpenrosen. Flora **159**, 258–285 (1970)

Schwarz, W. R.: Frost hardiness, resistance to heat, and CO_2 uptake in seedlings and mature stages of *Pinus cembra* L. In: Proc. 3d Forest Microclimate Symp., Kananaskis For. Exp. Stat., Alberta: 1969, pp. 137–145

Schwemmle, B., Lange, O. L.: Endogen-tagesperiodische Schwankungen der Hitzeresistenz bei *Kalanchoë blossfeldiana*. Planta **53**, 134–144 (1959)

Schwenke, H.: Untersuchungen zur Temperaturresistenz mariner Algen der westlichen Ostsee. I. Das Resistenzverhalten von Tiefenrotalgen bei ökologischen und nichtökologischen Temperaturen. Kieler Meeresforschung **15**, 34–50 (1959)

Scrutton, M. C.; Utter, M. F.: Pyruvate carboxylase. III. Some physical and chemical properties of the highly purified enzyme. J. Biol. Chem. **240**, 1–9 (1965)

Segal, E., Dehnel, P. A.: Acclimation of oxygen consumption to temperature in the American cockroach *(Periplaneta americana)*. Biol. Bull. **111**, 53–61 (1956)

Segal, H. L., Matsuzawa, T., Haider, M., Abraham, G. Y.: What determine the half-life of proteins *in vivo*? Some experiences with alanine aminotransferase of rat tissues. Biochem. Biophys. Res. Commun. **36**, 764–770 (1969)

Segal, H. L., Winkler, J. R., Miyagi, M. P.: Relationship between degradation rates of proteins *in vivo* and their susceptibility to lysosomal proteases. J. Biol. Chem. **249**, 6364–6365 (1974)

Semikhatova, O. A.: On the temperature dependence of respiration of high-mountain plants of the Eastern Pamirs. Tr. Bot. Inst. im. V. L. Komarova Akad. Nauk SSSR, ser. **4**, 91–112 (1959)

Seravin, L. N., Skoblo, I. I., Ossipoff, D. W.: The influence of thermal adaptation on the heat stability of enzymes of *Paramecium caudatum*. In: Teploustoichivost kletok zhivotnykh (Collection of works (Sbornik) Heat Resistance of Cells of Animals) Moscow, Leningrad: Izd. Nauka 1965, pp. 161–170

Serebrovsky, A.: Effect of temperature on *Paramecium caudatum*. On the problem of thermal method of analysis of biological phenomena. Uch. zap. Mosc. Univer. Tr. biol. lab. **1**, 345–445 (1916)

Sergeeva, E. P.: Changes in thermostability of isopods as a function of the environmental temperature. Ref. nauchn. rabot Inst. biol morya **1**, 148–150 (1969)

Setlow, R. B., Setlow, J. K.: Effects of radiation on polynucleotides. Ann. Rev. Biophys. Bioengin. **1**, 293–346 (1972)

Sevrova, O. K.: On the formation of an induced heat stability of plants. Tr. Tsentr. Sibirskogo Bot. sada **7**, 127–135 (1964)

Shapiro, B. M., Ginsburg, A.: Effects of specific divalent cations on some physical and chemical properties of glutamine synthetase from *Escherichia coli*. Taut and relaxed enzyme forms. Biochem. **7**, 2153–2167 (1968)

Shaw, M. K., Ingraham, J. L.: Fatty acid composition of *Escherichia coli* as a possible controlling factor of the minimal growth temperature. J. Bacteriol. **90**, 141–146 (1965)
Shcherbakova, A. M.: Registration of the experimental increase in plant cell thermostability by a tetrazolium test. Tsitologia **11**, 1467–1470 (1969)
Shcherbakova, A. M.: The influence of thermal hardening on extractability and heat resistance of watersoluble proteins of wheat leaves. Dokl. Akad. Nauk SSSR **199**, 727–729 (1971 a)
Shcherbakova, A. M.: The heat resistance of acid phosphatase from cold-hardened winter wheat leaves. Tsitologia **13**, 484–490 (1971 b)
Shcherbakova, A. M.: Heat resistance of glucose-6-phosphate dehydrogenase from leaves of heat and cold-hardened winter wheat. Dokl. Akad. Nauk SSSR **205**, 993–996 (1972)
Shcherbakova, A. M., Feldman, N. L., Shukhtina, H. G.: The changes in the thermostability of some proteins after the heat hardening of wheat leaves of different age. Tsitologia **15**, 391–398 (1973)
Shen, P. Y., Coles, J. L., Foote, J. L., Stenesh, J.: Fatty acid distribution in mesophilic and thermophilic strains of the genus *Bacillus*. J. Bacteriol. **103**, 479–481 (1970)
Shirley, H. L.: Lethal high temperatures for conifers and the cooling effect of transpiration. J. Agr. Res. **53**, 239–258 (1936)
Shkolnikova, M. D., Shterman, L. Ya.: Influence of heating on the rate of protoplasmic streaming and cellular resistance of plants. In: The Cell and Environmental Temperature, Moscow, Leningrad: Izd. Nauka 1964, pp. 200–204
Shkorbatov, G. L., Kudryavtseva, G. S.: Variation of tissue heat and cold-resistance in fishes, as dependent on the temperature of environment. Dokl. Akad. Nauk SSSR **156**, 452–454 (1964)
Shkorbatov, G. L., Salo, Z. T.: Physiological variation of fish populations as related to the temperature of their habitat. Dokl. Akad. Nauk SSSR **159**, 678–680 (1959)
Shlyakhter, N. A.: The influence of preliminary heating of the frog muscle on its resistance of the injurious effects of high temperature and various chemical agents. Tsitologia **1**, 692–698 (1959)
Shlyakhter, N. A.: The thermostability of frog muscles in different seasons. Tsitologia **3**, 95–100 (1961)
Shlyakhter, T. A.: A change in the staining of spinal ganglia of grass and lake frogs at the action of high temperature. In: Voprosy tsitologii i protistologii (Collection of works (Sbornik) Problems of Cytology and Protistology) Moscow, Leningrad: Izd. Akad. Nauk SSSR 1960, pp. 117–120
Shukhtina, H. G.: Seasonal variations of the heat stability of cells in some plants of the Khibiny. Bot. Zh. **47**, 100–105 (1962)
Shukhtina, H. G.: The influence of repeated heat-hardening on thermostability of plant cells. In: Tsitologicheskiye osnovy prisposobleniya rastenii k faktoram sredy (Collection of works (Sbornik) Cytological Aspects of Adaptation of Plants to the Environmental Factors) Moscow, Leningrad: Izd. Nauka 1964, pp. 26–30
Shukhtina, H. G.: The effect of ambient temperature on the heat stability of leaf cells of *Catalpa speciosa* Worder and of some other plants. Bot. Zh. **50**, 1310–1317 (1965)
Shukhtina, H. G., Yazkulyev, A.: Artificial heat hardening of plant cells with different initial levels of their heat resistance. Tsitologia **10**, 88–94 (1968)
Shukuya, R., Schwert, G. W.: Glutamic acid decarboxylase. III. The inactivation of the enzyme at low temperatures. J. Biol. Chem. **235**, 1658–1661 (1960)
Shuster, C. W., Doudoroff, M.: A cold-sensitive $\alpha(-)\beta$-hydroxybutyric acid dehydrogenase from *Rhodospirillum rubrum*. J. Biol. Chem. **237**, 603–607 (1962)
Sidler, W., Zuber, H.: Neutral protease with different thermostability from a facultative strain of *B. stearothermophilus* grown at 40° and 50°. FEBS Letters **25**, 292–294 (1972)
Sie, E. H., Sobotka, H., Baker, H.: Factors inducing mesophilic bacteria to grow at 55° C. Biochem. Biophys. Res. Commun. **3**, 205–209 (1961 a)
Sie, E. H., Sobotka, H., Baker, H.: Factor converting mesophilic into thermophilic microorganisms. Nature **192**, 86–87 (1961 b)
Simons, E. R., Schneider, E. G., Blout, E. R.: Thermal effects on the circular dichroism spectra of ribonuclease A and of ribonuclease S-protein, J. Biol. Chem. **244**, 4023–4026 (1969)

Sinensky, M.: Temperature control of phospholipid biosynthesis in *Escherichia coli*. J. Bacteriol. **106**, 449–455 (1971)
Sinensky, M.: Homeoviscous adaptation—a homeostatic process that regulates the viscosity of membrane lipids in *Escherichia coli*. Proc. Nat. Acad. Sci. USA **71**, 522–525 (1974)
Singleton, R. J., Kimmel, J. R., Amelunxen, R. E.: The amino acid composition and other properties of thermostable glyceraldehyde-3-phosphate dehydrogenase from *Bacillus stearothermophilus*. J. Biol. Chem. **244**, 1623–1630 (1969)
Sipos, T., Merkel, J. R.: Temperature-dependent activation of trypsin by calcium. Biochem. Biophys. Res. Commun. **31**, 522–527 (1968)
Skholl, E. D.: On one of the mechanisms of temperature adaptation of isolated mammalian cells. Tsitologia **4**, 562–565 (1962)
Skholl, E. D.: Increase in the resistance of isolated mammalian tissues to temperature. In: Problemy tsitoekologii zhivotnykh (Collection of works (Sbornik) Problems of Cytoecology of Animals) Moscow, Leningrad: Izd. Akad. Nauk SSR 1963a, pp. 220–229
Skholl, E. D.: Heat resistance of muscles in connection with the temperature gradient of the body in mammals. In: Problemy tsitoekologii zhivotnykh (Collection of works (Sbornik) Problems of Cytoecology of Animals) Moscow, Leningrad: Izd. Akad. Nauk SSSR 1963b, pp. 163–168
Skholl, E. D.: Seasonal changes in thermostability of muscles and actomyosin of *Citellus pygmalus* Pall. In: Teploustoichivost kletok zhivotnykh (Collection of works (Sbornik) Heat Resistance of Cells of Animals) Moscow, Leningrad: Izd. Nauka 1965, pp. 106–114
Skholl, E. D.: A comparative study of muscle thermostability of rodents. In: Izmenchivost teploustoichivosti kletok zhivotnykh v onto- i filogeneze (Collection of works (Sbornik) Variability in Cellular Heat Resistance of Animals in Ontogenesis and Phylogenesis) Leningrad: Izd. Nauka 1967, pp. 158–169
Skholl, E. D.: Seasonal changes in the heat stability of muscles and of muscle models in *Microtus* sp. Tsitologia **12**, 1020–1027 (1970)
Skholl, E. D.: A correlation between an initial level of thermostability of isolated ciliary epithelium of the mussel and changes of this indicator under the influence of a thermal treatment. Ekologia **6**, 69–73 (1971)
Skinner, F. A.: The limits of microbial existence. Proc. Roy. Soc. Ser. **B, 171**, 77–89 (1968)
Skulachev, V. P.: Accumulation of Energy in the Cell. Moscow: Izd. Nauka 1969
Skulachev, V. P.: Transformation of Energy in Biomembranes. Moscow: Izd. Nauka 1972
Sleptsova, L. A.: Adenosine triphosphatase activity and heat resistance of the rat actomyosin extracted from muscles with different iodine number of lipids. In: Problemy tsitoekologii zhivotnykh (Collection of works (Sbornik) Problems of Cytoecology of Animals) Moscow, Leningrad: Izd. Akad. Nauk SSSR 1963, pp. 217–219
Sluyterman, L. A., de Graaf, J. M.: The activity of papain in the crystalline state. Biochim. Biophys. Acta **171**, 277–287 (1969)
Smith, C. L.: Thermostability of some mitochondrial enzymes of lower vertebrates. I. General survey. Comp. Biochem. Physiol. **44**, 779–788 (1973a)
Smith, C. L.: Thermostability of some mitochondrial enzymes of lower vertebrates. II. Freshwater teleosts. Comp. Biochem. Physiol. **44**, 789–801 (1973b)
Somero, G. N.: Pyruvate-kinase variants of the Alaskan kingcrab. Evidence for a temperature-dependent interconversion between two forms having distinct and adaptive kinetic properties. Biochem. J. **114**, 237–241 (1969a)
Somero, G. N.: Enzyme mechanisms of temperature compensation: immediate and evolutionary effects of temperature on enzymes of aquatic poikilotherms. Amer. Naturalist. **103**, 517–530 (1969b)
Somero, G. N.: Temperature as a selective factor in protein evolution: the adaptational strategy of "compromise". J. Exp. Zool. **194**, 175–188 (1975a)
Somero, G. N.: The roles of isozymes in adaptation to varying temperatures. In: Isozymes II. Physiological function. (ed. P. D. Bayer) New York, San Francisco: 221–234 (1975b)
Somero, G. N., Doyle, D.: Temperature and rates of protein degradation in the fish *Gillichthys mirabilis*. Comp. Biochem. Physiol. **46**, 463–474 (1973)

Somero, G. N., Giese, A. C., Wohlschlag, D. E.: Cold adaptation of the antarctic fish *Trematomus bernacchii*. Comp. Biochem. Physiol. **26**, 229–233 (1968)

Somero, G. N., Hochachka, P. W.: The effect of temperature on catalytic and regulatory functions of pyruvate kinases of rainbow trout and the Antarctic fish *Trematomus bernacchii*. Biochem. J. **110**, 395–400 (1968)

Somero, G. N., Hochachka, P. W.: Isoenzymes and short-term temperature compensation in poikilotherms: activation of lactate dehydrogenase isoenzymes by temperature decreases. Nature **223**, 194–195 (1969)

Somero, G. N., Hochachka, P. W.: Biochemical adaptation to the environment. Amer. Zool. **11**, 159–167 (1971)

Somero, G. N., Johansen, K.: Temperature effects on enzymes from homeothermic and heterothermic tissues of the harbour seal. Com. Biochem. Biophys. **34**, 131–163 (1970)

Somero, G. N., de Vries, A. L.: Temperature tolerance of some Antarctic fishes. Science **157**, 257–258 (1967)

Sopina, V. A.: The role of the nucleus and cytoplasm in the inheritance of the resistance to the injuring action of ethyl alcohol and elevated temperature in *Amoeba*. Author's abstract of dissertation. Leningrad: 1968a

Sopina, V. A.: Interclonal differences in thermostability observed in *Amoeba* proteins. Tsitologia **10**, 207–217 (1968b)

Sorokin, C.: Kinetic studies of temperature effects on the cellular level. Biochim. biophys. Acta **38**, 197–204 (1960)

Sosfenov, N. I.: Development of a method of an X-ray small-angle scattering in the absolute scale of intensities and its application to investigation of proteins. Author's abstract of dissertation. Moscow: 1972

Spichtin, H., Verzör, F.: Calcium as stabilizing factor of the collagen macromolecule. Experientia **25**, 9–11 (1969)

Spilburg, C. A., Bethune, J. L., Vallee, B. L.: The physical state dependence of carboxypeptidase A_x and A_y kinetics. Proc. Nat. Acad. Sci. USA **71**, 3922–3926 (1974)

Spirin, A. S.: Ribonucleic Acids. Composition, Structure and Biological Role. Bakhovskiye chteniya XIX. Moscow: Izd. Nauka 1964

Spirin, A. S.: On the "masked" form of the messenger ribonucleic acid. Zh. evol. biokh. fiziol. **2**, 285–292 (1966)

Stabrovskaya, V. I., Braun, A. D.: Changes in ATPase activity and tinctorial properties of actomyosin at the action of urea on isolated protein and on the muscle. Tsitologia **11**, 201–209 (1969)

Stark, E., Tetrault, P. A.: Isolation of bacterial cell-free, starch saccharifying enzymes from the medium at 70° C. J. Bacteriol. **62**, 247–249 (1951)

Steinberg, D., Vaughan, M.: Observation on intracellular protein catabolism studies *in vitro*. Arch. Biochem. Biophys. **65**, 93–105 (1956)

Stellwagen, E., Cronlund, M. M., Barnes, L. D.: A thermostable enolase from the extreme thermophile *Thermus aquaticus* YT-1. Biochem. **12**, 1552–1559 (1973)

Stenesh, J., Holazo, A. A.: Studies of the ribosomal ribonucleic acid from mesophilic and thermophilic bacteria. Biochim. biophys. Acta **138**, 286–295 (1967)

Stenesh, J., Roe, B. A., Snyder, T. L.: Studies of the deoxyribonucleic acid from mesophilic and thermophilic bacteria. Biochim. biophys. Acta **161**, 442–454 (1968)

Stenesh, J., Yang, C.: Characterization and stability of ribosomes from mesophilic and thermophilic bacteria. J. Bacteriol. **73**, 930–936 (1967)

Stepanov, A. S., Voronina, A. S.: Formation of stabilized informosome-like particles at physiological temperatures. Dokl. Akad. Nauk SSSR **203**, 1418–1421 (1972)

Stewart, B. T., Halvorson, H. O.: Studies on the spores of aerobic bacteria. II. The properties of an extracted heat-stable enzyme. Arch. Biochem. Biophys. **49**, 168–178 (1954)

Stier, T. J. B., Taylor, H. E.: Seasonal variation in behavior of the intact frog heart at high temperatures. J. Cell. Comp. Physiol. **14**, 309–312 (1939)

Stokes, J. L.: Heat-sensitive enzymes and enzyme synthesis in psychrophilic microorganisms. In: Molecular Mechanisms of Temperature Adaptation (ed. C. L. Prosser) Washington: Am. Ass. Advan. Sci. 1967, pp. 311–323

Strachitsky, K. I., Kologrivova, L. Yu.: Changes in the optical activity and enzymatic hydrolysis of proteins in the course of their denaturation. Biokhimiya **11**, 384–389 (1946)

Straub, F. B.: Formation of the secondary and tertiary structure of enzymes. Adv. Enzymol. **26**, 89–114 (1964)

Sugimoto, S., Nosoh, Y.: Thermal properties of fructose-1,6-diphosphate aldolase from thermophilic bacteria. Biochim. biophys. Acta **235**, 210–221 (1971)

Sugiura, M., Takanami, M.: Analysis of the 5′-terminal nucleotide sequences of ribonucleic acids. II. Comparison of the 5′-terminal nucleotide sequences of ribosomal RNA's from different organisms. Proc. Nat. Acad. Sci. USA **58**, 1595–1602 (1967)

Sukhanova, K. M.: Temperature adaptation in parasitic *Protozoa* of *Amphibians* (On the dependence of the thermostability of *Opalinida* and *Intestinal infusoria* on various species of hosts). Tsitologia **1**, 587–600 (1959)

Sukhanova, K. M.: Temperature adaptations in parasitic *Protozoa* of *Rana ridibunda* and *Rana temporaria*. Dokl. Akad. Nauk SSSR **139**, 252–255 (1961)

Sukhanova, K. M.: Temperature adaptations of endoparasitic Protozoa from some species of poikilothermal animals. Zool. Zh. **16**, 1306–1316 (1962a)

Sukhanova, K. M.: Temperature adaptations in *Opalina ranarum* Ehrenberg. (Opalinidae) during its life cycle. Tsitologia **4**, 644–651 (1962b)

Sukhanova, K. M.: Studies on temperature adaptations and morphological peculiarities in some species of the infusoria *Astomata* from oligochaetes of the eastern murman. In: Morfologiya i fiziologiya prosteishikh (Collection of works (Sbornik) Morphology and Physiology of Protozoa) Moscow, Leningrad: Izd. Nauka 1963, pp. 75–91

Sukhanova, K. M.: Temperature adaptations in *Protozoa*. Leningrad: Izd. Nauka 1968

Sukhanova, K. M., Posnanskaya, T. M.: A study of the effect of sub-zero temperatures on some endoparasitic *Protozoa* of *Amphibians*. In: Reaktsiya kletok i ikh belkovykh komponentov yf ekstremalniye vozdeistviya (Collection of works (Sbornik) Reactions of Cells and Their Protein Components on the Action of Extreme Factors) Moscow, Leningrad: Izd. Nauka 1966, pp. 101–108

Sulkowski, E., Laskowski, M.: Protection of micrococcal nuclease against thermal inactivation. J. Biol. Chem. **243**, 651–655 (1968)

Sullivan, C. Y., Kinbacher, E. J.: Thermal stability of fraction 1 protein from heat-hardened *Phaseolus acutifolius* Gray, 'Tepary Buff'. Crop Sci. **7**, 241–244 (1967)

Sultzer, B. M.: Oxidative activity of psychrophilic and mesophilic bacteria on saturated fatty acids. J. Bacteriol. **82**, 492–497 (1961)

Sumner, J. E., Morgan, E. D., Evans, H. C.: The effect of growth temperature on the fatty acid composition of fungi in the order *Mucorales*. Can. J. Microbiol. **15**, 515–520 (1969)

Sussman, A. S.: A comparison of the properties of two forms of tyrosinase from *Neurospora crassa*. Arch. Biochem. Biophys. **95**, 407–415 (1961)

Suzdalskaya, I. P., Kiro, M. B.: The influence of temperature changes on excitability of retractors of *Emys orbicularis* L. maintained under different thermal conditions. In: Problemy tsitoekologii zhivotnykh (Collection of works (Sbornik) Problems of Cytoecology of Animals) Moscow, Leningrad: Izd. Akad. Nauk SSSR 1963, pp. 93–101

Suzuki, D. T., Piternick, L. K., Hayashi, S., Tarasoff, M., Baillic, D., Erasmus, U.: Temperature-sensitive mutations in *Drosophila melanogaster*. I. Relative frequencies among X-ray and chemically induced sex linked recessive lethals and semilethals. Proc. Nat. Acad. Sci. USA **57**, 907–912 (1967)

Svinkin, V. B.: Thermostability of spermatozoa in *Rana temporaria* L. and *Rana ridibunda* Pall. Tsitologia **1**, 580–586 (1959)

Svinkin, V. B.: Heat resistance of spermatozoa of certain *Unio*. Dokl. Akad. Nauk SSSR **139**, 1227–1230 (1961)

Svinkin, V. B.: Adaptation of *Rana ridibunda* Pall. to its environment in hot springs. Zh. obshch. biol. **23**, 155–157 (1962a)

Svinkin, V. B.: Thermoresistance of egg-cells of *Rana temporaria* L. at early stages of cleavage. Dokl. Akad. Nauk SSSR **145**, 913–916 (1962b)

Swick, R. W.: Measurement of protein turnover in rat liver. J. Biol. Chem. **231**, 751–764 (1958)

Szabolcsi, G., Biszku, E., Szörenyi, E.: Comparative studies on d-glyceraldehyde-3-phosphate dehydrogenase. VII. Studies on the digestibility of the enzyme isolated from various mammals. Biochim. biophys. Acta **35**, 237–241 (1959)

Szabuniewicz, B.: Über die Anpassung des *Paramecium caudatum* an höhere Temperaturen und über die Vererbung dieser Anpassung. Z. Induct. Abstamm. und Vererbungslehre **52**, 414–432 (1929)

Szer, W.: Cell-free protein synthesis at 0°. An activating factor from ribosomes of a psychrophilic microorganism. Biochim. biophys. Acta **213**, 159–170 (1970)

Tabidze, D. D.: Thermostability of the musculocutaneous sack in two Eisenia species. Tsitologia **14**, 261–262 (1972)

Tai, P. C., Jackson, H.: Apparent transformation of growth-temperature character in *Micrococcus cryophilus*. Can. J. Microbiol. **15**, 1119–1120 (1969)

Takahashi, T., Tanaka, T.: Physico-chemical studies on the skin and leather of marine animals. VIII. On the heat stability of fish skin and leather. Bull. Jap. Soc. Sci. Fish. **19**, 603–610 (1953)

Takano, T., Swanson, R., Kallai, O. B., Dickerson, R. E.: Conformational changes upon reduction of cytochrome C. Cold. Spr. Harb. Symp. Quant. Biol. **36**, 397–404 (1972)

Takemori, S., Furuya, E., Suzuki, H., Kotagiri, M.: Stabilization of enzyme activity by an organic solvent. Nature **215**, 417–419 (1967)

Taliev, D. N.: *Cottoidei* from the Baikal Lake. (ed. D. V. Nalivkin) Moscow, Leningrad: Izd. Akad. Nauk SSSR, 1955

Tanaka, M., Haniu, M., Yasunobu, K. T., Himes, R. H., Akagi, J. M.: The primary structure of the *Clostridium thermosaccharolyticum* ferredoxin, a heat-stable ferredoxin. J. Biol. Chem. **218**, 5215–5217 (1973)

Tanford, C.: Protein denaturation. Adv. Protein Chem. **23**, 121–282 (1968)

Taniuchi, H., Anfinsen, C. B., Sodja, A.: Nuclease-I: an active derivative of stphylococcal nuclease composed of two noncovalently bonded peptide fragments. Proc. Nat. Acad. Sci. USA **58**, 1235–1242 (1967)

Taniuchi, H., Moravek, L., Anfinsen, C. B.: Ligand-induced resistance of staphylococcal nuclease and α-chymotrypsin, and thermolysin. J. Biol. Chem. **244**, 4600–4606 (1969)

Tartakovsky, A. D., Pashkova, L, V., Pinaev, G. P., Zhirmunsky, A. V., Vorob'ev, V. I.: Thermostability and amino acid composition of myosins from *Rana ridibunda* and *R. temporaria*. In: Ref. nauchn. rabot Inst. biol. morya (Collection of Works of the Inst. Marine Biol.) Vladivostok: 1969, pp. 156–158

Tartakovsky, A. D., Pashkova, L. V., Pinaev, G. P., Zhirmunsky, A. V., Vorob'ev, V. I.: Thermostability and amino acid composition of the DEAE—Sephadex A-50 pyrified myosins from the lake and grass frogs. Tsitologia **15**, 855–861 (1973)

Tasaki, I. L., Carnay, R., Sandlin, R., Watanabe, A.: Fluorescence changes during conduction in nerves stained with acridine orange. Science **163**, 683–685 (1969)

Tasaki, I., Watanabe, A., Sandlin, R., Carnay, L.: Changes in fluorescence, turbidity and birefringence associated with nerve excitation. Proc. Nat. Acad. Sci. USA **61**, 883–888 (1968)

Tashian, R. E., Osborne, W. R. A.: Some genetic and molecular aspects of enzyme adaptations. In: Physiological adaptation to the environment. (ed. F. J. Vernberg) New York: Intext Educational Publishers 1975, 37–49

Terentyev, P. V.: The Frog. Moscow: Izd. Sovetskaya Nauka, 1950

Terroine, E. F., Hatterer, C., Rochrig, P.: Les acides gras des phosphatides chez les poikilothermes, les vegetaux supérieurs et les microorganismes. Bull. Soc. Chem. Biol. **12**, 682–702 (1930)

Theorell, H., Chance, B., Yonetani, T.: Effect of crystallization upon the reactivity of horse liver alcohol dehydrogenase. J. Mol. Biol. **17**, 513–524 (1966)

Thompson, P. J., Thompson, T. L.: Some characteristics of a purified heat stable aldolase. J. Bacteriol. **84**, 694–700 (1962)

Thörner, W.: Untersuchungen über Wärmeerregung und Wärmelähmung und den Erscheinungskomplex der „Gewöhnung" bei der letzteren. Z. allgem. Physiol. **18**, 226–276 (1919)

Thörner, W.: Leitungsverlangsamung und Verringerung des Stoffumsatzes als Grundlage der scheinbaren „Gewöhnung" der wärmegelähmt gewesenen Nerven. Pflüg. Arch. ges. Physiol. **195**, 602–616 (1922)

Tieszen, L. L., Helgager, J. A.: Genetic and physiological adaptation in the Hill reaction of *Deschampsia caespitosa*. Nature **219**, 1066–1067 (1968)

Tischler, G.: On some problems of cytotaxonomy and cytoecology. J. Indian Bot. Soc. **16**, 165–169 (1937)

Tonomura, Y., Kanasava, T., Sekiya, K.: Phosphorylation and conformational rearrangement of myosin in its interaction with adenosine triphosphate acid. In: Molekulyarnaya biologiya. Problemy i perspectivy (Collection of works (Sbornik) Molecular Biology. Problems and Perspectives) Moscow: Izd. Nauka 1964, pp. 213–226

Towers, N.R., Raison, J.K., Kellerman, G.M., Linnane, A.W.: Effects of temperature—induced phase changes in membranes on protein synthesis by bound ribosomes. Biochim. biophys. Acta **287**, 301–311 (1972)

Tranquillini, W.: The physiology of plants at high altitudes. Ann. Rev. Plant Physiol. **15**, 345–362 (1964)

Treharne, K.J., Cooper, J.P.: Effect of temperature on the activity of carboxylases in tropical and temperate *Gramineae*. J. Exp. Bot. **20**, 170–175 (1969)

Treharne, K.J., Eaglis, C.F.: Effect of temperature on photosynthetic activity of climatic races of *Dactylis glomerata* L. Photosynthetica **4**, 107–117 (1970)

Trewavas, A.: Determination of the rates of protein synthesis and degradation in *Lemna minor*. Plant Physiol. **49**, 40–46 (1972a)

Trewavas, A.: Control of the protein turnover rates in *Lemna minor*. Plant Physiol. **49**, 47–51 (1972b)

Troitsky, G.V., Zavyalov, V.P., Kiryukhin, I.F.: Reversible conformational transitions of proteins in the region of physiological temperatures. Biokhimiya **36**, 1107–1114 (1971)

Tschudy, D.P., Marver, H.S., Collins, A.: A model for calculating messenger RNA half-life: short lived messenger RNA in the induction of mammalian δ-amino-levulinic acid synthetase. Biochem. Biophys. Res. Commun. **21**, 480–487 (1965)

Tsukuda, H., Ohsawa, W.: Effects of acclimation temperature on the composition and thermostability of tissue proteins in the goldfish (*Carassius auratus* L.) Annotat. Zool. Jap. **44**, 90–98 (1971)

Tsvelyov, N.N.: Some problems of the evolution of *Poaceae*. Bot. Zh. **54**, 361–373 (1969)

Tsyperovich, A.S.: Studies on denaturation and stabilization of globular proteins. Author's abstract of dissertation. Kiev: 1954

Tsyperovich, A.S.: On the mechanism of protein denaturation. Denaturative stabilization of globular proteins. Biokhimiya **21**, 203–209 (1956)

Tsyperovich, A.S., Loseva, A.L.: On the mechanism of protein denaturation. Properties of a globular protein stabilized by an action of a denaturing factor. Biokhimiya **21**, 546–556 (1956)

Tsyperovich, A.S., Loseva, A.L.: Stabilization of pepsin, trypsin and chymotrypsin by amino acids. Ukr. biokhim Zh. **32**, 25–43 (1960)

Tumanov, I.I.: Physiological Basis of Winter-Resistance of Cultivated Plants. Moscow, Leningrad: Selkhogiz 1940

Tysdal, H.M.: Determination of hardiness in *Alfalfa* varieties by their enzymatic responses. J. Agric. Res. **48**, 219–240 (1934)

Ullrich, H., Heber, U.: Über das Denaturieren pflanzlicher Eiweiße durch Ausfrieren und seine Verhinderung. Ein Beitrag zur Klärung der Frostresistenz bei Pflanzen. Planta **51**, 399–413 (1958)

Ullrich, H., Heber, U.: Ursachen der Frostresistenz bei Winterweizen. IV. Das Verhalten von Fermenten und Fermentsystemen gegenüber tiefer Temperatur. Planta **57**, 370–390 (1961)

Upadhyay, J., Stokes, J.L.: Temperature-sensitive formic hydrogenase in a psychrophilic bacterium. J. Bacteriol. **85**, 177–185 (1963a)

Upadhyay, J., Stokes, J.L.: Temperature-sensitive hydrogenase and hydrogenase synthesis in a psychrophilic bacterium. J. Bacteriol. **86**, 992–998 (1963b)

Ushakov, B.P.: Heat resistance of the somatic muscular system in the terrestrial animals as related with the conditions of existence. Zool. Zh. **34**, 578–587 (1955)

Ushakov, B.P.: Muscle thermostability of *Crustaceans* in relation to environmental conditions of a species. Izv. Akad. Nauk SSSR, ser. biol. 67–75 (1956a)

Ushakov, B.P.: Thermostability of the muscles of the *Mytilus*-species and *Lerches* in relation to the environmental temperature of a species. Zool. Zh. **35**, 953–964 (1956b)

Ushakov, B.P.: On the conservatism of protoplasm proteins of the species of poikilothermal animals. Zool. Zh. **37**, 693–706 (1958)

Ushakov, B.P.: Cellular physiology and the problem of species in zoology. Tsitologia **1**, 541–565 (1959a)

Ushakov, B. P.: Heat resistance of tissues as a specific character of poikilothermic animals. Zool. Zh. **38**, 1292–1302 (1959 b)

Ushakov, B. P.: Cytophysiological analysis of adaptation of reptiles to high temperatures of the desert. In: Voprosy tsitologii i obshchei fiziologii (Collection of works (Sbornik) Problems of Cytology and General Physiology) Moscow, Leningrad: Izd. Akad. Nauk SSSR 1960 a, pp. 355–367

Ushakov, B. P.: Thermostability of various tissues of frogs in relation to the temperature of environment characteristic of the species. In: Problemy tsitologii i protistologii (Collection of works (Sbornik) Problems of Cytology and Protistology) Moscow, Leningrad: Izd. Nauka 1960 b, pp. 84–89

Ushakov, B. P.: On some disputable problems of cytoecology. Tsitologia **3**, 455–466 (1961)

Ushakov, B. P.: A cytophysiological analysis of the intraspecific differentiation of *Phrynocephalus helioscopus*. Dokl. Akad. Nauk SSSR **144**, 1178–1180 (1962)

Ushakov, B. P.: On the classification of animal and plant adaptations and the role of cytoecology in solving the problem of adaptation. In: Problemy tsitoekologii zhivotnykh (Collection of works (Sbornik) Problems of Cytoecology of Animals) Moscow, Leningrad: Izd. Akad. Nauk SSSR 1963 a, pp. 5–20

Ushakov, B. P.: Cytophysiological analysis of intraspecies divergence of *Rana ridibunda* Pall. In: Problemy tsitoekologii zhivotnykh (Collection of works (Sbornik) Problems of Cytoecology of Animals) Moscow, Leningrad: Izd. Akad. Nauk SSSR 1963 b, pp. 145–157

Ushakov, B. P.: Changes of cellular heat resistance in ontogenesis and problem of conservatism of cells of higher poikilothermal animals. In: Problemy Tsitoekologii zhivotnykh (Collection of works (Sbornik) Problems of Cytoecology of Animals) Moscow, Leningrad: Izd. Akad. Nauk SSSR 1963 c, pp. 21–42

Ushakov, B. P.: Changes in heat resistance of muscle tissue of Reptiles in relation to seasons and the reproduction cycle. In: Problemy tsitoekologii zhivotnykh (Collection of works (Sbornik) Problems of Cytoecology of Animals) Moscow, Leningrad: Izd. Akad. Nauk SSSR 1963 d, pp. 51–61

Ushakov, B. P.: An analysis of thermostability of cells and proteins of poikilothermal animals in relation to the problem of species. Author's abstract of dissertation. Leningrad: 1964 a

Ushakov, B. P.: Thermostability of cells and proteins of poikilotherms and its significance in speciation. Physiol. Rev. **44**, 518–560 (1964 b)

Ushakov, B. P.: The mechanism of heat injury of cells. In: Teploustoichivost kletok zhivotnykh (Collection of works (Sbornik) Heat Resistance of Cells of Animals) Moscow, Leningrad: Izd. Nauka 1965 a, pp. 5–54

Ushakov, B. P.: The problem of related protein changes in the process of speciation. Tsitologia **7**, 467–480 (1965 b)

Ushakov, B. P., Amosova, I. S.: Changes in thermostability of isolated muscle tissue from different individuals of frogs produced by a preliminary heating. Ekologia **2**, 15–20 (1972)

Ushakov, B. P., Chernokozheva, I. S.: Changes in thermostability level of the muscle tissue of *Rana temporaria* tadpoles as a result of thermal action of spermatozoa. Tsitologia **5**, 238–241 (1963)

Ushakov, B. P., Darevsky, I. S.: A comparison between heat resistance in muscle fibers and temperature responses in two sympatric species of Lizards, living in semi-desert. Dokl. Akad. Nauk SSSR **128**, 833–835 (1959)

Ushakov, B. P., Gasteva, S. V.: The temperature thermoanesthesia coefficient of somatic muscles. Dokl. Akad. Nauk SSSR **88**, 1071–1074 (1953)

Ushakov, B. P., Glushankova, M. A.: On the lack of a definite correlation between the iodine number of protoplasma lipids and the cell thermostability. Tsitologia **3**, 707–710 (1961)

Ushakov, B. P., Glushankova, M. A.: Thermostability of the muscle tissue and proteins of *Bombina bombina* (L.) during the acclimation to cold. Tsitologia **12**, 510–515 (1970)

Ushakov, B. P., Glushankova, M. A., Salmenkova, E. A., Chernokozheva, I. S.: A correlation between an original level of cellular and protein thermostability and the direction of a phenotypic shift produced by a temperature adaptation of tadpoles. Ekologia **3**, 9–18 (1971)

Ushakov, B. P., Kusakina, A. A.: On lability and conservatism of the animal cell adaptation as discovered at the level of proteins. Tsitologia **2**, 428–441 (1960)

Ushakov, B. P., Pashkova, I. M.: The dynamics of individual changes in thermostability of muscle tissue in the process of temperature acclimation of *Asellu aquaticus* L. Zh. obshch. biol. **33**, 387–396 (1972)

Ushakov, B. P., Pashkova, I. M., Chernokozheva, I. S.: Changes in thermostability of an organism and muscle tissue of tadpoles during thermal acclimation visualized as a stabilizing adaptation. Dokl. Akad. Nauk SSSR **203**, 935–938 (1972)

Ushakov, B. P., Sleptsova, L. A.: Changes in muscle thermostability of Leeches at the thermal acclimation. Tsitologia **10**, 259–262 (1968)

Ushakov, B. P., Vinogradova, A. N., Kusakina, A. A.: Cytophysiological analysis of intraspecific differentiation of *Coreganus autumnalis* and *Thymalus arcticus* of the Baikal Lake. Zh. obshch. biol. **23**, 56–63 (1962)

Ushakov, B. P., Zander, N. V.: Adaptation of muscle fibers from warm-watered habitats *Rana ridibunda* to the temperature factor. Biofizika **6**, 322–327 (1961)

Ushakov, V. B.: A possible cause of thermal death of skeletal muscles. Tsitologia **5**, 204–211 (1963)

Ushakov, V. B.: On the cause underlying the thermal death of skeletal muscles in poikilothermal animals. Dokl. Akad. Nauk SSSR **155**, 1178–1181 (1964)

Ushakov, V. B.: An analysis of the causes of thermal death of skeletal muscles. Fiziol. Zh. SSSR **51**, 388–394 (1965a)

Ushakov, V. B.: The thermostability of excitable and contractile systems of a muscle fiber. In: Teploustoichivost kletok zhivotnykh (Collection of works (Sbornik) Heat Resistance of Cells of Animals) Moscow, Leningrad: Izd. Nauka 1965b, pp. 55–60

Ushakov, V. B.: The thermostability of cholinoreceptive system of skeletal phasic and tonic muscles of the frog. In: Teploustoichivost kletok zhivotnykh (Collection of works (Sbornik) Heat Resistance of Cells of Animals) Moscow, Leningrad: Izd. Nauka 1965c, pp. 70–75

Ushakov, V. B.: A comparison of the thermostability of muscles and fibrils of *Rana temporaria*. Tsitologia **8**, 96–99 (1966)

Ushakov, V. B., Vasilyeva, V. V.: Photometric studies of thermal death of frog skeletal muscles. In: Biofizika kletki (Collection of works (Sbornik) The Biophysics of the Cell) Moscow: Izd. Nauka 1965, pp. 131–139

Ushakov, V. B., Vasilyeva, V. V., Nikolaeva, E. N.: After effect of heating of phasic muscle fibers of the frog upon soluble proteins, actomyosin adenosine triphosphatase and contractile properties of fibrils. Tsitologia **13**, 311–318 (1971)

Ushatinskaya, R. S.: Diapause of insects and its modifications. Zh. obshch. biol. **34**, 194–215 (1973)

Vainman, G. M.: Seasonal changes in thermostability of the spermatozoa of *Rana temporaria*. Tsitologia **3**, 398–401 (1966)

Vallee, B. L., Stein, E. A., Summerwell, W. N., Fischer, E. H.: Metal content of α-amylases of various origins. J. Biol. Chem. **234**, 2901–2905 (1959)

Vanhumbeeck, J., Lurquin, P.: Purification and some properties of leucyl-tRNA synthetase from *Bacillus stearothermophilus*. Eur. J. Biochem. **10**, 213–218 (1969)

Vasilyeva, V. S., Glushankova, M. A., Zhirmunsky, A. V., Kosyuk, G. N.: Thermostability of cells and muscle aldolase from the sea urchin of the *Strongylocentrotus* genus and bivalve molluscs of the *Pectiniae* family in reference to the habitat conditions. In: Referaty nauchnykh rabot Instituta biologii morya (Abstr. Inst. Marine Biol.) Vladivostok: 1969, pp. 46–48

Vasquez, C., Parisi, M., Robertis, E.: Fine structure of ultrathin artificial membranes. I. Changes by acetylcholine addition in lipid proteolipid membranes. J. Membrane Biol. **6**, 353–367 (1971)

Vasyanin, S. I.: Thermostability of the muscle tissue of some species of birds of the finch family. Tsitologia **2**, 483–485 (1960)

Venetianer, P., Straub, F. B.: Enzymic formation of the disulfide bridges of ribonuclease. Acta Physiol. Acad. Sci. Hung. **24**, 41–53 (1963a)

Venetianer, P., Straub, F. B.: The enzymic reactivation of reduced ribonuclease. Biochim. Biophys. Acta **67**, 166–168 (1963b)

Vengerova, T. I., Rokhlin, O. V., Lezlin, R. S.: Retention of complete antigenic activity of light chains of rat immunoglobulins after their splitting into halves. Immunochem. **9**, 413–420 (1972)

Vernberg, F. J.: Comparative physiology: latidudinal effects on physiological properties of animal populations. Ann. Rev. Physiol. **24**, 517–546 (1962)

Vernberg, F. J., Schleiper, C., Schneider, D. E.: The influence of temperature and salinity of ciliary activity of excised gill tissue of molluscs from North Carolina. Comp. Biochem. Physiol. **8**, 271–285 (1963)

Vernon, H. M.: Heat rigor in cold-blooded animals. J. Physiology **24**, 238–287 (1899)

Veronese, F. M., Boccu', E., Fontana, A.: Denaturation of thermophilic and mesophilic 6-phosphogluconate dehydrogenase by 8 M urea. Intern. J. Peptide Protein Res. **7**, 341–343 (1975)

Viktorov, S. B.: Cytoecological studies and their significance for phytocoenology. Usp. sovr. biol. **13**, 172–174 (1940)

Vinnikov, Ya. A.: Sensory reception cytology, molecular mechanisms and evolution. In: Molecular Biology, Biochemistry and Biophysics. Berlin, Heidelberg, New York: Springer 1974, p. 392

Vinogradova, A. N.: The thermostability of succinate dehydrogenase of the *Rana temporaria* and *R. ridibunda* muscles and the changes in the enzyme activity during acclimation. Tsitologia **3**, 595–598 (1961)

Vinogradova, A. N.: Heat resistance and optimal temperature of adenosine triphosphatase activity of actomyosin in *Carcinus maenas* (L.), *Hyas araneus hoeki* Bir., *Roja clavata* L. and *Roja radiata* Donov. from the Black and Barents Seas. In: Problemy tsitoekologii zhivotnykh (Collection of works (Sbornik) Problems of Cytoecology of Animals) Moscow, Leningrad: Izd. Akad. Nauk SSSR 1963 a, pp. 189–194

Vinogradova, A. N.: Investigation of heat resistance of actomyosin of *Testudo horsfieldi* Gray and *Emys orbicularis* (L.) in connection with the problem of adaptation to living in deserts. In: Problemy tsitoekologii zhivotnykh (Collection of works (Sbornik) Problems of Cytoecology of Animals (Moscow, Leningrad: Izd. Akad. Nauk SSSR 1963 b, pp. 195–198

Vinogradova, A. N.: Thermostability and temperature optimum of ATPase activity of actomyosin of two species of marine *Mollusca*. Tsitologia **5**, 246–249 (1963 c)

Vinogradova, A. N.: Studies on actomyosin thermostability of frogs in relation to seasonal and experimentally induced decrease of muscle thermostability. In: Teploustoichivost kletok zhivotnykh (Collection of works (Sbornik) Heat Resistance of Cells of Animals) Moscow, Leningrad: Izd. Nauka 1965 a, pp. 115–118

Vinogradova, A. N.: Thermostability of actomyosin and myosin in two species of frogs. In: Teploustoichivost kletok zhivotnykh (Collection of works (Sbornik) Heat Resistance of Cells of Animals) Moscow, Leningrad: Izd. Nauka 1965 b, pp. 186–192

Vinogradova, A. N.: Ecological analysis of thermostability of lactate dehydrogenase from poikilothermal animals. Ekologia **5**, 60–67 (1970)

Vinogradova, A. N., Kusakina, A. A.: Heat resistance of protoplasmic proteins in representatives of different populations of *Rana ridibunda* Pall. In: Problemy tsitoekologii zhivotnykh (Collection of works (Sbornik) Problems of Cytoecology of Animals) Moscow, Leningrad: Izd. Akad. Nauk SSSR 1963, pp. 158–162

Viswanatha, T., Liener, J. E.: Utilization of native and denatured proteins by *Tetrahymena pyriformis* W. Arch. Biochem. Biophys. **56**, 222–229 (1955)

Vitvitsky, V. N.: Influence of methyl orange dye on the tryptic hydrolysis and heat denaturation of serum albumin. Molek. Biol. **3**, 678–682 (1969)

Vitvitsky, V. N.: Distinctions in the resistance of serum albumin to heating and to trypsin, observed in two frog species differing as to their thermophylic habit. Dokl. Akad. Nauk SSSR **194**, 950–952 (1970)

Vogel, W.: Über die Hitze- und Kälteresistenz von*Zoothamnium hiketes* Precht. *(Ciliata, Peritricha)*. Z. wiss. Zool. **173**, 344–378 (1966)

Volkenshtein, M. V.: The Physics of Enzymes. Moscow: Izd. Nauka 1967

Volkenshtein, M. V.: Electronic conformational interactions in biological systems. Izv. Akad. Nauk SSSR, biol. ser. **6**, 805–818 (1971)

Volotovsky, I. D., Konev, S. V.: On the connection between the conformation and UV-luminescence of proteins. Biofizika **12**, 200–205 (1967)

Volpe, E. P.: Embryonic temperature adaptations and relationships in toads. Physiol. Zool. **26**, 344–354 (1953).

Volpe, E. P.: Embryonic temperature tolerance and rate of development in *Bufo valliceps*. Physiol. Zool. **30**, 164–176 (1957)

Vorobyev, V. I.: Contractility. In: Rukovodstvo po tsitologii (A Handbook of Cytology) Moscow, Leningrad: Izd. Nauka 1966, pp. 67–91)
Vroman, H. E., Brown, J. R. C.: Effect of temperature on the activity of succinic dehydrogenase from livers of rats and frogs. J. Cell. Comp. Physiol. **61**, 129–131 (1963)
Waddington, C. H.: The evolution of adaptations. Endeavour **12**, 134–139 (1953)
Wagenbreth, D.: Das Auftreten von zwei Letalstufen bei Hitzeeinwirkung auf Pappelblätter. Flora **156**, 116–126 (1965)
Walker, H. W., Matches, J. R., Ayres, J. C.: Chemical composition and heat resistance of some aerobic bacterial spores. J. Bacteriol. **82**, 960–966 (1961)
Waugh, D. F.: Protein-protein interactions. Adv. Prot. Chem. **9**, 325–437 (1954)
Weiss, P. A.: A cell is not island entire of itself. Perspectives Biol. Med. **14**, 182–205 (1971)
Wernick, A., Künnemann, H.: Der Einfluß der Temperatur auf die Substrataffinität der Laktat-dehydrogenase aus Fischen. Marine Biol. **18**, 32–36 (1973)
West, M., Mooney, H. A.: Photosynthetic characteristics of three species of sagebrush as related to their distribution patterns in the White Mountains of California. Amer. Midland Naturalist **88**, 479–484 (1972)
Westra, A., Dewey, W. C.: Variation in sensitivity to heat shock or X-rays during the cell-cycle of Chinese hamster cells *in vitro*. Intern. J. Radiat. Biol. **20**, 197 (1971)
Wilson, A. C., Kaplan, N. O., Levine, L., Pesce, A., Reichlin, M., Allison, W. S.: Evolution of lactic dehydrogenases. Fed. Proc. **23**, pt. I, 1258–1266 (1964)
Wilson, F. R., Whitt, G. S., Prosser, C. L.: Lactate dehydrogenase and malate dehydrogenase isozyme patterns in tissues of temperature-acclimated goldfish (*Carassius auratus* L.) Com. Biochem. Physiol. **46 B**, 105–116 (1973)
Wilson, G., Fox, C. F.: Biogenesis of microbial transport systems: evidence for coupled incorporation of newly synthesized lipids and proteins into membrane. J. Mol. Biol. **55**, 49–60 (1971)
Wise, J., Swanson, A., Halvorson, H. O.: Dipicolinic acid-less mutants of *Bacillus cereus*. J. Bacteriol. **94**, 2075–2076 (1967)
Yagi, K., Ozawa, J.: Complex formation of apoenzyme, coenzyme and substrate of α-amino acid oxidase. IV. Changes in physico-chemical properties of the protein. Biochim. biophys. Acta **62**, 397–401 (1962)
Yakovleva, V. I.: Isoenzymes. In: Uspekhi biologicheskoi khimii (Collection of works (Sbornik) Progress of biological chemistry) **9**, 55–94 (1968)
Yakovleva, V. I., Gubnitsky, L. S.: The effect of heating on catalytic and allosteric properties of glutamate dehydrogenase in bovine liver. Dokl. Akad. Nauk SSSR **190**, 231–234 (1970)
Yakovleva, V. I., Gubnitsky, L. S.: Catalytic activity and allosteric properties of beef liver glutamate dehydrogenase after heat alteration *in situ*. Biokhimia **36**, 572–579 (1971)
Yakovleva, V. I., Gubnitsky, L. S.: Determination of the catalytic activity and allosteric properties of glutamate dehydrogenase after a thermal treatment of the liver tissues taken from two frog species. Tsitologia **14**, 721–730 (1972)
Yakovleva, V. I., Gubnitsky, L. S.: Thermal stability of the catalytic activity and allosteric properties of beef liver crystalline glutamate dehydrogenase. Biokhimia **38**, 1163–1168 (1973)
Yamada, S.: Über die Wirkung höherer Temperaturen auf sympathische Kaltblüternerven. Pflüg. Arch. ges. Physiol. **202**, 73–87 (1924)
Yamanaka, T., Mizushima, H., Nozaki, M., Horio, T., Okunuki, K.: Studies on cytrochrome C. III. Determination of "native" mammalian heart muscle cytochrome C and its physiological properties. J. Biochem. (Tokyo) **46**, 121–132 (1959)
Yarwood, C. E.: Acquired tolerance of leaves to heat. Science **134**, 941–942 (1961)
Yarwood, C. E.: Acquired sensitivity of leaves to heat. Plant Physiol. **37**, suppl. 70 (1962)
Yarwood, C. E.: Sensitization of leaves to heat. Adv. Frontiers Plant Sci. **7**, 195–203 (1963)
Yarwood, C. E.: Adaptation and sensitization of bean leaves to heat. Phytopathology **54**, 936–940 (1964 a)
Yarwood, C. E.: Thermophylaxis in bean rust. Nature **203**, 426–427 (1964 b)
Yarwood, C. E.: Adaptation of plants an plant pathogens to heat. In: Molecular Mechanisms of Temperature Adaptation, (ed. C. L. Prosser) Washington: Am. Ass. Advan. Sci. 1967, pp. 75–89

Yarwood, C. E., Holm, E. W.: Heat adaptation in a rust and a virus. Phytopathology **52**, 709–712 (1962)
Yazkulyev, A.: The increase in cell thermostability of *Aristida karelini* (Trin. et Rupr.) Roshev. and *Arundo donax* L. under influence of environmental temperatures in natural conditions. In: Tsitologicheskiye osnovy prisposobleniya rastenii k faktoram sredy (Collection of works (Sbornik) Cytological Aspects of Adaptation of Plants to the Environmental Factors) Moscow, Leningrad: Izd. Nauka 1964a, pp. 3–25
Yazkulyev, A.: Cell thermostability of some grasses growing in Turkmenia. In: Tsitologicheskiye osnovy prisposobleniya rastenii k faktoram sredy [Collection of works (Sbornik) Cytological Aspects of Adaptation of Plants to the Environmental Factors]. Moscow, Leningrad: Izd. Nauka 1964b, pp. 88–94
Yazkulyev, A.: On differences in the resistance of cells to heating, high hydrostatic pressure and ethanol in closely related species of plants differing in their periods of vegetation. Tsitologia **11**, 180–188 (1969)
Yazkulyev, A.: Seasonal and diurnal variations of the cell thermostability, photosynthesis and water dificit of the leaf in *Aristida karelinii*. Bot. Zh. **55**, 938–945 (1970)
Yoshida, M.: Allosteric nature of thermostable phosphofructokinase from an extreme thermophilic bacterium. Biochemistry **11**, 1087–1093 (1972)
Yoshida, M., Oshima, T.: The thermostable allosteric nature of fructose-1,6-diphosphatase from an extreme thermophile. Biochem. Biophys. Res. Commun. **45**, 495–500 (1971)
Yoshida, M., Oshima, T., Imahori, K.: The thermostable allosteric enzyme: phosphofructokinase from an extreme thermophile. Biochem. Biophys. Res. Commun. **43**, 36–39 (1971)
Yoshida, M., Oshima, T., Imahori, K.: Fructose-1,6-disphosphatase of an extreme thermophile. J. Biochem. (Tokyo) **74**, 1183–1191 (1973)
Yudin, A. L., Sopina, V. A.: On the role of nucleus and cytoplasm in the inheritance of some characters in Amoebae experiment on transfer of nuclei. Acta Protozool. **8**, 1–39 (1970)
Yuen Wan Shing, Akagi, J. M., Himes, R. H.: Thermolabile triose phosphate isomerase in a psychrophilic *Clostridium*. J. Bacteriol. **109**, 1325–1329 (1972)
Yutani, K., Yutani, A., Isemura, T.: Accelerating effect of proteins on the renaturation of denatured bacterial α-amylase. J. Biochem. (Tokyo). **62**, 576–583 (1967)
Zaalishvili, M. M.: Physico-chemical basis of muscle activity. Tbilisi: Izd. Metsniersba 1971
Zavadskaya, I. G.: On the rate of increase of thermostability of plant cells after a short preliminary exposure to high temperature. Bot. Zh. **48**, 755–758 (1963a)
Zavadskaya, I. G.: The influence of high temperature on the viscosity of cytoplasm of plant cells. Tsitologia **5**, 151–158 (1963b)
Zavadskaya, I. G.: Changes in carbohydrate content of plants under heat hardening. In: The Cell and Environmental Temperature. Moscow, Leningrad: Izd. Nauka 1964, pp. 124–125
Zavadskaya, I. G., Denko, E. I.: The effect of dehydration on thermostability of plant cells. Bot. Zh. **51**, 696–705 (1966)
Zavadskaya, I. G., Denko, E. I.: The effect of insufficient water supply on the stability of leaf cells of certain plants of the Pamirs. Bot. Zh. **53**, 795–805 (1968)
Zavadskaya, I. G., Shukhtina, H. G.: The influence of dehydration and superoptimal temperatures of leaf cell thermoresistance of drought resistant barley. Tsitologia **13**, 1304–1307 (1971)
Zavodszky, P., Abaturov, L. B., Varshavsky, Y. M.: Structure of glyceraldehyde-3-phosphate dehydrogenase and its alteration by coenzyme binding. Acta biochim. biophys. Acad. Sci. Hung. **1**, 389–402 (1966)
Zeikus, J. G., Brock, T. D.: Protein synthesis at high temperatures: aminoacylation of tRNA. Biochim. biophys. Acta **228**, 736–745 (1971)
Zeikus, J. G., Taylor, M. W., Brock, T. D.: Thermal stability of ribosomes and RNA from *Thermus aquaticus*. Biochim. biophys. Acta **204**, 512–520 (1970)
Zernov, S. A., Schmalhauzen, O. I.: On the limits of life at below-zero temperatures. Dokl. Akad. Nauk SSSR **44**, 84–85 (1944)
Zhestyanikov, V. D.: Radioresistance of thermoresistant *Escherichia coli* strains. In: Reaktsiya kletok na ekstremalnye vozdeistviya [Collection of works (Sbornik) Reaction of Cells to Extreme Influences]. Moscow, Leningrad: Izd. Akad. Nauk SSSR 1963, pp. 135–142

Zhestyanikov, V. D.: Thermoresistance of *Escherichia coli* cultivated at high temperatures. In: Tsitologicheskiye osnovy prisposobleniya rastenii r faktoram sredy [Collection of works (Sbornik) Cytological Aspects of Adaptation of Plants to the Environmental Factors]. Moscow, Leningrad: Izd. Nauka 1964, pp. 31–45

Zhestyanikov, V. D.: Recovery and Radioresistance of the Cell. Leningrad: Izd. Nauka 1968

Zhilenko, T. P., Vasilyeva, V. S.: Thermostability of the sea urchin muscles during quiescence and reproduction. In: Referaty nauchnykh rabot Instituta biologii morya (Collection Works Inst. Marine Biol.). Vladivostok: 1969, pp. 63–78

Zhirmunsky, A. V.: Thermostability of *Actinia* and their ciliary epithelium under natural conditions and at experimental changes of temperature of the medium. Tsitologia **1**, 270–276 (1959)

Zhirmunsky, A. V.: A study of temperature adaptations of invertebrates from the South China Sea. Tsitologia **2**, 675–690 (1960a)

Zhirmunsky, A. V.: Sensitivity of the Black Sea *Mytilus* and its ciliary epithelium to elevated temperatures. Dokl. Akad. Nauk SSSR **133**, 683–685 (1960b)

Zhirmunsky, A. V.: Thermostability of the ciliary epithelium cells of two ascidias, *Tethyum aurantium* (Pollos) and *T. robetzi* Drasch, in connection with thermal conditions of their habitat. Tsitologia **5**, 227–230 (1963)

Zhirmunsky, A. V.: A comparative study of cellular thermostability of marine invertebrates in relation to their geographical distribution and ecology. In: The Cell and Environmental Temperature. Moscow, Leningrad: Idz. Nauka 1964, pp. 142–150

Zhirmunsky, A. V.: A comparative study of cell thermostability of molluscs from the White Sea in relation to the vertical distribution of species and the history of fauna formation. Zh. obshch. biol. **30**, 686–703 (1969a)

Zhirmunsky, A. V.: Cellular thermostability and distribution of benthic animals in the upper marine zones. Author's abstract of dissertation. Leningrad: 1971

Zhirmunsky, A. V.: Vertical distribution and cellular heat resistance of benthic animals from the Possyet Bay (Japan Sea). Helgol. wiss. Meeresuntersuch. **24**, 247–255 (1973)

Zhirmunsky, A. V., Chu Li-chun: Thermostability of the ciliary epithelium of the *Nerita* genus molluscs in connection with thermal conditions of the habitat. Tsitologia **2**, 479–482 (1960)

Zhirmunsky, A. V., Pisareva, L. N.: Thermostability of the tissues of some sea animals living at various depths. In: Problemy tsitologii i protistologii [Collection of works (Sbornik) Problems of Cytology and Protistology]. Moscow, Leningrad: Izd. Akad. Nauk SSSR 1960a, pp. 112–116

Zhirmunsky, A. V., Pisareva, L. N.: A study of heat sensitivity in certain marine invertebrates and of their tissues. Dokl. Akad. Nauk SSSR **133**, 957–959 (1960b)

Zhirmunsky, A. V., Schlyachter, T. A.: Heat resistance of the organisms of frogs and of their cells with experimental changes in the temperature of environment. In: Problemy tsitologii zhivotnykh [Collection of works (Sbornik) Problems of Cytoecology of Animals]. Moscow, Leningrad: Izd. Akad. Nauk SSSR 1963, pp. 78–86

Zhirmunsky, A. V., Tsu Li-tsun: Adaptation of tropical molluscs *Nerita* and *Donax* to habitat temperature. In: Collection of Deposited Works: Molluscs, Moscow, Leningrad: 1964, pp. 62–77

Zhukov, E. K.: On the thermal parabiosis of the nerve related to the calorimetric modification of the parabiotic area. Tr. Leningr. Obshch. estestvoispytatelei (Collection of Works of the Leningrad Soc. Natural.) Leningrad: 1935, pp. 407–428

Zilber, L. A.: Sugar bacillar vaccines (AD-vaccines). Zh. mikrobiol. epidemiol. i. immunobiol. **16**, 6–22 (1936)

Zito, R., Antonioni, E., Wyman, J.: The effect of oxygenation on the rate of digestion of human hemoglobins by carboxypeptidases. J. Biol. Chem. **239**, 1804–1808 (1964)

Zondag, H. A.: Lactate dehydrogenase isozymes: lability at low temperature. Science **142**, 965–967 (1963)

Zydowo, M., Makarewicz, W., Umiastowski, J.: Temperature dependence of AMP deamination catalysed by muscle extracts from homeothermic and poikilothermic animals. Acta biochim. pol. **12**, 319–325 (1965)

Subject Index

Ecological Studies

Analysis and Synthesis

Editors: W. D. Billings, F. Golley, O. L. Lange, J. S. Olson

Vol. 1: **Analysis of Temperature Forest Ecosystems**
Editor: D. E. Reichle
1st corrected reprint
91 figs. XII, 304 pages. 1973

Vol. 3: **The Biology of the Indian Ocean**
Editor: B. Zeitzschel in cooperation with S. A. Gerlach
286 figs. XIII, 549 pages. 1973

Vol. 4: **Physical Aspects of Soil Water and Salts in Ecosystems**
Editors: A. Hadas, D. Swartzendruber, P. E. Rijtema, M. Fuchs, B. Yaron
221 figs., 61 tables. XVI, 460 pages. 1973

Vol. 5: **Arid Zone Irrigation**
Editors: B. Yaron, E. Danfors, Y. Vaadia
181 figs. X, 434 pages. 1973

Vol. 6: K. Stern, L. Roche
Genetics of Forest Ecosystems
70 figs. X, 330 pages. 1974

Vol. 7: **Mediterranean Type Ecosystems**
Origin and Structure
Editors: F. di Castri, H. A. Mooney
88 figs. XII, 405 pages. 1973

Vol. 8: **Phenology and Seasonality Modeling**
Editor: H. Lieth
120 figs. XV, 444 pages. 1974

Vol. 9: B. Slavik
Methods of Studying Plant Water Relations
181 figs. XVIII, 449 pages. 1974
Distribution Rights for the Socialist Countries: Academia Publishing House of the Czechoslovak Acàdemy of Sciences, Prague

Vol. 10: **Coupling of Land and Water Systems**
Editor: A. D. Hasler
95 figs. XVII, 309 pages. 1975

Vol. 11: **Tropical Ecological Systems**
Trends in Terrestrial and Aquatic Research
Editors: F. B. Golley, E. Medina
131 figs. XV, 398 pages. 1975

Vol. 12: **Perspectives of Biophysical Ecology**
Editors: D. M. Gates, R. B. Schmerl
215 figs. XIII, 609 pages. 1975

Vol. 13: **Epidemics of Plant Diseases**
Mathematical Analysis and Modeling
Editor: J. Kranz
46 figs. X, 170 pages. 1974

Vols. 14—20 see next page

Springer-Verlag
Berlin
Heidelberg
New York